MARINE TAILINGS DISPOSAL

MARINE TAILINGS DISPOSAL

Edited by
DEREK V. ELLIS

ANN ARBOR SCIENCE
THE BUTTERWORTH GROUP

FOREWORD

A cynical public might well be forgiven for viewing the ocean disposal of mine tailings as a perverse extension of the ocean-as-a-garbage-can idea. The findings presented in this book provide clear grounds for a more optimistic assessment. We already know a great deal about the science of inshore environments, even those with complex embayments. Physical oceanography occupies a leadership position. While existing tailings disposal systems have provided some unpleasant surprises, these have been relatively minor when one considers the newness of the technology. The special value of the present volume is that it constitutes an important contribution to an exponential learning curve which will rapidly improve the predictability of future tailings disposal systems. Contributions from the fields of sediment and water column chemistry, biological community structure and dynamics, along with innovative engineering design will be especially important for improved predictability of new disposal systems and eventual public acceptance.

J. E. McInerney, Chairman
Technical Panel, Kitsault Tailings Discharge Assessment
for the Minister of Fisheries and Oceans
British Columbia

PREFACE

This book represents the proceedings of a Symposium on Marine Tailings Disposal held at Ketchikan, Alaska, on March 22–23, 1982. The chapters follow the sequence of the symposium papers, although authors and chairmen have had the opportunity to modify their verbal presentations. Some of the questions and answers from the sessions have been incorporated where they provide extra information and conclusions relevant to the theme. Introductions have been inserted at the beginning of each section to assist in continuity. Two of the presentations at the symposium have not been prepared for publication.

It will be asked: "Why such a symposium on mining waste disposal in Ketchikan?" This does not mean "Why such a symposium?" There is no question that the need for one existed. The disposal of mine tailings to the sea is a controversial environmental management action with both proponents and objectors either at specific sites, or in principle. The question really is "Why at Ketchikan?"

The answer is simple. Some 50 miles to the east of Ketchikan there is a massive molybdenum ore-body known as Quartz Hill (see Section I frontispiece). U.S. Borax & Chemical Corporation located the ore-body in 1974, and as the prospect size was confirmed, started appropriate environmental impact assessment in 1978. The symposium at Ketchikan is a joint response of government regulatory agencies and the mining company to assemble the most recent environmental information from other marine disposing mines and the Quartz Hill site and make it available to the public, to the industry and to the many government agencies who must make decisions on the numerous permits that will be granted if the mine eventually comes into operation.

This book is the public record of the symposium and the discussions between scientists, and between scientists and concerned Alaskan residents. I hope that it will serve as a reference book for those with a continuing interest not only in the Quartz Hill site, but also the general topic of marine tailings disposal. I have attempted to ensure that the

vii

chapters, although providing highly technical information in places, are nevertheless understandable to the nonspecialist. He or she might have to struggle a little with technicalities, since there is no avoiding them, but nothing as complex as an environmental engineering project can be understood without some effort.

The words "tailings" and "tailing" are treated as synonyms in this book. This follows the mixed use at the symposium. Perhaps it reflects an increasing spectrum of interest in mining developments outside the traditional mining community. Units also have not been standardized, but have been kept as used by authors.

Much of the information provided in this book comes from existing mines, and of these, Island Copper Mine of Utah Mines Ltd. is featured frequently. There is a reason for this. Island Copper Mine was developed and went into operation in the early 1970s when environmental regulatory action such as the U.S. National Environmental Policy Act (NEPA) and Canadian equivalents were being legislated and implemented. Island Copper was the first marine discharging mine to come into operation with a comprehensive environmental monitoring and impact assessment program. It now has a decade of data banked and impacts assessed. Other information presented at the symposium comes from the Canadian mine at Kitsault, but there are other marine discharging mines on the coasts of Canada and other countries. From the frontispiece in Section I, it can be seen that both Kitsault and Island Copper in the province of British Columbia are sufficiently near Quartz Hill (35 and 400 miles distant, respectively) that they are climatically and topographically quite similar.

Derek V. Ellis

ACKNOWLEDGMENTS

The symposium was sponsored jointly by:

- U.S. Forest Service, Ketchikan Area;
- U.S. Environmntal Protection Agency, Region 10, Seattle, Washington;
- Alaska Department of Environmental Conservation, Juneau, Alaska; and
- U.S. Borax & Chemical Corporation, Los Angeles, California.

I must also acknowledge the cooperation of the authors, the symposium chairman, Dr. I. S. Bengelsdorf, the organizer, Dr. K. Kitasaki, and the guidance of Ann Arbor Science Publishers, as we brought this book to publication. In addition, I am very grateful to the commitment of my secretary, Gretchen Moyer, in managing all aspects of the editorial office; to Bonnie Oldershaw, Laurel Carr and Katharine Ellis for typing; and to Elizabeth and Jurgen Campolin of Technigraphics for drafting. Photographs were provided by chapter authors or the editor.

Derek V. Ellis is a Professor of Biology at the University of Victoria in Victoria, British Columbia, Canada. He received his PhD and MSc from McGill University, Montreal, with theses in the area of arctic marine ecology, and his BSc with Honors in Zoology from the University of Edinburgh in Scotland. During his PhD investigations he held a one-year National Research Council of Canada Graduate Student Fellowship at the University of Copenhagen. He spent six years as an Associate Scientist with the Fisheries Research Board of Canada at the Pacific Biological Station in Nanaimo, British Columbia, and a year as Assistant Professor at the University of Manatoba in Winnipeg.

Since 1970 Dr. Ellis has been involved in environmental impact assessment (EIA) and research on problems of waste disposal from mines, pulp and paper mills, log storage areas, and cities. Since then he has consulted for government and industry on mining and other impacts at Kitsault, Tasu, the Philippines, Argentina, Greenland, Alaska and the Canadian Northwest Territories. He teaches an undergraduate course in Marine Field Biology, a graduate student course in EIA, and has presented short courses in EIA attended by engineers, physicians and administrators. Dr. Ellis is the editor and a co-author of *Pacific Salmon: Management for People*, and has published many articles in professional journals and chapters in books on his research and developments in the marine environmental sciences. His particular field of fundamental research has been on the distribution, biology and species associations of coastal benthos.

CONTENTS

xi

Section II: Case Studies

Section III: Regulatory Action

Section IV: Quartz Hill

Modern tailings outfalls. The older (right), now a standby, discharges to its seawater mix tank, whereas the newer design (left) pipes the tailings underwater to the mix tank.

SECTION I

ENGINEERING AND SCIENTIFIC PRINCIPLES

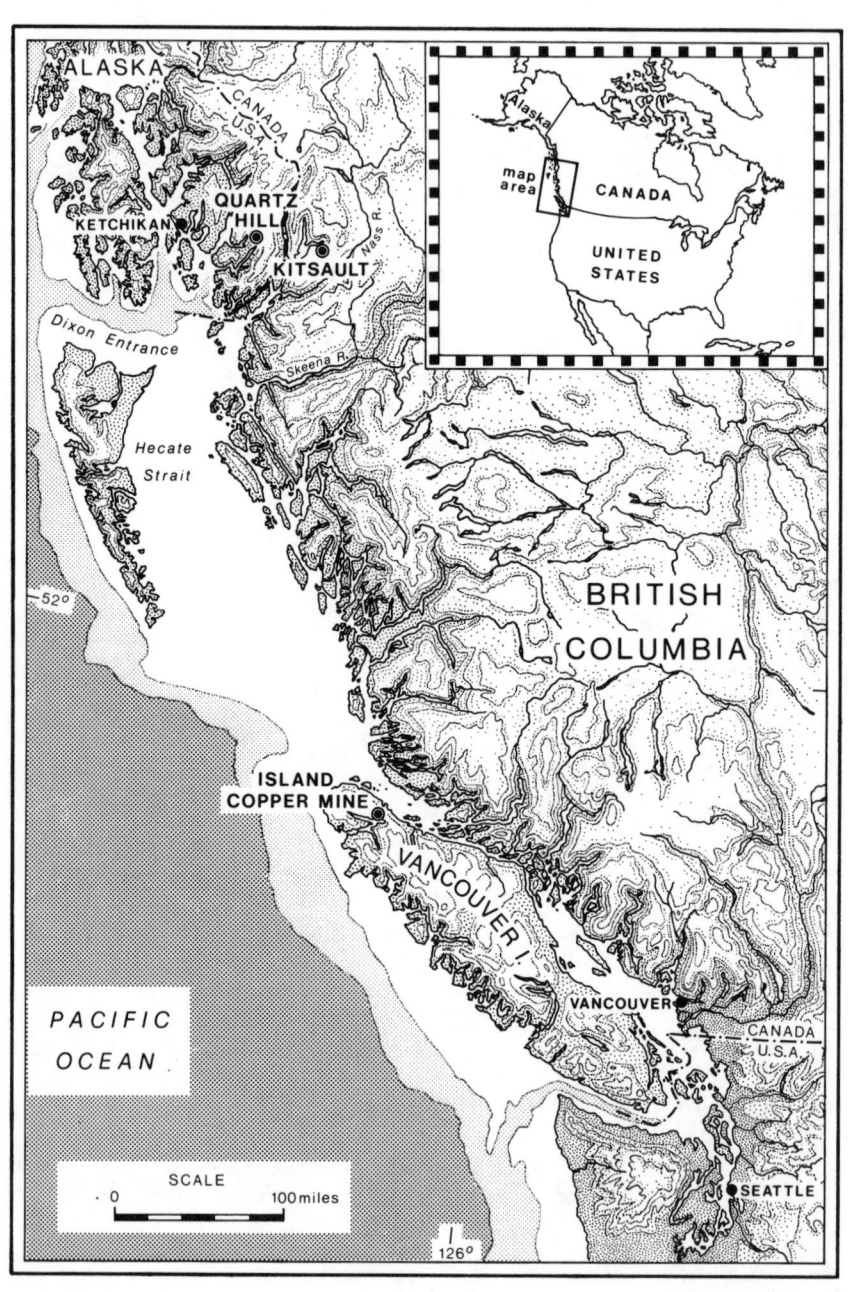

Section I Frontispiece. The coast near Ketchikan, Alaska.

INTRODUCTION

ENGINEERING AND SCIENTIFIC PRINCIPLES

Any industrial development must be based on a sound understanding of engineering principles by the design and construction engineers. Nowadays any major development with the potential to impact its surrounding environment must be designed by engineers associating with environmental scientists, who in turn must have a sound understanding of environmental management principles. Prevention is better than cure. It is far easier to mitigate potential impacts before concrete is poured. During operations the surrounds need monitoring, so that the concepts of quality control can be applied to the environment, i.e. information fed back by ecologically trained scientists and technicians so that production staff can fine-tune operations to maintain environmental quality, as well as production.

Section I provides chapters on engineering and scientific principles underlying the development of a mine with the possibility of disposing its tailings to the sea. First there is a chapter on tailings, their engineering and the way in which decisions are made on how to dispose of massive amounts. Then follows a chapter on the physical aspects of tailings, what they are and how they behave when introduced into seawater in large amounts. Basic fjord oceanography is summarized to lead to an understanding of how deep and near-surface currents and waves can affect massive and quickly introduced silt particles differently from the natural, slower but still massive injections of river silt. (Most mine wastes discharged to the sea presently utilize fjords rather than open coasts, where the oceanography is somewhat different.)

A chapter on the important topic of metal contamination in the sea was scheduled for the symposium, but due to illness the participant was unable either to attend or prepare his material. This topic cannot be eliminated from a mine waste disposal review, and a brief introduction to some recent information and concepts has been provided as Chapter 4.

Finally in this section, a chapter on the actual behavior of tailings in a fjord environment at a marine discharging mine bridges the gap from the fundamentals of Section I to the case studies of Section II. This chapter is perhaps the most innovative of the book, and provides information on a topic never before intensely investigated. Its results are surprising: tailings do not simply form an apron of sediments obliterating all benthic life before it. Instead, the discharged tailings may form a density current snaking its way between levees. This information gives considerable hope that consequent engineering of a tailings outfall can minimize the physical impact of tailings disposal to the sea.

CHAPTER 1

TAILINGS DISPOSAL IN RUGGED,
HIGH PRECIPITATION ENVIRONMENTS:
AN OVERVIEW AND COMPARATIVE ASSESSMENT

J.A. Caldwell
Steffen Robertson and Kirsten
Suite 500 - 1281 West Georgia St.
Vancouver, B.C.
Canada V6E 3J7

J.D. Welsh
Steffen Robertson and Kirsten
7510 W. Mississippi - Suite 210
Lakewood, Colorado 80226
U.S.A.

INTRODUCTION

This chapter presents an overview of the disposal of tailings. Particular attention is given to land disposal in impoundments constructed of the tailings or borrow material in high precipitation, rugged environments. Considered are:

*tailings disposal options;
*techniques for constructing tailings impoundments;
*design considerations for safe, economic impoundments;
*construction considerations and problems;
*operating considerations and problems; and
*approaches to impoundment reclamation.

A procedure for site selection of a tailings impoundment is described. The procedure has been used in many cases to select the best site for a particular mining project. As an exercise to compare different tailings disposal options, the procedure is used in this paper to rank disposal of tailings in the following hypothetical situations: (a) marine disposal; (b) flat arid terrain disposal; (c) wet rugged terrain disposal; (d) in-pit disposal; and (e) disposal of dry tailings. From this exercise, the advantages and disadvantages of the different options may be rigorously derived and rationally evaluated, and in particular marine disposal compared to land disposal in rugged, high precipitation environments.

THE DESIGN OF TAILINGS DEPOSITION FACILITIES

Tailings System Criteria

A tailings disposal system must:

- be compatible with the other aspects of mining and milling;
- provide for the safe storage of the tailings during and after operation;
- satisfy environmental, legal, and regulatory requirements;
- be economically feasible to construct and operate; and
- be able to be stabilized economically at closure to provide long-term, low maintenance containment integrity.

Elements of a Tailings Disposal System

The basic elements common to most tailings disposal systems are:

- a place to dispose of the tailings;
- a way to transport the tailings from the mill to the disposal area;
- a method to contain or impound the tailings;
- a method to distribute the tailings;
- a water management system; and
- a pollution control and reclamation method.

Storage Location Options

There are three options available for the location of tailings storage. These are:

- marine and lake disposal;
- mine backfill; and

˙surface disposal.

Marine Disposal

Tailings may be deposited into the sea, an inlet, or an inland lake. Such options are site-specific and are limited to operations within economical transport distances of such water bodies. There are few places where rivers may be used to transport tailings to the sea. The chemical nature of the tailings may preclude ocean or lake deposition. The impact of tailings deposition on the marine environment must be acceptable, and suitable engineering is required to keep impacts to as low a level as possible. Some coastal conditions may not be conducive to marine disposal; shallow productive coastlines and those with strong currents may be unacceptable deposition locations.

Marine disposal of tailings has several advantages. Large volumes may be disposed of without creating a prominent on-land impoundment. Placing tailings in the ocean closely duplicates the earth's natural processes of erosion. Such deposition reduces the risk of surface water contamination and eliminates dust, visual impact and the potential for long-term instability.

Mine Backfill

Mining may be in an open pit or underground. Tailings disposal in an open pit that is being mined is usually not possible. Only if successive pits are mined at an operation may mined-out pits be used for tailings deposition. Example 1 (see p. 55) illustrates such a case. At some strip mines backfilling with tailings may be done. In rugged terrain, where only one pit is to be dug, in-pit disposal is impossible.

An underground mine may be backfilled as mining proceeds. Usually only the coarse fraction of the tailings is used for backfill. Because of this and because of bulking (increase of volume in mining and grinding), fine tailings still have to be deposited on land. Thus mine backfill cannot be a complete solution to tailings deposition.

Surface Disposal

There are few mines where tailings may be disposed of below water or in the mine. At most mines tailings are disposed of on land in tailings impoundments. There are many different types of tailings impoundments and numerous ways of disposing of the tailings. However, before details of a tailings disposal system are worked out, a site for the impoundment must be selected. Thus the first consideration in surface disposal of tailings is the selection of a suitable site.

Tailings Impoundment Site Selection

Selection of a site for a tailings impoundment may be done by the method described by Robertson and Moss (1). The method involves a number of steps, summarized as follows:

1. Regional screening. Examine on topographic maps an area within a 10 km to 50 km radius of the site. Areas obviously unsuitable for tailings impoundments are eliminated. Some factors that may eliminate an area are: (a) topography too steep; (b) access too difficult; (c) sensitive ecological areas; (d) important land use zones; (e) too large a catchment area; (f) groundwater discharge area; and (g) unsuitable geology or mineralization.
2. Site identification. Identify all possible impoundment sites. List capacity, embankment height, distance from mine and other salient characteristics that might affect tailings disposal.
3. Fatal flaw analysis. Eliminate sites flawed by characteristics sufficiently unfavorable or severe that they preclude use of the site. Some such factors are: (a) unacceptable visual impact; (b) land use or ecological factors (such as critical fish or wildlife habitat); (c) archaeologically important sites; (d) too exposed to winds, flood plain, or active faults; (e) too small; and (f) too costly to develop.
4. Investigate sites. Visit the sites and gather as much data as possible about such factors as visibility, land use, meteorology, geology, soils, vegetation and groundwater. Formulate for each a conceptual design of an impoundment and hence estimate costs of construction operation and reclamation. Tabulate operational constraints or difficulties.
5. Qualitative evaluation and ranking. For defined factors, rate each site subjectively as very good, good, moderate, poor or very poor. Robertson and

Moss (1) list many factors; some may only be appropriate in particular cases.

6. Semi-quantitative evaluation and ranking. To each of the subjective ratings, assign a number - one for very good through 5 for very poor. Average ratings in similar categories such as operational, cost or environmental impact. Assign suitable weightings to each category; then add and determine a final numerical rating. The ranking of sites is according to these ratings.

7. Detailed investigation. The highest ranking sites are further investigated. This may involve field investigations, drilling, further design, costing and environmental studies and impact analysis. A minimum of two and preferably three sites should be studied in detail. Fig. 1 shows the system as adapted for use to rank and select a tailings impoundment site for a project on the Queen Charlotte Islands, B.C.

Types of Surface Impoundments

Tailings disposal impoundments may be classified with respect to the topography or the method used to construct the embankment, as is shown in Fig. 2.

Ring dike impoundments are best suited to flat or gently sloping topography where containment is required around the entire perimeter. They have a high ratio of embankment volume to total contained volume. This increases the amount of work in building the embankment, which may be of tailings, borrow material or mine waste rock. An advantage of the ring dike impoundment is that only the impoundment acts as catchment; hence precipitation inflow is kept low. Example 2 (p. 55) describes a ring dike impoundment.

Side-hill impoundments are normally used in sloping terrain. Only a part of the perimeter is enclosed with embankments. Such an impoundment has a larger drainage catchment area than ring impoundments. They can be located on a wide range of natural slope angles, and hence are adaptable to many sites. Example 3 (p. 57) describes a side-hill impoundment.

The third type of impoundment is the valley impoundment. Blocking a valley provides a large basin for tailings storage. A valley impoundment offers the lowest ratio of embankment volume to containment volume. In steep terrain, valley impoundments may be the only choice available for tailings storage. The principal disadvantage of a valley impoundment is the large drainage area above

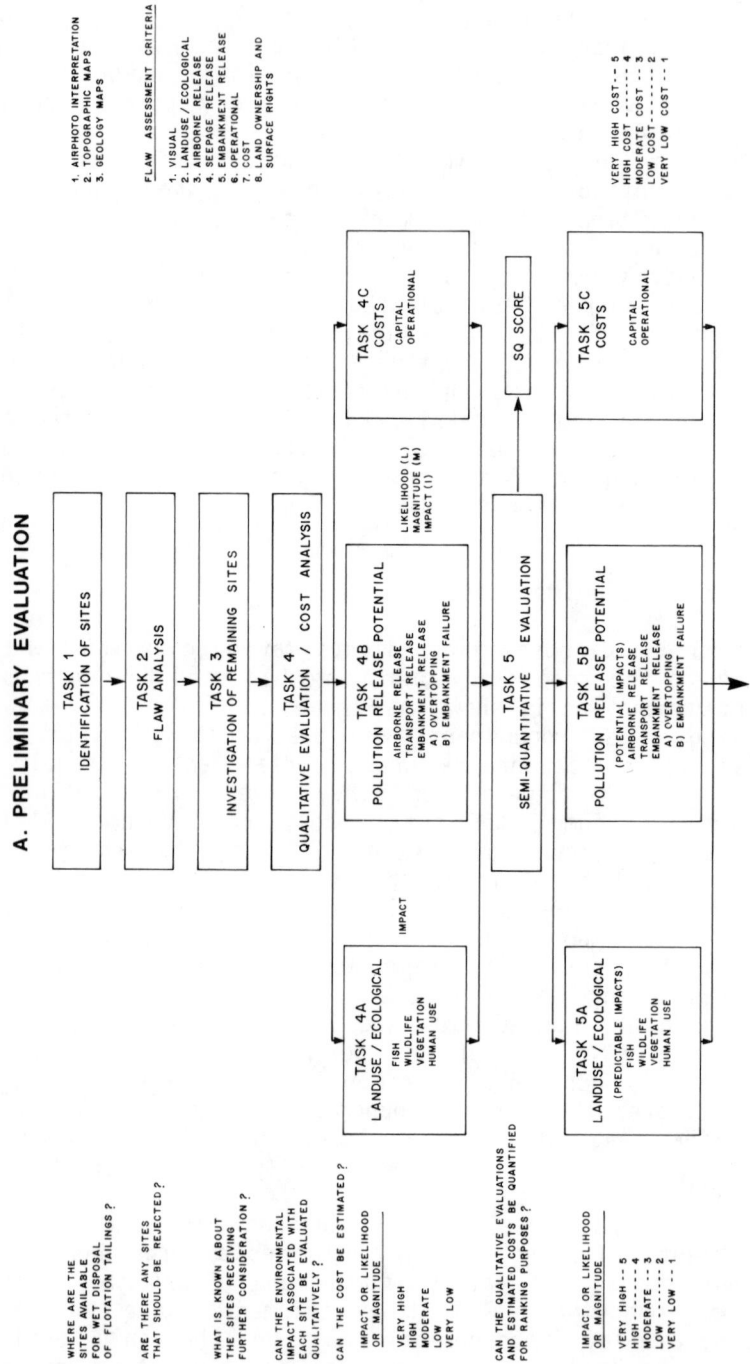

A. PRELIMINARY EVALUATION

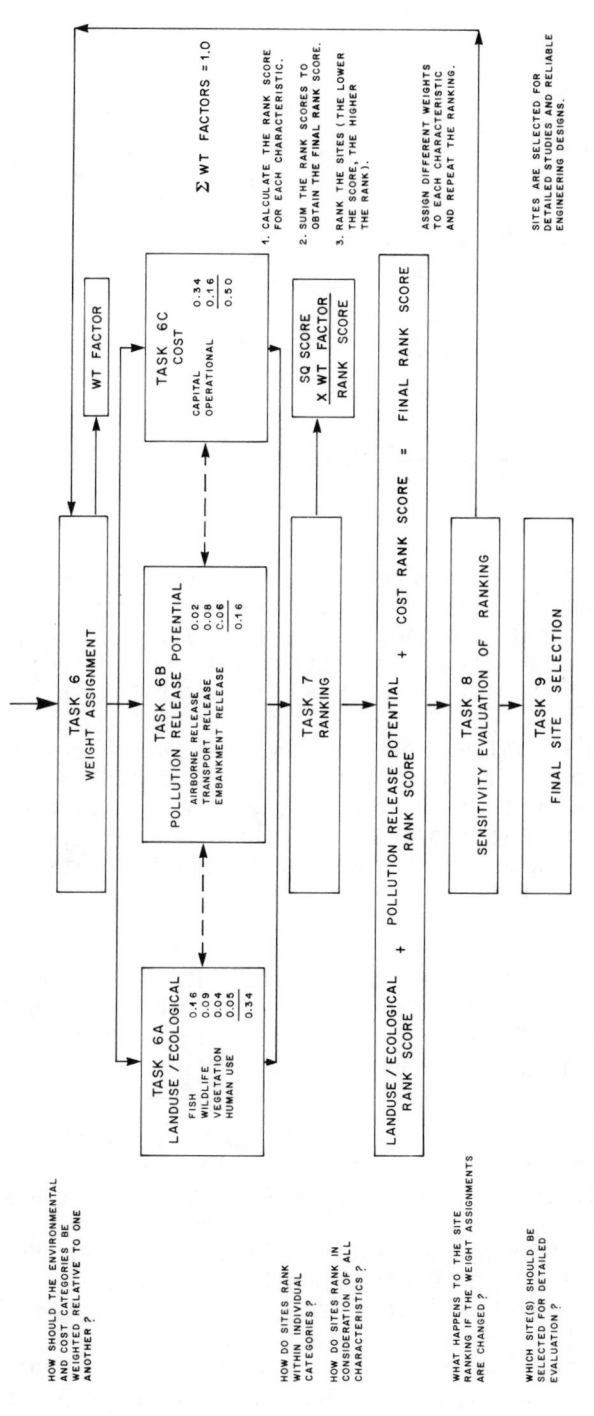

Figure 1. Use of site selection method (2).

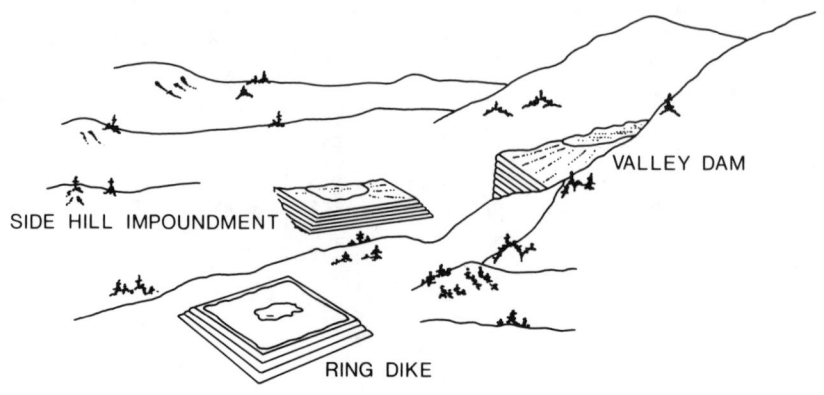

Figure 2. Impoundment location types.

the impoundment. This drainage may provide flood potential and a continual inflow of excess water if not diverted. Because of natural drainage patterns, valley impoundments are also more difficult to stabilize and decommission after mining. A valley impoundment is described in Example 4 (p.58).

Because a valley impoundment is most suited to tailings disposal in rugged terrain a detailed assessment of such impoundments is included as Table I. The table is taken from a recent publication (3) of the International Atomic Energy Agency on current practices and options for confinement of tailings.

Table I. Factors to be considered in assessing the location and use of a valley dam impoundment (from "Current Practices and Options for Confinement of Uranium Mill Tailings").

---

(1) Valley impoundments are chosen to maximize the use of natural topographic features in the area. The shape and locations of the impoundment basin are therefore determined by the availability of suitable sites within a reasonable distance from the mine. Because the characteristics of the site are fixed by nature there is usually little scope for varying the site location to avoid unfavourable features.

(2) Since site locations are dictated by the available natural features, distances from the ore bodies and

mines to the mill plant may be substantial. Short distances reduce the surface area that is affected by the overall mining and milling operations, with attendant reduced risk of contaminating the environment by spillage.

(3)   Material borrowed from the vicinity of the dam wall further increases the area of excavation and thus the environmental impact, unless the borrow area can be located within the tailings storage area.

(4)   Since valley dams are in topographically low areas, the visual impact is low.

(5)   Surface water run-off from the valley catchment is automatically collected in the impoundment area. Long-term seepage and erosion control require that such flow be diverted around the impoundment area. Where the catchment area is large in comparison with the impoundment area the failure potential associated with concentrated water flows can be relatively high. Additional maintenance and surveillance may therefore be necessary after shut-down of the mill.

(6)   As these impoundments are located in water courses, the capability exists for comparatively rapid transport of any spillage or seepage into the river system. The transport rate will depend on the catchment area flow conditions, the distance to the nearest river and the gradients applicable.

(7)   Since the valley forms three sides of the impoundment and only a relatively short dam wall is required to complete the impoundment, the impoundment basin is therefore cost-effective.

(8)   In most areas, for a given volume of tailings, a valley dam embankment will have to be relatively high, with subsequent high hydraulic gradients.

(9)   Because of the shape of the valley, tailings depths are shallow along the three natural sides and become progressively deeper towards the middle. The average tailings depth is therefore generally fairly low and the final surface area relatively high. This tends to increase the area of environmental impact, the surface area for seepage and wind erosion, and the volume of the cover required for stabilization.

(10)   The shape of the valley floor is often irregular and installation of any liner may be difficult and require considerable reshaping. Complex shapes also may render installation of a well-managed tailings facility more difficult and costly to operate. Where line discharges are used along the dam embankment, soft slimes zones usually form along the outer basin edges, creating placement difficulties and differential settlement problems for the cap.

(11) The form of the valley normally restricts seepage and surface flow to a known path or channel generally directed down and towards the centre line of the valley. Monitoring and measures to minimize or return seepage flows are therefore relatively simple.

(12) Owing to the shape of the valley floor the dam height varies over its length, resulting in differential loadings on the foundation and in the dams themselves. Care is required in dam design and construction to minimize differential settlement and possible resultant dam weakening.

(13) Deep alluvium deposits are often associated with valley floors. The horizontal permeability of such deposits is often high. This permits the rapid lateral migration of seepage water and makes water diversion channels and/or liners necessary.

(14) Valleys often occur as a result of some underlying weak geological feature, such as a fault or shear zone, which may also be a zone of higher permeability.

(15) Groundwater is usually shallowest in the valley floor area, and thus seepage from the impoundment has only a relatively short distance to travel before it joins the groundwater.

(16) For an open-pit mine the distance between the impoundment and the mine has a considerable effect on the cost of transporting overburden waste for either dyke building or covering and capping.

(17) The volume and cost of earthworks associated with covering a tailings impoundment usually exceed that required for embankment or dyke construction.

Types of Construction

The embankment to impound the tailings may be constructed of tailings, mine overburden or earth borrow materials. Fig. 3 shows the different forms of embankments and lists their advantages and disadvantages. An upstream construction method involves hydraulic deposition of tailings from spigotts. Example 1 (p. 55) describes an impoundment constructed by this method. The tailings are contained by peripheral dikes constructed of previously deposited tailings. Upstream construction is used at Climax Molybdenum mines in Colorado to construct embankments in excess of 120 m. However, the method is generally not favored for high dams in seismic risk areas, particularly as a result of the failure of several Chilean dams built by this method.

If the tailings are coarse enough to cyclone to

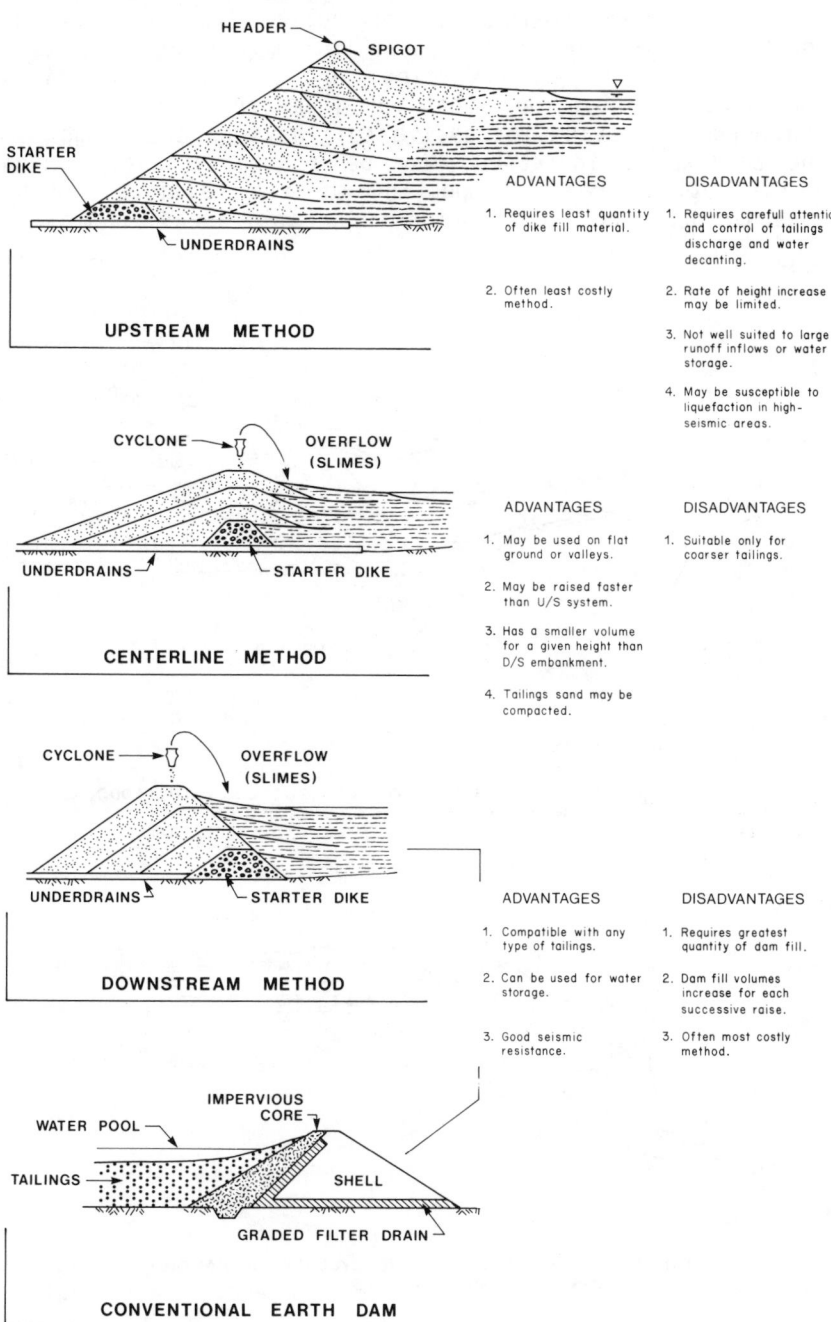

Figure 3. Types of embankments.

separate the coarser and finer fraction a centerline on-downstream embankment may be built. Fig. 4 shows two forms of centerline embankments. Both may be used in steep high valleys; good examples are the Brenda Mines impoundment in B.C, and that for Thompson Creek in Idaho. The need to provide successive spillways as the height of the embankment increases, may be a disadvantage in high precipitation areas. Compaction of the tailings may be done to preclude liquefaction in seismic areas. Example 2

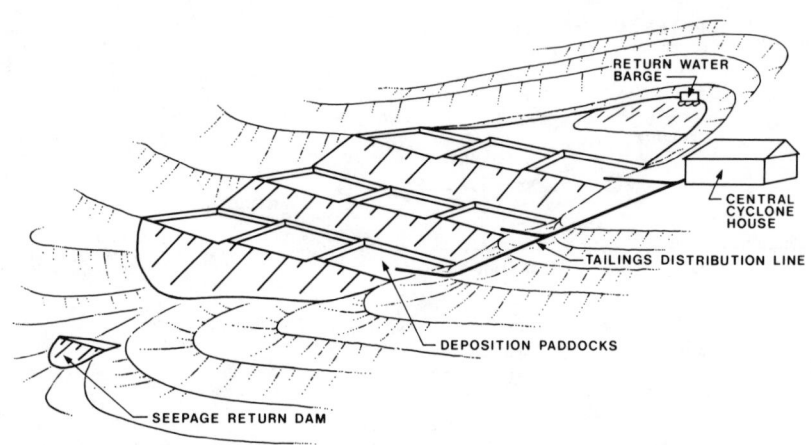

(a)   CENTERLINE EMBANKMENT CONSTRUCTION WITH PADDOCKS
AND CENTRALLY LOCATED CYCLONE STATION

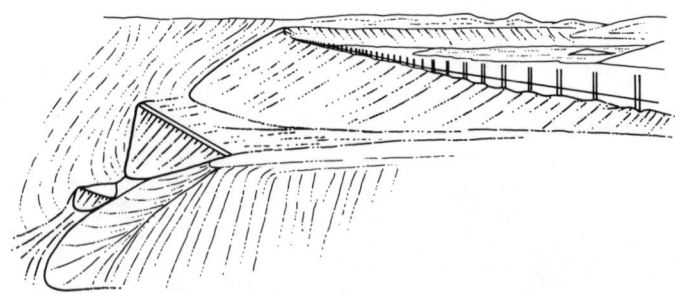

(b)   SCHEMATIC OF CENTERLINE CYCLONE OPERATION —
CYCLONES ON DAM

Figure 4. Examples of centerline construction.

(p. 55) describes an impoundment the embankment of which is constructed by the centerline method.

Downstream construction of embankments with tailings or borrowed soil or rock is favored for areas of high seismic activity or when the embankment will be high. If the tailings are too fine to be used to construct embankments, earth dam construction is necessary. The tailings may be too fine because of the grind required to free the ore or because the coarser fraction is removed to use as mine backfill. An embankment that can hold water can be constructed if a conventional dam is built. Spillways may be incorporated at successive stages of construction. The main disadvantage of downstream embankments is the large volume of construction material required and the associated cost. Example 5 (p. 59) describes an impoundment for which the embankment is constructed of rock and fill stripped from the area of the pit.

Tailings Deposition Options

Tailings may be deposited in an impoundment in one of three ways: dry, semi-dry or wet.

"Dry" disposal involves transporting and placing the tailings in an essentially dry state. Drying is done in the mill; tailings are transported by truck or conveyor to the impoundment and placed in layers. The pile is covered and reclaimed as tailings placement advances. As there is little or no water in the tailings, liners and other containment features may not be necessary. Example 3 (p.57) describes an impoundment where dry disposal is used.

"Semi-dry" disposal involves moving tailings as a slurry and using placement techniques and impoundment design features that cause the tailings to dry out once placed. This drying out results either from sun drying of unsubmerged tailings or controlled seepage of water from the deposited mass. Such designs do not involve liners or impermeable dikes or embankments; but they do provide for control of seepage and excess pool water. Potentially contaminated water is treated during the mining period. When deposition at the impoundment comes to an end the volume of water that could seep from the tailings is small and will continue to reduce as the tailings dries out.

"Wet" disposal involves depositing and containing the tailings in such a way that they are placed and are likely to remain wet over extended periods. The design of a wet tailings impoundment may involve liners and impermeable cores in embankments or containment dikes. Impermeable barriers are provided to contain and prevent movement to the surrounding environment of water that may be contaminated. Example 4 (p. 58) describes an impoundment where wet disposal is used.

Dry disposal is presently being used for disposal of coal and uranium mine tailings. Disposal costs are high (at least $3.00 per tonne) and the method is best suited to small operations in dry climates. Semi-dry disposal, too, is practical only where evaporation exceeds precipitation. Thus in areas of high precipitation wet disposal is the norm. If the tailings are a potential pollutant, impermeable barriers will be required.

Tailings Transport Systems

Tailings are usually conveyed as a slurry in a pipeline to the impoundment. Ideally the impoundment should be below the mill so that tailings slurry flow can be by gravity. If possible the line should be so arranged that tailings leaking or spilled will flow down overland or in a trench to the impoundment and not to the environment. In steep terrain or sensitive environmental areas the pipe conveying the tailings may be placed within a larger pipe. This can be expensive for high volume projects.

If the difference in elevation between the mill and impoundment is large, drop boxes may be used to dissipate the energy of the tailings. Ideally drop boxes too should be located so that spills flow or are washed to the impoundment.

If it is not possible, as frequently occurs in rugged terrain, to locate the mill above the impoundment, pumping of tailings will be required. Provision must be made to contain and deal with spills or leaks from pipes or to empty the pipe when pumping is inadvertently interrupted. The tailings may be thickened before pumping; this reduces pipeline sizes, the amount of water to be recycled, and the energy consumed in pumping.

Both for gravity and pumped slurry lines a major concern is achieving the correct slurry velocity. Too high a velocity leads to excessive pipe abrasion and wear and hence high maintenance costs. If the velocity is too low the solids will settle and the line will clog with sand. This can be a particular problem for upgrade pumping.

Pipelines in cold climates potentially subject to seismic disturbance require special design considerations. For example, that at the Climax Mine in Colorado, at an elevation of 3300 m, is designed for a temperature from $32^{o}C$ during August to $-51^{o}C$ in January. The system consists of 24,400 m of large diameter concrete and high density polyethylene pipe. All valves and extremities are heat traced and insulated. Freezing problems have not occurred.

Water Management

Most milling operations and tailings systems use large quantities of water. Water control and management is, therefore, a critical element in the tailings disposal system. Many physical failures have resulted from mismanaging water in the tailings impoundment. Failures due to mismanagement of water often result in water pollution, plant shutdowns, or emergency construction. Storing and handling water is often more troublesome for the operator than storing tailings. The higher the precipitation at the impoundment site the more crucial to the operation of the impoundment is a proper water management system. The higher the precipitation, the more difficult and costly water control will be.

Water Control Facilities

The most common water control facilities at a tailings impoundment are diversion trenches, spillways, decant towers, return water barges, and drains from beneath the impoundment or embankment. In dry environments diversion trenches and spillways are usually not needed. Where precipitation is high both will be required. The smaller the impoundment catchment area the better. A smaller catchment yields less inflow to the impoundment, hence the less the need, if any, for spillways and diversion trenches.

Diversion Trenches

Diversion trenches are constructed around the perimeter of an impoundment in order to reduce the accumulation of water on the impoundment. The trenches intercept surface runoff before it enters the impoundment and convey it downstream of the embankment. In high precipitation climates, it is common to employ such trenches. It is often necessary to install several trenches at different levels within the ultimate tailings impoundment. Staging ensures that a diversion trench is always close to the edge of the tailings water pool, and the portion of the catchment diverted is greatest. As the tailings level rises and drowns out the lower trench, a trench is constructed higher up.

Staged construction of diversion trenches must be integrated with embankment construction. Ideally, both trench and embankment construction are staged and the diversion trenches located just above the embankment crest. However, where embankments are constructed in one stage, pipes through the embankment must be provided, if several stages of diversion trenches are used.

Diversion trenches are generally sized to divert a medium-sized flood with a 50 to 200 year recurrence interval. This ensures that they have a sufficiently low probability of being overtopped during the operation of the mine but are not excessively large and expensive.

The advantages of diversion trenches are that they provide a convenient method for reducing water accumulation in the impoundment and are less costly than treating the excess water that would otherwise enter the impoundment.

The disadvantages of trenches, particularly in rugged terrain, are:

'trenches may be difficult and costly to construct;
'trenches generally increase the sediment load in the runoff; and
'routine maintenance is required and it may be difficult and costly.

Trenches in steep terrain where slopes are greater than 2 to 5% often require erosion protection by riprapping or lining unless they are cut into sound bedrock. Construction costs vary significantly depending on the size of the ditch, the nature of the material the ditch is cut into, the access that is available, the topography and the amount and type of lining that is required. A simple case of cutting a triangular ditch with a bulldozer at a slope flat enough to avoid erosion may cost on the order of $1.50 to $3.00 per meter for a 0.6 to 1.0 m depth. However, lined ditches of the same size in rugged terrain may approach $15.00 to $30.00 per meter or higher if substantial amounts of rock excavation are required.

Spillways and Decant Facilities

If the impoundment is in a high precipitation environment and its catchment area is large, a spillway may be required. If it is not possible to construct the impoundment embankment so as to ensure sufficient freeboard to contain at all times inflow from the probable maximum precipitation then a spillway must be built. The spillway is required to ensure that during extreme floods the impoundment can discharge water without endangering the integrity of the embankment. Nelson and Shepherd (4) note that in the long-term the most severe threat to a tailings embankment is flooding, and that the flood potential is most severe for valley embankments because of the size of the drainage they block.

The spillway may be designed not only to pass flood waters but also to control the operating water level in the tailings impoundment. In such a case the spillway usually takes the form of a decant tower. Flow from a spillway or decant tower used to control the operating water level may be treated before release to the environment.

Table II summarizes the types of spillways that may be used at a tailings impoundment and notes their advantages and disadvantages. Fig. 5 shows the schematics of the various types of spillways and decant towers.

Pipe outlets or decant towers are better suited to controlling the level of water in the impoundment than are spillways. They may be fitted with openings at specific levels, whereas the spillway level is fixed and water is kept at that level. Tailings deposition in the impoundment may have to be done under water. Pipe outlets and decant towers have certain disadvantages and are not generally favored, although they are used. They may block if not properly maintained. They may deteriorate and fail. In such cases they are difficult to repair. Other problems include: piping along the exterior of the decant pipe, pipe collapse, pulling apart of the joints, and riser breakage. Because the cost and difficulty of fixing and the risk of failure are high, a high factor of safety should be used for design of buried decant pipelines.

If the tailings embankment is to be constructed in several stages a separate spillway will be required for each. The expense of numerous spillways and the difficulties of construction scheduling may thus preclude the use of embankments constructed of tailings at impoundments where spillways are required. Indeed, if a natural saddle is used as a spillway, then even stage construction of an earth embankment may not be possible.

Barges

Because of the problems associated with decant towers as a way of controlling the operating water level on the impoundment, barges are often used. Barges float on the pool and the pumps mounted on them return water to the

Table II. Spillway Options.

Category	Type	Capital Cost	Potential for Staging with Embankment Construction	Impoundment Water Level Control	Other
Pressure flow	Decant tower or drop inlet spillway	High	Very good	Very good	Problems can be encountered due to blockage and damage caused by differential settlement, particularly in highly seismic areas.
	Discharge pipes	Low – medium	Limited. Construction of alternate outlet pipes at different levels is generally required.	Fair	Potential blockage problem. Tailings have to be discharged underwater if spillway is used for continuous discharge.
Open channel	Conventional	High	Good	Poor	Underwater discharge of tailings is required if spillway is used for continuous discharge.
	Channel type	Low – medium	Limited. Construction of alternative channels at different levels is generally required.		
	Natural saddle (if available)	Very good	None	Very poor	

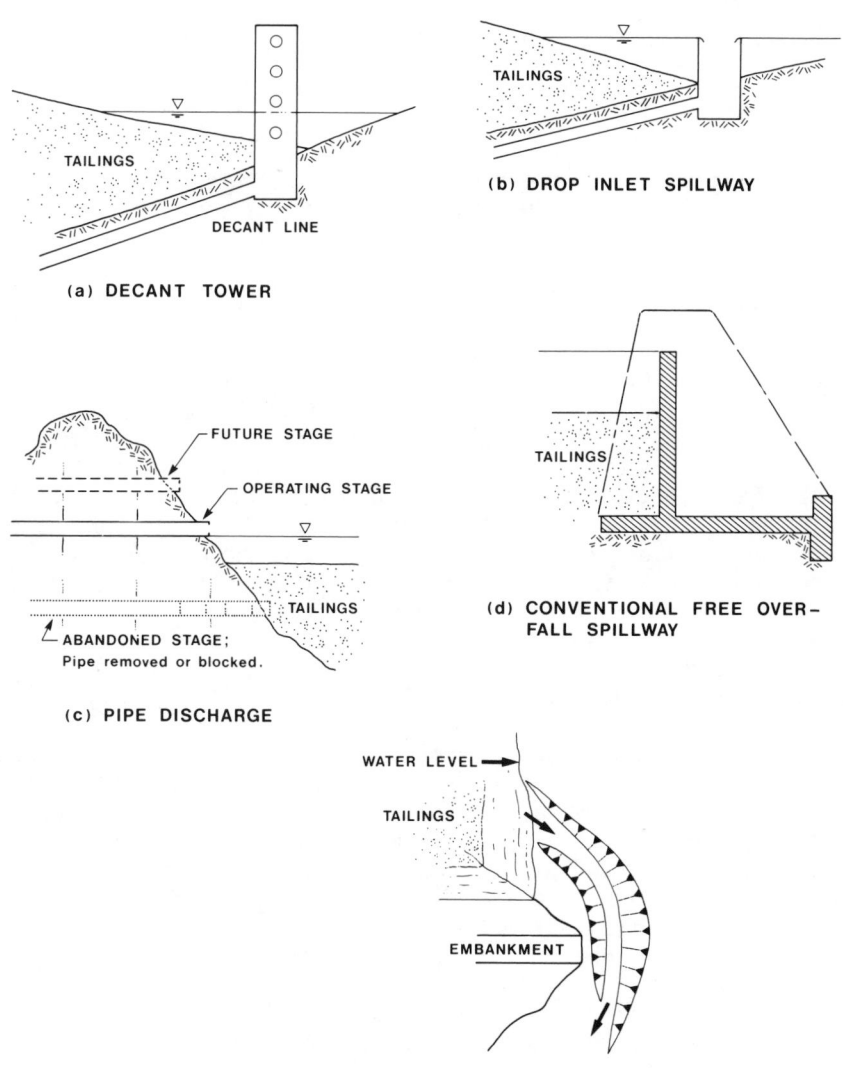

(a) DECANT TOWER

(b) DROP INLET SPILLWAY

(c) PIPE DISCHARGE

(d) CONVENTIONAL FREE OVER-FALL SPILLWAY

(e) CHANNEL SPILLWAY – PLAN VIEW

Figure 5. Spillway types.

mill for reuse or send it to treatment facilities for release to the environment. Barges on impoundments in cold climates have bubble systems to prevent them from being frozen in the ice that forms during winter on the impoundment. Barges may be used to pass flood waters. Limited pumping capacity and the potential for failure during flood inflow usually require that the impoundment be able to contain all flood inflow; the barge pumps can be used after the storm to pass the flood waters.

Underdrains

Underdrains are required to control the water pressures within the embankment and hence ensure the stability of the impoundment. They must be designed to prevent blocking or clogging, for if they do block, the resulting rise of the water level in the embankment could lead to its instability.

The quantity of water seeping from underdrains is usually small by comparison with quantities diverted, released over the spillway, or returned to the mill. Water from tailings impoundment or embankment underdrains may be high in potential contaminants. In such a case the water may be treated for release or pumped back to the tailings impoundment. Where long-term seepage from abandoned impounded tailings is undesirable, the embankment to contain the tailings should be constructed of suitable borrow and have a very low permeability core.

Tailings Impoundment Water Balances

Fig. 6 shows the components of a water balance model of a tailings impoundment. As shown, water flows into the system from a number of sources. It is stored within or on the tailings and is lost from the system by evaporation, seepage or discharge. The water balance system or model of an impoundment and mill can be analyzed as three interlinking cycles:

·the mill water cycle;
·the tailings impoundment mill cycle; and
·the tailings impoundment natural cycle.

The first cycle deals with the water input to the mill with the ore, from make-up water and by way of return flow from the impoundment. The tailings impoundment mill cycle deals with water input from the mill, the amount trapped in the deposited tailings and returned to the mill. The impoundment natural cycle deals with rainfall, runoff, evaporation and seepage. It is the only one of the three cycles that deals with stochastic, that is randomly variable, processes. Statistical methods are usually required to deal with this cycle.

Table III summarizes the results of water mass balance studies on three typical tailings impoundments which are either being designed or constructed in the U.S.

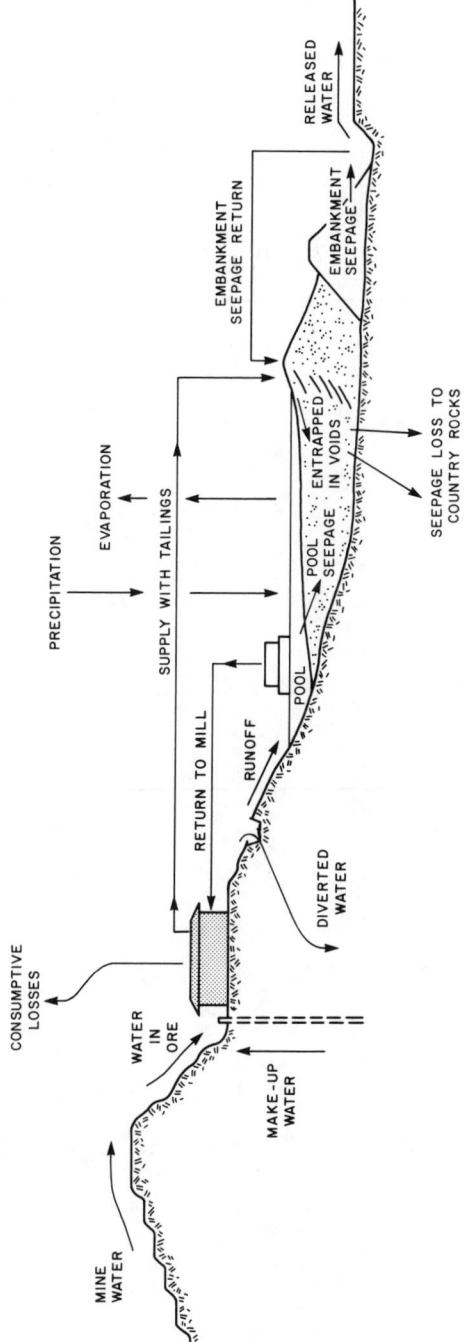

Figure 6. Water balance components.

Table III. Typical Average Annual Water Balances

Water Balance Item	Case		
	1. Wet	2. Moderate	3. Dry
1. Ore production - tonnes/year	127,000	1,600,000	8,300,000
2. Mill water - L/min			
2.1 Input			
ore	25	170	
fresh make-up	45	60	
tailings return	380	2,000	
2.2 Output			
tailings discharge	450	2,230	
2.3 Mill balance	0	0	
3. Tailings impoundment - mill cycle L/min			
3.1 Input			
tailings discharge	450	2,230	32,200
mine dewatering	7,660	0	3,200
3.2 Output			
trapped in tailings	70	600	3,700
return to mill	380	2,000	35,000
3.3 Mine cycle balance	7,660	-370	-3,300

Table III, continued

Water Balance Item	Case		
	1. Wet	2. Moderate	3. Dry
4. Tailings impoundment - natural cycle L/min			
4.1 Input			
precipitation	1,300	920	4,000
runoff	2,380	1,440	
4.2 Output			
diversion	2,300	650	0
evaporation	170	460	300
seepage	0	0	400
4.3 Natural cycle balance	1,210	1,250	3,300
5. Discharge from tailings impoundment L/min	8,870	880	0

The values listed represent the average annual water volumes over the life of the impoundment. Many of the values vary from year to year and show trends with time. Such details are not shown.

Case 1 is a very wet natural cycle as the precipitation at the impoundment is high. The mill cycle has a large surplus of water from mine dewatering activities; this is discharged into the impoundment. Consequently, a relatively large volume of 8870 L/min must be discharged. Even without the mine dewatering component, the impoundment generates excess water at the rate of 1210 L/min and 70 L/min per thousand tonnes.

Case 2 represents a slightly drier natural cycle. The mill cycle has a slight deficit balance (370 L/min) which means the mill removes more water from the impoundment than it contributes to surface storage. Overall, there is a slight surplus of 0.5 L/min per thousand tonnes which must be discharged.

Case 3 represents a relatively dry natural cycle with a small surplus coupled with a mill cycle which has a slight deficit balance. In this case, a zero total balance has been achieved and the system is, in fact, a closed circuit.

Water balances such as those shown in Table III can be evaluated very easily using regional hydrologic and climatic data, preliminary information on the mill process, and time averaged impoundment and catchment areas.

As the more detailed studies proceed, the mass balances computational process becomes correspondingly more detailed. The tailings impoundment system can be divided into subcomponents such as the main tailings impoundment, the tailings embankment and the seepage return dam. Computer models are used to perform stepwise solutions to the mass balance equations, thus simulating a time sequence that shows how the tailings impoundment develops.

The Stability and Failure of Tailings Impoundments

To assess the stability of a tailings impoundment, possible causes and modes of failure must be identified. The factor of safety against such failure occurring must be calculated and the consequences of failure, if it were to occur, evaluated.

Embankment instability is a major potential cause of failure of impoundments. Embankment failure may occur because of:

- inadequate embankment material or foundation strength;
- high water table or water pressures in the embankment;
- liquefaction of the soils due to earthquake; or
- embankment erosion from flood overtopping, spigotting errors, pipeline failure or runoff from high intensity rainfall.

Spillways, diversion trenches, and decant lines may be too small to cope with inflow from extreme precipitation, and thus may be a cause of impoundment failure. Structural failure of spillways and decant lines may also lead to failure. For example, a decant line may fail as a result of piping along the outside of the pipe, breaking of the pipe from load or settlement, tower failure, or spreading at the pipe joints.

The factor of safety against failure by these modes or causes may be done by standard engineering analytical methods. The majority of large failures have occurred on embankments which had little or no engineering. Most significant failures can be prevented with proper design, diligent construction, observation and monitoring, and providing adequate supervision.

Many regulations and agencies require an assessment of the consequences of failure of an impoundment. One must, by law, ask and answer what would happen if complete failure of the tailings impoundment were to occur. Thus one must say that if a large impoundment in a steep valley were to fail, in a short while much of the tailings would flow into the inlet, and in the longer term the high rainfall of the site would wash the remainder into the sea. There is a growing body of knowledge about the flow of tailings from breached impoundments; thus it is now possible to predict how far and in what manner tailings would flow from a failed impoundment. If necessary, the design can take account of such predictions.

Some jurisdictions permit an assessment of the probability of failure rather than of the consequences. This too can be done with the advances now being made in probability and its application to dams and tailings impoundments. In many instances the probability of failure may be so low as to be difficult to comprehend, interpret or act on.

THE CONSTRUCTION AND OPERATION OF TAILINGS IMPOUNDMENTS

Construction Stages

The construction of a tailings system usually involves the following phases:

- construction of Stage I facilities;
- continuous operational construction and placement of tailings, or construction of subsequent stages; and
- construction of permanent stabilization for closure.

Initial Construction

Construction of Stage I facilities is usually started one or two years in advance of tailings deposition. This phase is commonly performed by independent contractors. Initial construction normally involves the construction of underdrains, a starter embankment, tailings delivery and reclaim water pipelines, a seepage recovery and return system, and a water control system which may include decant lines, spillways, or interceptor ditches.

Common Construction Problems

During initial construction, common problems include providing water diversion and turbidity control, quality control of underdrain construction, material gradation and moisture control from borrow areas, and scheduling materials and activities to avoid conflicts or delays.

Weather is usually a serious constraint. Water diversion plans must accommodate a reasonable design storm during construction, especially in the case of starter embankments constructed across natural stream channels. If it is not practical to include temporary culverts through the embankment, it may be necessary to provide a standby pumping system or temporary spillways to protect the embankment from overtopping.

Initial Tailings Deposition

The most critical period for a tailings embankment is usually the start-up period. Extreme care must be exercised during start-up to avoid washout or blinding underdrains with tailings slimes. In the case of a new mill, start-up is often complicated by variable mill output, inexperienced personnel and a variable tailing gradation.

Valley impoundments often have a rapid rate of rise during the first few months of operation. This may dictate many pipeline raises, additions to monitoring access pipes, reclaim barge raises, and underdrain extensions. Seepage through underdrains may be large during initial deposition.

Continuous Tailings Deposition

The construction of a large tailings impoundment is a major undertaking. The scale of operation, the high capital investment and the consequences of such liability for accidental discharge or failure, demand application of the best principles of sound engineering. Management tools such as computerized water and mass balance models assist, but ultimately it is the skill, experience and intuition of the operator which assures success.

Continuous tailings embankment construction involves these elements:

- a thoroughly designed system and disposal plan from initial disposal to closure;
- an operating plan which explains the operational objectives of the disposal system to the mine staff; and
- a systematic inspection, monitoring, reporting, and documentation system.

Slurry Distribution within the Impoundment

The distribution of the tailings within the impoundment is an important element in impoundment construction. The solids must be placed in a manner which is compatible with the embankment construction method and the water management system.

For most upstream construction techniques, a header and spigots are used to distribute the tailings. As the embankment rises, the spigot pipes are extended from the header until the head losses in the small diameter pipes become too great to continue deposition, so the header pipe is raised and spigotting continues. The control of the water pool is done by spigotting at various points.

Sand distribution is the prime concern in downstream or centerline methods. The cyclone underflow may be too dewatered to flow to the desired position. The sand may be repulped by adding water or must be moved mechanically. In high precipitation areas erosion of the sand may be a major problem.

Impoundments in high precipitation environments, in a steep valley and with an embankment formed of borrow material, may involve subaqueous deposition of the tailings. The spillway is likely to be some height above the tailings, which will accordingly be covered with water. Discharge may be done from a single point. In this case tailings may rise above the water. The beach above the water will be flat, but will be steep where it meets the water. Distribution of tailings into the deep pool that forms may be a problem. A ring delivery system is preferable. Discharge can be done at a number of places around the impoundment and can usually be done below water. If the water freezes subaqueous discharge can continue.

Common Operational Problems

Table IV provides a list of operating problems which have been experienced in several operations. The list does not contain all possible problems nor can the remedial actions indicated represent all possible solutions. Proper engineering should accompany any remedial measure taken.

When embankments are built using tailings as a construction material, most of the day-to-day problems are usually associated with placing material of proper gradation in the correct location. Cyclone operation and maintenance to obtain sands of adequate permeability demand constant attention.

The Cost of Tailings Disposal

Tailings disposal does not produce revenue for a project; thus the aim is to keep disposal costs as low as possible. The cost of tailings disposal varies greatly; for some large scale operations the cost can be as low as $0.05 per tonne, whereas the disposal of small volumes of highly contaminated tailings can cost as much as $15.00 per tonne.

The three major cost items in tailings disposal relate to design and construction, operation and reclamation. Table V gives capital (construction) and operating costs for a range of disposal schemes. Accurate data for reclamation costs for these projects are not available. Estimated reclamation costs range from $.20 to $1.00 per tonne.

Marine disposal costs are likely to be less than those for on-land or mine backfill disposal. The high capital costs of the outfall and the operating cost of environmental monitoring of the waters into which the tailings are discharged, are the major cost items of marine disposal.

Fig. 7 is a comparison of the cost of embankment construction by different methods. Upstream construction requires the least volume of material to construct an embankment. A downstream embankment has a volume at least six times as much as an upstream embankment. If the embankments are constructed of tailings, the ratio of embankment cost is similar to the ratio of embankment volumes. If the downstream embankment is constructed of imported borrow or fill, placement costs could be three to four times that of embankments made of tailings. Thus if an upstream embankment of tailings is taken as the base case, an embankment of compacted soil or rock could cost upwards of twenty times the base case.

Table IV. Common Operating Problems

Description	Potential Cause	Possible Remedial Action
I. Tailings delivery system		
A. Line blockage/ sanding	Insufficient gradient, mill output fluctuations.	Regrade pipeline, remove or pressurize drops, install pumps, change pulp density, change pipe size, change pipeline material.
B. Excessive line wear	Too much gradient (velocity).	Install energy dissipating drops or orifice plates, line pipeline, regrade line.
C. Wear in vertical drops	Impingement of tailings stream on side of drop.	Install conveyor belt baffle, incline drop, install wear resistant lining.
D. Burping and uneven spigot discharge	Entrained air with tailings stream, plugging of spigot pipes at pinch valves.	Install air vent downstream of last drop, open first spigot on embankment, operate pinch valves with jaws vertical.

Table IV, continued

Description	Potential Cause	Possible Remedial Action
II. Cyclone system		
A. Over-steepening	Loss of water content as slurry fans out.	Cut trenches down face with small dozer to prevent fanning, add dilution water to underflow, move mechanically.
B. Erosion trenches	Improper cyclone adjustment. Too much water in underflow. Overflow not free flowing.	Adjust or replace cyclones apex, decrease number of cyclones to increase pressure.
C. Downstream surface sliding in sheets	Underflow contains too many fines, flow lines parallel to surface.	Adjust or replace cyclones apex, disturb low permeability layers mechanically, clean up underflow.
III. Embankment		
A. Crest settlement	Normal consolidation of foundation or embankment. Embankment instability.	Add material in low spot; add toe berm, flatten slope, decrease phreatic surface, move water pool farther from crest, monitor.

B. Differential settlement and cracking normal to embankment.	Uneven settlement of foundation.	Move water pool away from embankment, cracks in tailing should self heal, monitor, decrease rate of rise.
C. Traverse cracking parallel to embankment	Differential consolidation of coarse and fine tailings.	Monitor, cracks should be self healing.
	On downstream face, could be indication of slide scarp.	Add toe berm, flatten slope, decrease phreatic surface, move pool away from crest, horizontal drains, monitor, check underdrains.
D. Toe bulging	Embankment instability.	As above.
E. High phreatic surface (water table in dike)	Pool too close to crest, blocked underdrains, excessive rate of rise, deposition too long in one area.	Move pool away from crest, install vertical sand drain, install well point dewatering system, deposit on another impoundment, flatten slope to provide stability, monitor.
F. Seepage exiting face	Same as above, melting of ice lens, layering.	Same as above; if seepage is dirty or carrying solids install inverse filter.

Table IV, continued

Description	Potential Cause	Possible Remedial Action
IV. Impoundment		
A. Filling faster than anticipated	Lower than predicted in-place tailings density.	Reduce area covered by water pool to promote consolidation by evaporation, reduce rate of rise, install vertical drains, redesign embankment if necessary to provide flood protection.
B. Dust	Wind drying and pickup of cohesionless particles.	More frequent wetting by moving discharge point, install irrigation system, add crusting agent to slurry or to dry surfaces, flood impoundment by increasing pool size.
C. Pooling of fines near crest	Fanning of coarse particles forming deltas blocking normal sheet flow, uneven spigotting.	Operate spigots or cyclones in panels, move discharge points more frequently, install training dikes with dozer.
D. Depression in beach	Melting of ice lens, foundation subsidence, material loss due to piping into decant pipe or foundation.	Move pool away from depression, perform geotechnical investigation through tailings, survey and monitor, check for mine workings.

E. Too much water on impoundment	Flood, wet season, large drainage area, changed water balance.	Decrease mill make-up intake and increase pumping from impoundment, install diversion ditches, treat and release, build separate storage reservoir for temporary surpluses.
V. Rock underdrains		
A. Not flowing	Blinded by slimes cover, no phreatic surface present at particular location, blocked or sheared drain.	Install vertical sand drain into underdrain (carefully), abandon, install horizontal well from downstream embankment face, monitor.
B. Seepage carrying fines	Graded filter disrupted or improperly installed.	Starting at farthest practical point upstream of embankment, drill and pressure grout a section of drain. Monitor. If seepage does not clear up, repeat until seepage is clear or drain is completely grouted.

Table IV, continued

Description	Potential Cause	Possible Remedial Action
VI. Monitoring		
A. Sudden increase in piezometric level at one location	Reference elevation error, arrival of a wetting front, actual rise as indicated, operator error, equipment malfunction.	Retake reading and check reference elevation, read more frequently, replace instrument.
B. Embankment movement indicated by surveying	Raising crest is often associated with downstream strains, downward movement may be normal consolidation, foundation or embankment failure, survey point disturbance, survey error.	Compare with previous readings, triangulate reference points, recheck on more frequent basis, inspect downstream embankment.

Table V. Waste Disposal Costs

Case	Costs	Metal
TAILINGS		
Case 1		
Capital	$7.00 - 13.00	Silver and Gold
Operating	.17 - 3.40	3.4×10^6 tonnes
Case 2		
Initial capital		
:wet	$ .23 - .80	
:dry	.35 - .70	Uranium
Total capital		
:wet	2.55 - 3.60	21×10^6 tonnes
:dry	2.30 - 4.55	
Case 3		
Capital	$ .64 - .77	Cu/Zn
Operating	.08 - .09	32×10^6 tonnes
Reclamation	.32 - .34	
Case 4		
Capital	$ .85 - 1.05 (max. 4.44)	Gold
Operating	.15 - .18 (max. .22)	26×10^6 tonnes
Case 5		
Capital	$ .05 - .10	Cu/Mo
Operating	approx. .05	300×10^6 tonnes

Table V, continued

Case	Costs	Metal
ROCK		
Case 6		
Capital	$ .20 - .44	
Operating	.24 - .60	100×10^6 tonnes

ENVIRONMENTAL CONTROL AND RECLAMATION

Pollution Control

Control of pollution from a tailings impoundment involves precluding or reducing to as low a level as practicable release from the impoundment of water or tailings. Water may be released from an impoundment as a result of:

    ˙planned discharge;
    ˙inadvertent spills;
    ˙overflow of spillways during floods; and
    ˙seepage.

Tailings may be released from an impoundment as a result of:

    ˙wind (airborne release);
    ˙erosion; and
    ˙failure of the containing embankment.

In a rugged environment of high precipitation, all of these are possible, although wind dispersion of the tailings is unlikely except during occasional dry periods.

Water Quality Requirements for Release

Water stored in a tailings impoundment is generally of a lower quality than the natural surface and subsurface waters.
If continuous discharge of water from the tailings

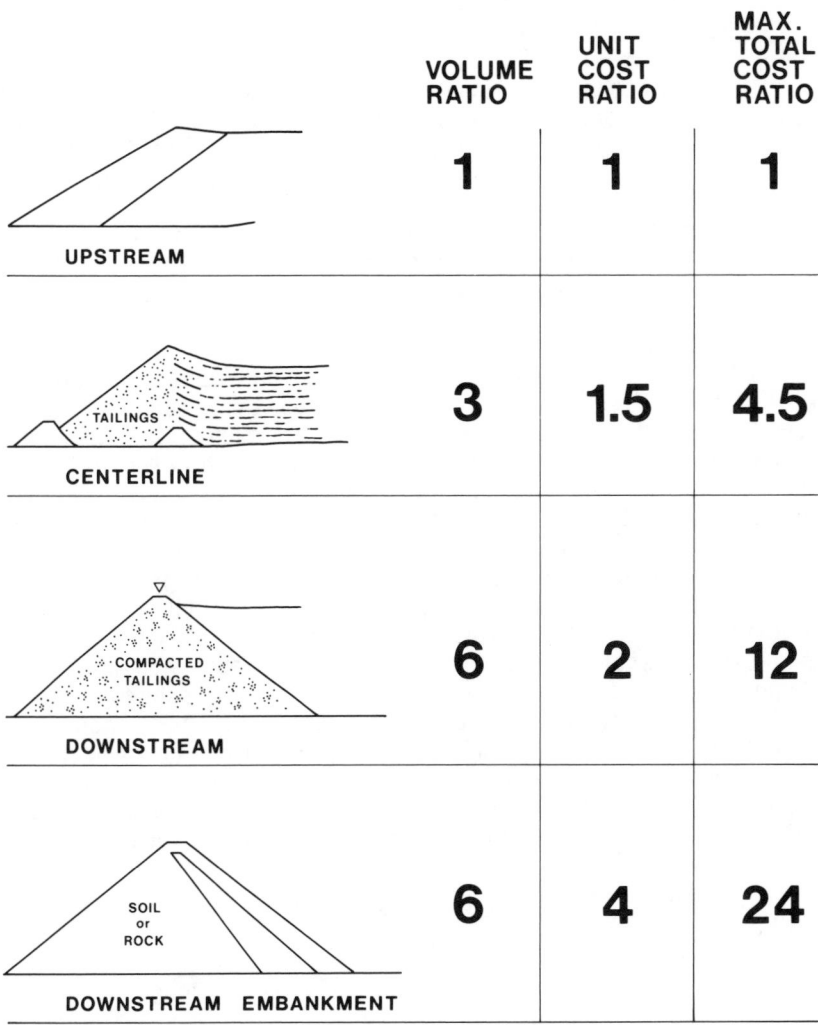

	VOLUME RATIO	UNIT COST RATIO	MAX. TOTAL COST RATIO
UPSTREAM	1	1	1
CENTERLINE	3	1.5	4.5
DOWNSTREAM	6	2	12
DOWNSTREAM EMBANKMENT	6	4	24

Figure 7. Embankment cost comparison.

impoundment to a natural drainage system is done, state and federal regulations require that the quality of the water meet the more stringent of the two following criteria.

1. Effluent limitations guidelines: EPA guidelines require that effluent quality does not exceed the concentrations of the parameters listed in Table VI. Values are based on Best Practicable Treatment

(BPT) technology.

2. Stream water quality standards: These are generally set by state agencies and indicate the criteria for stream or surface water into which the effluent may be discharged. Stream water quality should not exceed these criteria once the effluent is added and dilution occurs. Where actual stream quality is worse than the applicable standards, before the effluent is added, the state generally requires that adding the effluent to the stream should not further degrade the stream quality.

It is necessary to apply for, and obtain a National Pollutant Discharge Elimination System (NPDES) permit before any discharge to the environment can be done. Usually, with the mine wastes associated with base and heavy metals milling, water treatment is required before discharging the water. In issuing the NPDES permit, the state agency will generally incorporate monitoring and reporting requirements and indicate the degree to which stream flow dilution will be allowed, and the process for controlling this dilution.

Many states are prepared to accept non-continuous type discharges, such as emergency discharge during extreme flood events, provided they do not occur more frequently than a specified return period, which can vary from 100 years upwards. The philosophy behind this approach is that during such extreme flood events, there will be a substantial amount of water flowing in the

Table VI. Effluent Limitations (EPA Guidelines)

Effluent Characteristic	Maximum for any one day	Average of daily values for 30 consecutive days shall not exceed
	Milligrams per liter	
TSS	30	20
Cu	.30	.15
Zn	1.5	.75
Pb	.5	.3
Hg	.002	.001
pH	within the range 6.0 - 9.0	

TSS = Total Suspended Solids

natural water course, thereby providing adequate dilution.

Regulations for seepage into the groundwater from the impoundment are generally less well-defined than for surface water. Agencies often indicate that they will accept certain maximum seepage rates (10^{-6} to 10^{-8} cm/sec). In other cases, they may insist that the mine guarantee that no pollution of the groundwater will occur.

If natural ground conditions do not provide sufficient protection, engineering measures, such as the placement of clay or membrane type liners within the impoundment, downstream cutoff walls or slurry trenches, and downstream collection wells, can be used to deal with any potential problems. When designing such measures, it is important to consider the chemical attenuation of the constituents in the wastewater and the dilution of the effluent by the groundwater flow under the impoundment. These aspects usually contribute significantly to reducing the impact on groundwater quality. As they are generally difficult to predict with accuracy, it is important to devote funds to obtain field data and undertake analyses to firm up these aspects during the baseline studies.

Reclamation of Tailings Impoundment

When mining stops and all tailings are deposited the impoundment must be reclaimed. Table VII lists some objectives that have been adopted in tailings impoundment reclamation. It is seldom possible to achieve all of them. A tailings impoundment becomes a new topographic form in the environment. A well reclaimed impoundment will have a shape similar to the natural landforms of its surroundings. And to the extent that this is not possible, the impoundment should at least respond to the natural processes of landforming in an acceptable way.

A water dam, when it is old, can be breached, its contents discharged and what was, will no longer be. This is not possible with a tailings impoundment; it cannot be breached, its contents cannot be discharged, and what is deposited must remain where placed. Thus engineering solutions for tailings impoundments should provide for permanent retention of the tailings. Advanced engineering design can incorporate qualitative consideration of the effect of predictable geomorphological and climatological processes on the integrity of the impoundment system. Typically, this may be no longer than 10,000 years. Geomorphological processes over that period are, however, such that even the most soundly engineered impoundment system cannot be expected to ensure complete retention of the tailings. This is particularly true of impoundments in

steep valleys where great volumes of runoff from high
precipitation will flow over the impoundment. Ultimately
the processes of valley formation will seek to reestablish
a valley, and the tailings will become sediment in the
ocean.

Table VII. Tailings Impoundment Reclamation Objectives.

A reclaimed tailings impoundment ideally should:

(1) provide physical and environmental stability;
(2) return the land to productive status;
(3) fit post-mining land use objectives;
(4) comply with state and federal regulations;
(5) meet local environmental quality and safety standards;
(6) rehabilitate disturbed lands and enhance the visual
    character of the area;
(7) assure protection of water quality in the watershed
    and adjacent land;
(8) control erosion and sediment runoff;
(9) prevent seepage loss;
(10) be cost effective;
(11) provide for the establishment of a permanent
    vegetative cover of high diversity requiring little
    or no maintenance and which is capable of
    self-regeneration; and
(12) ensure the safety of future land users.

To accomplish successful tailings impoundment
reclamation, comprehensive preplanning is needed. Site the
impoundment to avoid steep slopes and large watersheds. If
there is only one option that is technically and
economically feasible, the luxury of choice of sites does
not exist.

Flexibility does exist in determining the surface
features such as relief, drainage, soils, plant cover, and
aspects that remain after mining is complete. To avoid
long-term post-mining sediment problems, these features
should be designed and installed in a manner compatible
with the off-site drainage system, local climatic
conditions, and the physical and chemical properties of
the construction materials.

The tailings impoundment must be designed to promote
the greatest possible surface runoff and to reduce
infiltration. An important factor to be considered in the
design of the drainage system for the tailings impoundment
is the location of stable areas on which to dispose of

surface runoff collection. These include: (a) well vegetated and relatively flat areas that will tend to slow, spread and filter the water discharge and runoff; (b) well armored (riprap) rough and partially vegetated surfaces that are highly resistant to erosion and dissipate flow energy by spreading and slowing runoff; and (c) natural drainage ways that can handle increased flow with little or no increase in channel erosion.

A second erosion and sediment control principle is to expose the smallest practical area of land for the shortest possible time. If possible, the reclamation of embankment surfaces should be accomplished soon after construction. Valley fill embankments should be stabilized at the end of construction of each bench, beginning at the toe and progressing upwards. Staging the revegetation operation in this manner provides a place to trap sediment from higher, unprotected areas, as well as prevent rainfall and runoff from removing soil from lower sections of the slope.

The area of land exposed to erosion is determined by the geometry of the embankment. The vertical interval between benches and the bench widths for lateral drainage must be considered.

Approaches to reclamation of tailings in southeast Alaska are strongly affected by basic environmental factors including climate, soil parent material, and native biota. The particular problems to be faced in tailings revegetation are acidity, heavy metals, and in the case of some recent gold-extraction operations, cyanide. Each of these problems requires case by case analysis to determine the most effective and economic course of action.

Acids residual from an extraction process can be leached from the surface with supplemental irrigation or, in locations such as southeast Alaska, by the abundant natural precipitation. Acids produced by weathering of pyrite can also be leached but may be balanced by the addition of lime; this may be expensive at sites distant from sources of lime. Acid tolerant species such as articgrass or bluejoint reedgrass may be used in Alaska.

Stabilization of tailings containing potentially toxic heavy metals such as copper, lead, or zinc is possible with conventional reclamation techniques. Highly efficient modern extraction techniques result in low concentrations of heavy metals, and this facilitates reclamation. On tailings from older processes or where the heavy metal is not one of the materials extracted, revegetation may involve metal-tolerant varieties such as Merlin red fescue. Organic top dressings such as sewage sludge have been used to ameliorate the heavy metals problem; organic molecules from the sludge complex to prevent plant uptake of the heavy metals, at least temporarily. Top dressing with waste rock of low metal content or, preferably, with topsoil, is probably the most effective long-term solution.

COMPARISON OF TAILINGS DISPOSAL OPTIONS

General

In this section, the differences between alternative tailings disposal methods are discussed. Emphasis is placed on the differences between marine disposal and land disposal in rugged, seismic, high rainfall terrain. For comparison, reference is made to in-pit tailings disposal and impoundments on flat terrain and disposal of "dry tailings" in rolling country.

There are probably no mining projects where the tailings engineer will have to choose between disposal in the sea, in steep valleys, or on flat land, either by wet or dry disposal techniques. However, there are projects where the choice is between disposal in the sea or steep valleys; and there are projects where the choice is between disposal in a steep valley or on flat ground. When disposal is on land, there is always a choice between dry or wet disposal. Hence, although the comparisons made in this paper are hypothetical, they are not entirely without practical application as some of the examples described illustrate.

The method used to compare disposal methods is based on a site selection procedure used in numerous cases by Steffen Robertson and Kirsten and associated firms. This method is described in the first part of this paper.

Comparison of Disposal Options

The tailings impoundment site selection method described above involves a qualitative evaluation of each site. For a list of defined factors, five categories are defined ranging from very good to poor (or very low to very high impact). A similar method is used to compare alternative tailings disposal options. For the list of factors in Tables VIII and IX, the impact of each disposal method may be evaluated. The considerations leading to an evaluation of poor, good, etc. is based on that described by Robertson and Moss (1). A typical set is shown on Table X.

Table VIII. Environmental Impact Comparison of Tailings Disposal Options.

Environmental Factors	Deep Water Disposal	Flat Terrain	Rugged Terrain, High Rainfall, Seismic Area	In-pit Disposal	Dry Disposal in Rolling Country, Moderate Rainfall
Visibility	Very low	Very high	Moderate	Low	Moderate
Land use/ecology					
Area of disturbance	Very high	High	Moderate	Low	High
Long-term effect on:					
vegetation	Very low	Moderate	Moderate	Very low	Moderate
wildlife	Moderate	Moderate	Moderate	Very low	Moderate
human use	Very low	Moderate	Moderate	Very low	Moderate
Airborne release potential	Very low	High	High	Moderate	Very high
Seepage release potential	Very high	High	Low	Low	Very low

Table VIII, continued

Environmental Factors	Deep Water Disposal	Flat Terrain	Rugged Terrain, High Rainfall, Seismic Area	In-pit Disposal	Dry Disposal in Rolling Country, Moderate Rainfall
Surface release pollution characteristics					
Concentrated flow	Very low	Low	High	Very low	Low
Erosion	Very low	Moderate	High	Very low	High
Foundation failure	Very low	Low	Moderate	Very low	Low
Liquefaction	Very low	Low	Very high	Very low	Very low
Semi-qualitative evaluation					
Total	21	38	41	17	31
Average	1.9	3.5	3.7	1.6	2.8

Table IX. Cost and Operational Comparison of Tailings Disposal Options.

Environmental Factors	Deep Water Disposal	Flat Terrain	Rugged Terrain, High Rainfall, Seismic Area	In-pit Disposal	Dry Disposal in Rolling Country, Moderate Rainfall
Operational storage capacity	Very good	Moderate	Good	Poor	Moderate
Need for construction materials	Very low	Moderate	Very high	Very low	Low
Expansion capacity	Very good	Good	Moderate	Poor	Good
Surface drainage control requirements	Very low	Moderate	Very high	Moderate	Moderate
Operating difficulties	Low	Moderate	High	Very high	Moderate
Semi-qualitative evaluation					
Total	6	14	19	17	13
Average	1.2	2.8	3.8	3.4	2.6

Table X. Typical Evaluation Criteria.

Element	Description	Very high(5)	High (4)	Moderate (3)	Low (2)	Very low (1)
1. Aesthetic features						
1.1 Prominence	Topographic locations and angle through which the site is visible	High in the landscape with no screening and visible through 360°	Moderately high in the landscape with little screening	Moderately high in the landscape but well screened, or low in the landscape with little screening	Low in the landscape and well screened	Essentially not visible from a short distance
1.2 Visibility	Location relative to public facilities such as cities, roads or parks	Visible from towns or cities	Visible from highly used highways or parks	Visible from moderately used roads or public facilities	Visible from low use public facilities	Essentially no public visibility
1.3 Scenic value	Identified as an important area by agencies or by public use	Official designation	High use	Moderate scenic value	Low scenic value	No scenic value

2. Land use and ecological factors					
2.1 Recreational areas	Type of use, intensity of use, and proximity to site (mi.)	Very highly used camping, fishing, picnic areas	Highly used areas	Moderately used areas	Low use areas	No use within influence of site
2.2 Historic or archaeological sites					
2.2.1 Historic value	Designated or potential historic sites	Designated historic site	Known historic site	Old structures present	Isolated old structures present	No old structures
2.2.2 Archaeologic value	Designated archaeologic site, or found to contain sites of archaeologic value	Designated archaeologic site	Locations of archaeologic interest on site	Potential for archaeologic sites	Low potential for archaeologic sites	No archaeologic sites suspected

Table X, continued

Element	Description	Very high(5)	High (4)	Moderate (3)	Low (2)	Very low (1)
2.3 Operational conditions						
2.3.1 Technical feasibility	Evaluation of difficulties of developing any disposal plan for the site	Requires innovative and untried technology	Requires appreciable innovative application of established technology	Requires moderate innovative application of established technology	Requires little innovative application of established technology	All proven standard technology
2.3.2 Availability of construction materials	Evaluation of soils and rock near site for construction of liner, embankment and reclamation cover	None available; very long haulage necessary	Long haulage necessary	Moderate haulage necessary	All materials available nearby	All materials on site
2.3.3 Snowdrift areas	Identification of areas of snow accumulation as a constraint	Deep snowdrifting (over 30 ft.)	Accumulation greater than average snowfall	Moderate snow accumulation equal to average snowfall	Accumulation less than average snowfall	No significant accumulation of snow

There is room for argument about the definition of ranking considerations and about the assignment of a particular scale to each. Those given reflect the subjective opinion of the authors. Others may wish to change definitions and scales and hence test the sensitivity of the evaluation system to different opinions or judgment. As done in Tables VIII and IX a system is established for comparing marine tailings disposal to disposal in areas of high precipitation and rugged topography, or any of the other disposal options considered. Table VIII considers primarily on-land environmental impacts. Table IX compares operational and cost considerations.

In order to obtain a semi-quantitative evaluation, hence ranking, a number is assigned to each factor; 1 for very good or very low impact through 5 for very poor or very high impact. This has been done for Tables VIII and IX. Totals and averages are given. No weighting is done, except to the extent that all factors are considered equal. Indeed, the ratings for the two categories considered are not added but are viewed separately. For the scales given, it is clear that marine disposal ranks ahead of on-land disposal in areas of high precipitation and rugged topography. To the extent that such numbers have meaning, marine disposal may be ranked higher by a factor of two to three.

Examination of the tables and the associated ranking criteria list enables an assessment and identification to be made of those areas where marine disposal is better or worse than on-land disposal. Land disposal results in an entity that is visible, which does impact large land areas and can have both short and long-term effects on wildlife, vegetation and the potential for human use. By comparison tailings disposed of in the marine environment are not visible and only the pipeline and outlet corridor may be said to affect the land environment, and that area is small relative to the area of an impoundment.

Airborne release of tailings from a land impoundment can occur, but is most unlikely for a marine disposal system. Seepage release is a function of bedrock; in a valley bedrock is likely to be competent and release small. Seepage from marine disposed tailings to land groundwater is not possible. Similarly, surface release of tailings from a valley impoundment can occur. Tailings in an inlet on the sea is where the ultimate landforming processes will seek to place tailings disposed of on land.

Table IX, comparing the operational and cost considerations of tailings disposal systems, shows that marine disposal usually provides a greater potential storage and expansion capacity than most land disposal options. The need for construction materials for marine disposal is negligible by comparison with land disposal systems. No surface drainage problems, except perhaps those associated with the pipeline to the outfall, occur for marine disposal. This is, however, a major problem with impoundments in high precipitation areas and steep valleys. Operation of on-land impoundments is inevitably more troublesome and difficult than marine disposal. The final numerical ratings are a measure, however imprecise, of the extent of the advantages and disadvantages of the various systems.

CONCLUSIONS

The comparative assessment made in this paper of different tailings disposal options is simple and certainly not adequate to substantiate a conclusion that in all cases, marine disposal is preferable to on-land disposal. Each case should be considered in detail on its merits. A complete assessment of the impact of different disposal options is required for the site-specific circumstances. The methods used here may be appropriate, but a particular environment, regulations or public attitude may dictate a need for other or more sophisticated approaches to the matter.

Many more sophisticated methods of choosing options involving multiple objectives and assessing conflicting preferences exist (5). To use them requires time, money and the considered input of specialists and laymen. Such an approach is not warranted for our simple comparisons; they may be required for real life situations.

REFERENCES

1. Robertson, A.M. and A.S.E. Moss. "Site Selection and Optimization for Mill Sites and Tailings Impoundments", Asian Mining Conference (1981).
2. Arrangement by Thurber Consultants Ltd., Geotechnical Engineers.
3. International Atomic Energy Agency. "Current Practices and Options for Confinement of Uranium Mill Tailings", Technical Report Series No. 209 (1981).
4. Nelson, J. and J.J. Kane. "Failure of the Church Rock Tailings Dam", Proc. of the Third Symposium on Uranium Mill Tailings Management (Colorado State University, 1980), pp. 505-511.
5. Keeney, R. Siting Energy Facilities (NY: Academic Press, 1980).

EXAMPLE ONE: IN-PIT TAILINGS SCHEME

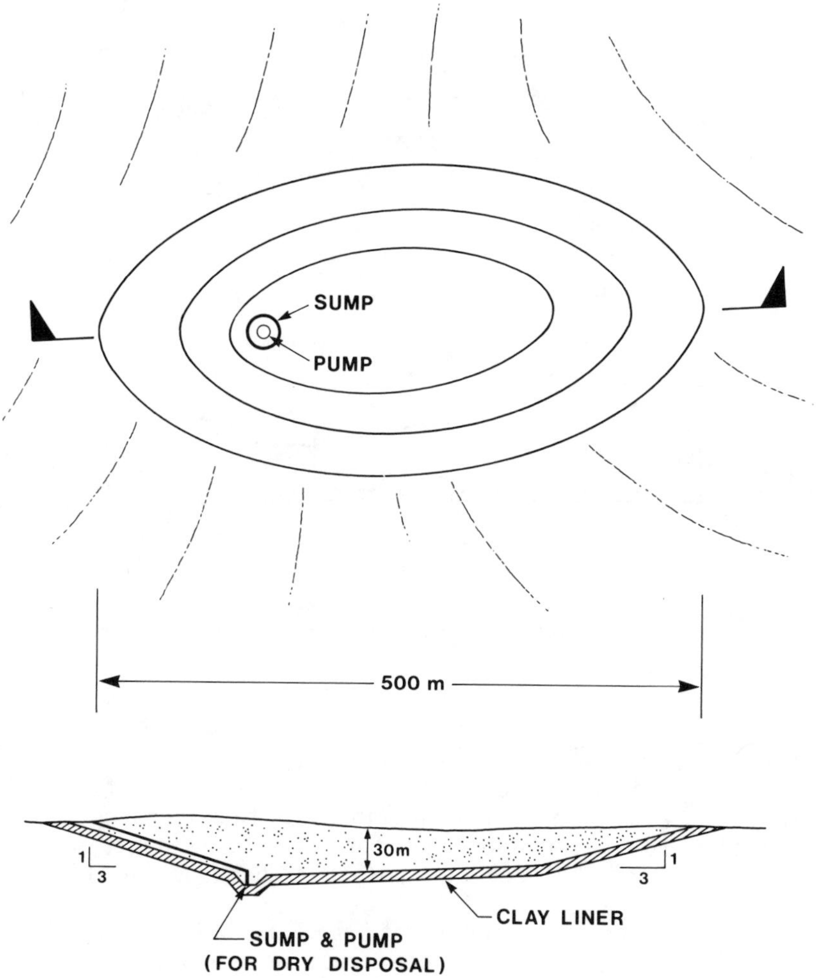

Figure E-1. In-pit disposal.

EXAMPLE TWO: A TAILINGS IMPOUNDMENT IN A FLATTISH TERRAIN

Fig. E-2 shows the plan and construction details of a tailings impoundment constructed in flat terrain in a low rainfall, high evaporation environment. The bedrock is high strength, moderately jointed norite. Soil cover is on average about 1 m of high plasticity silty clay of very stiff consistency. Peak undrained strength is high (+ 150 kPa) but residual effective strength is low ($\phi = 15^{\circ}$). The groundwater table is usually at the clay and bedrock

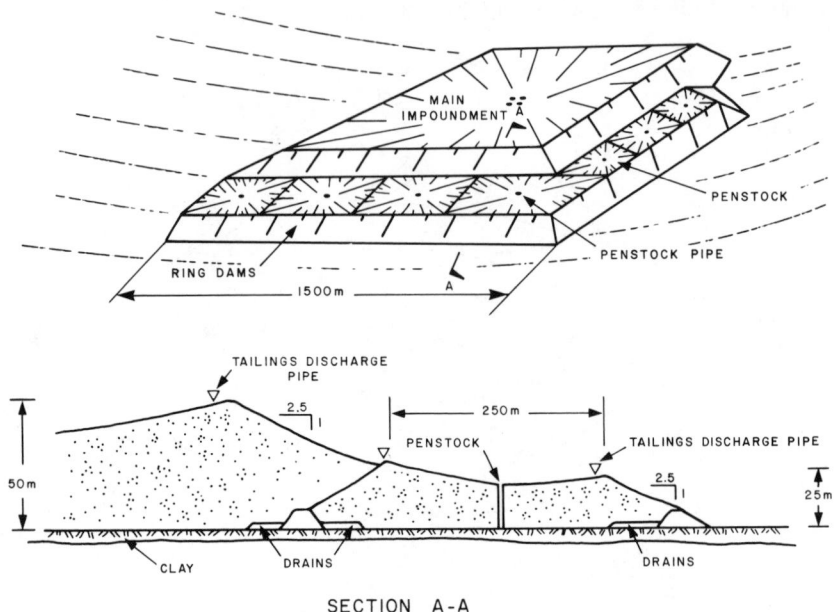

SECTION A-A

Figure E-2. Upstream spigotted tailings impoundment.

interface. Seismicity is insignificant.

The tailings grade from a medium sand to fine silt. They contain potential contaminants, thus construction on the clay to preserve groundwater quality is done. A closed circuit water balance is possible because of the arid climate.

In order to construct to a height of 50 m, essentially by upstream construction methods, hydraulic fill dams are used in addition to conventional upstream discharge. In this way the overall flat slopes required for stability in the clays are created.

Construction of a surround dam starts with a 1 m high wall built of in situ clays. Tailings are discharged behind the wall from spigots. Segregation occurs along the beach which forms; coarser tailings near the perimeter, and finer tailings near the penstocks or decant towers which are used to remove water.

Once the surround dams reach a suitable height, discharge of tailings into the main impoundment begins. This too is done by conventional spigotting.

EXAMPLE THREE: DRY DISPOSAL TAILINGS IMPOUNDMENT

In order to reduce as far as practicable seepage of contaminants from tailings a dry disposal impoundment as shown in Fig. E-3 is used. Underlying bedrock is sound and a clay liner is installed over in situ sandy silts before tailings deposition.

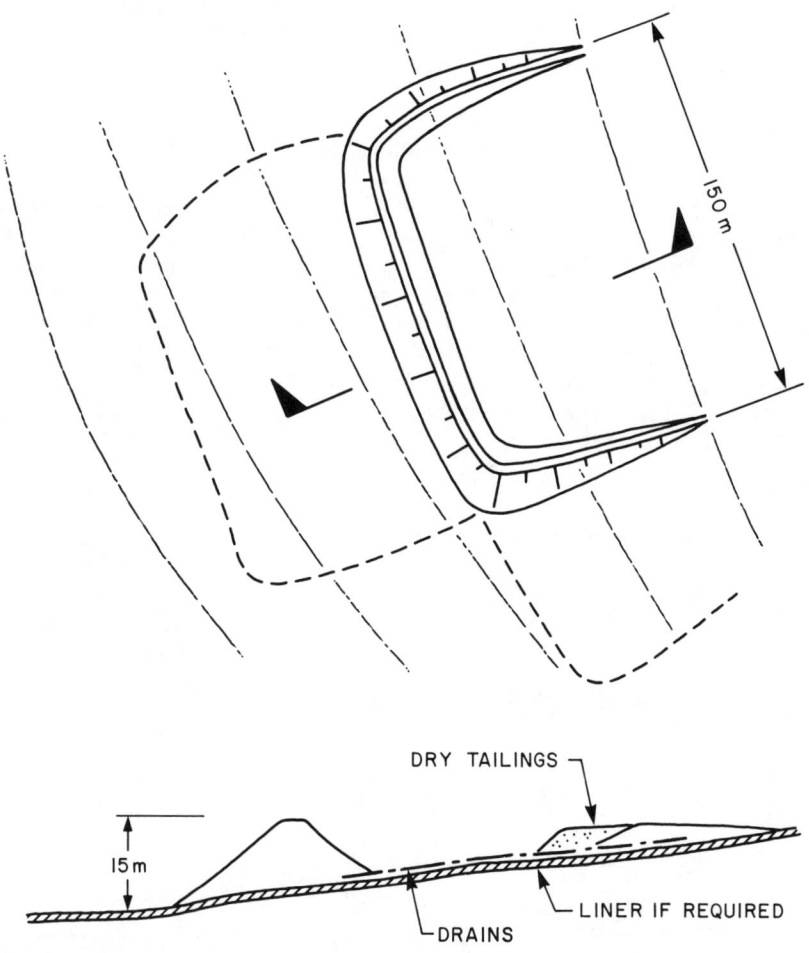

Figure E-3. Dry tailings disposal facility layout.

The tailings are dried to a moisture content of 15% in the mill with belt filters. At the mill it is loaded onto trucks, for transport to the impoundment.

The impoundment consists of paddocks or cells formed by dikes of pit rock and overburden. The dikes are built to a maximum height of 15 m. Copious drains are installed to expedite drainage, hence further in situ drying of the tailings occurs. The tailings are dumped from the trucks on the edge of the dikes into the cells. Distribution of the tailings is by front-end loaders and bulldozers.

Once a cell is full it is covered with a layer of rock and low permeability soil.

EXAMPLE FOUR: A TAILINGS IMPOUNDMENT IN A STEEP VALLEY

Fig. E-4 shows the plan and cross section of a tailings impoundment designed for a steep narrow valley. Bedrock in the area is a competent quartz monzonite with few fractures or joints and an hydraulic conductivity of the order of 10^{-8} m/sec. Filling the base of the valley to a depth of 20 m is a deposit of medium dense alluvial sands with a hydraulic conductivity of 10^{-5} m/sec. A tongue of low strength ($c' = 0$, $\emptyset' = 20^{o}$) silty clay underlies a part of the sands. The design earthquake

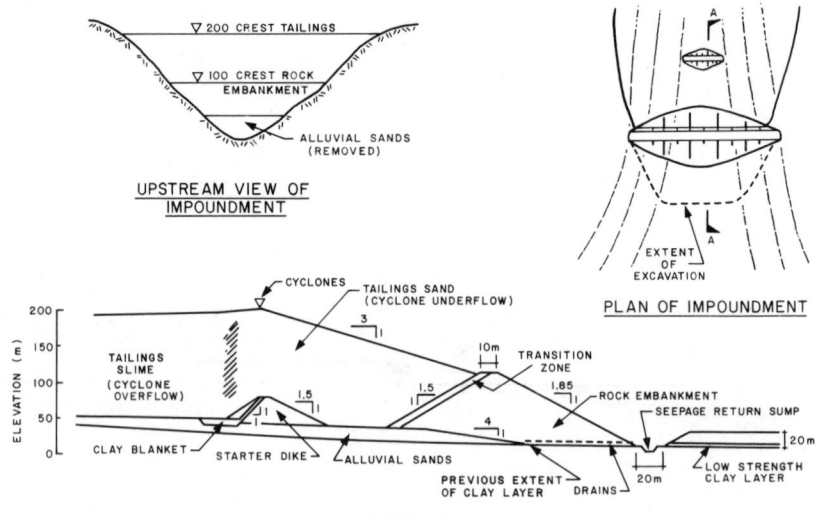

Figure E-4. Centerline cycloned embankment.

acceleration at the site is 0.1 g.

The climate is dry: annual precipitation is 300 mm and evaporation is 750 mm. Temperatures seldom fall below freezing and winds are moderate due to shielding by neighboring hills.

The tailings grade from a medium sand to a fine silt and are suitable for cycloning. Chemically they do not give rise to any potential ground or surface water contamination.

The toe embankment is constructed of mine waste rock. Part of the alluvial material is removed both to construct an upper starter dike and to remove the underlying low strength clay. Dewatering of the sand before excavation is required. A transition zone of screened rock is placed on the upstream side of the rock embankment and a clay blanket and liner on the upstream side of the starter dike.

Cyclones along the starter dike separate the tailings into coarse and fine fractions. The coarser sands are used to build the sand embankment by centerline techniques. Drains beneath the cyclone underflow zone, the starter dike and the rock embankment prevent the buildup of a phreatic line. A series of diversion ditches plus the very small catchment area of the impoundment control long-term buildup of excess water on the impoundment. Sufficient freeboard for extreme precipitation is provided at all times.

EXAMPLE FIVE: TAILINGS IMPOUNDMENT IN A
HIGH PRECIPITATION ENVIRONMENT

Fig. E-5 shows the layout and cross sections of an impoundment built in a high precipitation environment. Bedrock is competent andesite and basalt. Generally the upper 3 - 10 m of the bedrock is fractured such that its hydraulic conductivity is about 10^{-5} m/sec. Bedrock is overlain by 1 - 10 m of medium dense to dense glacial till.

The climate at the impoundment site is wet and generally cold. Regulations require that no process water be discharged from the site, hence the impoundment must be run as a closed circuit.

Two embankments are built successively to contain the tailings. The embankments are built of mine overburden till and waste rock. To make the best use of the materials available according to the optimum pit mining schedule, the upper embankment is constructed primarily of tills, and the lower embankment mainly of waste rock.

The tailings are fine grained and potentially acid generating. Complete containment of the tailings is

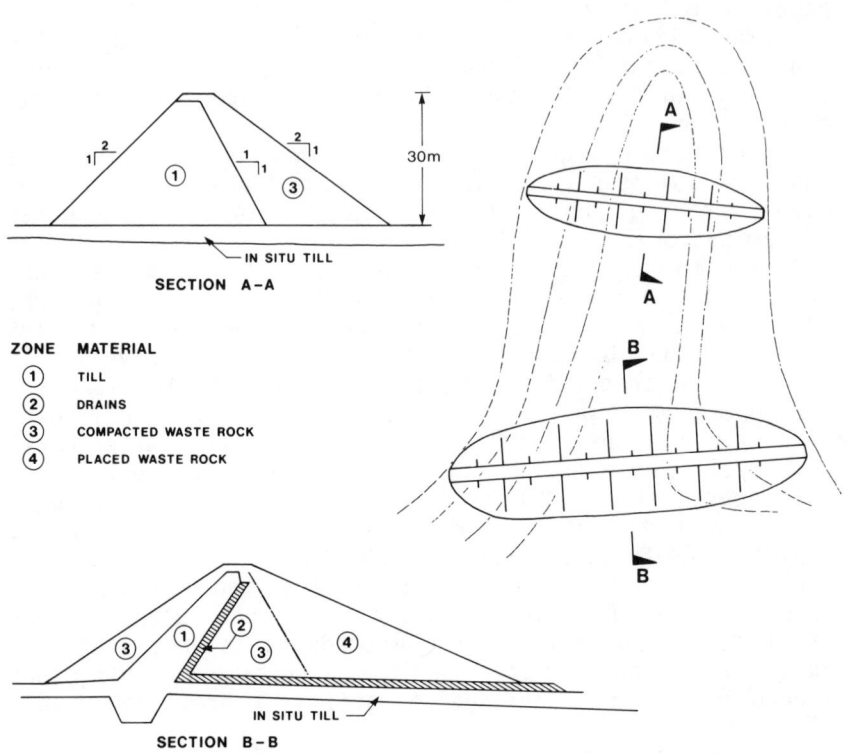

SECTION A-A

ZONE	MATERIAL
①	TILL
②	DRAINS
③	COMPACTED WASTE ROCK
④	PLACED WASTE ROCK

SECTION B-B

Figure E-5. Earth embankment for tailings impoundments.

required. The existing in situ tills, enchanced where necessary, constitute the impoundment liner, and the cores of the embankments are designed to be watertight.

Tailings deposition is by ring discharge, and under water. Water is returned to the mill from a barge. The closed circuit water balance is achieved by the successive construction of the embankments and filling of the impoundments. Excess water accumulating on the impoundments will, at reclamation, be treated and discharged to the mined out pit.

Reclamation of the impoundment will involve covering the tailings with 1-3 m of rock and a layer of till, and establishing suitable vegetation on a soil layer.

DISCUSSION

Question: I was intrigued by some of the disposal problems, in particular the plans for quite elaborate means of getting water away from the land disposal system, including those concrete structures. What are the long-term implications of that? Once the mine has gone, who is going to look after the disposal system?

Answer: The tailings impoundment in the long-term must be regarded as a new landform. You are asking how one can ensure that a landform in the long-term operates without maintenance. You must design your facility, so that at the end reclamation which precludes as far as possible the use of pipes and so on, can be achieved. Water control must not rely on unnatural man-made type facilities, but must blend with the environment. You try to reestablish natural stream channels or create stream channels which approximate those which nature in its wisdom designed. Spillways would have to be located and placed in natural areas. With Mt. Tolman, for example, which we consider the ultimate spillway for long-term control, we managed to find an old spillway that had been used or at least formed during the glaciation. The valley had been blocked off by ice almost at the area where we wanted our impoundment to be. Water had flowed over a spillway which had been cut, as it were, into the rock. We were able to take advantage of the natural feature, something that had existed for a long, long time. That is not necessarily possible everywhere. At two other impoundments that I am working with at the moment we have been able to find natural features which, in the long-term, will be usable as a spillway. They need a little enhancement with engineering input.

Question: The final disposition of the tailings could be an area that might even enhance the seabed or some other aspect of storage, rather than saying that through cost benefit planning we will dispose in a fashion that may not be immediately detrimental to the environment, but won't enhance it either. Do you have such fundamentals in your final decision process in addition to cost benefit planning?

Answer: When I put forward the various criteria, cost was only one of them. You can't look at something like a new landform, in the long-term, simply as a cost benefit. How long do you go on costing it for? When do you cut off your costs? You must try to create something that is natural.

I think a designer must take into account all things and obviously in a project such as the one being considered here, cost must be one of them. But creating the best system for the environment is also essential and must be integrated. Whatever system is better, whether it be land disposal or marine disposal, that which best approximates the natural processes is a matter for discussion.

Question: Protection of renewable resources should be a primary axiomatic concern of all mining industry, and I don't see that this is so as yet. It might be, but I am not aware of it.

Answer: Let me put it this way. Since I grew up in a mining camp a long while ago, the attitudes have changed considerably. Miners have a reputation, but I think they take protection of renewable resources into account much more these days than they ever did in the past.

Question: Since you have put up some matrices to relate environmental factors to different kinds of developments, I really have to raise an issue with you. I think you made the point that the environmental factors and the ratings reflected a point of view. You didn't mention the problem of weighting those factors. Different people will weight those factors in different ways and come to a different set of numbers, totals and averages than you have. I think that you should point out that you have in fact weighted those factors equally. Others may weight them very differently. This will result in averages different from yours, so your figures could be slightly misleading.

Answer: I didn't weight them on purpose because it usually leads to an argument about the weighting which is greater than the argument that you haven't weighted them. There are methods of dealing with this, such as preference analysis. You try to relate the opinions of a large number of people. I wouldn't like to go into the theories of weighting, of multi-objective analysis and so on. I think quite a large advance has been made in enabling us to take those things into account, but they are complicated. I didn't want to present more than the basic essentials here.

CHAPTER 2

CHARACTERISTICS OF MILL TAILINGS
AND THEIR BEHAVIOR IN MARINE ENVIRONMENTS

G.W. Poling
    Department of Mining and Mineral Process Engineering
    The University of British Columbia
    Vancouver, B.C.
    Canada V6T 1W5

Characteristics of mill tailings most important to the determination of their behavior in marine environments are presented. Typical mining and mineral processing operations that produce mill tailings are described briefly. Criteria that should affect the choice between on-land impoundment and submarine disposal methods are discussed.

INTRODUCTION

Mining and mineral processing operations around the world face the problem of how best to dispose of mill tailing wastes. For hundreds of years, both terrestrial and submarine disposal methods have been employed. Mill tailings are normally composed of both solid-particulate and liquid phases. The main environmental concerns in disposing of these wastes are:

* aesthetics and land use conflicts;
* the safety and stability of any terrestrial impoundments both during and following cessation of mining operations;
* the possibilities of air and water pollution;
* the reclamation and rehabilitation of the disposal site; and
* possible alienation of residual contained mineral resources in the future.

63

Of these, most people consider the threat of water pollution to be the most significant.

Every ore-body is unique and can vary widely in both mineralogical (or chemical) composition and physical properties. As a result, each waste tailing material is also unique. In addition, each mine site and potential tailings disposal site is unique. Therefore, one cannot expect realistically to generalize about which is the "best" disposal method. Concern for minimizing environmental impacts, governmental regulations, and economic constraints demand careful evaluation of several possible disposal alternatives.

This paper describes important characteristics of mill tailings, how they are produced, what their chemical characteristics are, and what their sedimentation behaviors are. Criteria that should affect the choice between on-land and marine disposal systems will also be discussed. Examples are drawn from some existing marine disposal systems.

NATURE OF MILL TAILINGS

Mineral Processing Operations

Valuable minerals often comprise a very small proportion of ores fed to mineral processing plants. For example, low grade porphyry copper ore-bodies typically contain around 0.5% copper by weight. Often this copper is present in the form of small isolated crystallites (0.1 to 0.2 mm in diameter) of chalcopyrite ($CuFeS_2$) dispersed throughout a siliceous waste rock or gangue matrix. Since pure chalcopyrite is composed of 34.6% Cu, this typical low grade copper ore would contain approximately 1.4 weight percent of chalcopyrite or 28 lbs. of chalcopyrite per short ton of ore feed. Since chalcopyrite has a density of 4.2 gm/cm^3, while the siliceous gangue matrix density would be around 2.7 gm/cm^3, the percentage of chalcopyrite by volume is significantly less than the 1.4 weight percent mentioned above. In this particular example, the valuable chalcopyrite mineral would occupy only 0.90% of the total solid volume of the ore feed, i.e.:

$$\frac{1 \text{ part } CuFeS_2}{111 \text{ parts ore}} \text{ by volume.}$$

Primary molybdenite ores are even lower in grade than the typical copper ore described above. An ore containing

0.084% molybdenum in the form of molybdenite (MoS_2) would contain 0.14% MoS_2 mineral by weight but only 0.08% MoS_2 (of s.g. = 4.8) by volume, i.e.:

$$\frac{1 \text{ part } MoS_2}{1250 \text{ parts ore}} \text{ by volume.}$$

Mineral processing plants physically separate such ores into relatively pure, valuable mineral concentrates and waste tailings. To achieve such physical separations, the ore, which often arrives at the mill in lumps up to 1 m in diameter, first must be crushed and ground fine enough to "liberate" the valuable crystallites of contained chalcopyrite or molybdenite. Achieving adequate liberation of the sparse chalcopyrite or molybdenite crystallites might require the ore to be ground to an average particle size of .02 to .08 mm (20-80 μm). This ground material will be composed of particles ranging in size from less than 1 μm to perhaps 200 μm. Fig. 1 shows a typical particle size analysis of a crushed and ground ore from a porphyry copper operation in British Columbia (1).

Fig. 2 presents a highly simplified schematic flowsheet of a mineral processing plant designed to recover both copper and molybdenum from a low-grade porphyry type ore-body. This particular flowsheet depicts ore being derived from an open pit type mine. The ore is drilled and blasted before being loaded, usually with electric shovels, into large trucks (up to 450 tons each). At this stage the largest ore fragments would typically be up to 1 m in diameter.

In order to liberate the valuable copper and molybdenite minerals the ore must first be crushed and ground until it has the consistency of a fine powder. In the flowsheet in Fig. 2, this is achieved first in a gyratory crusher (primary crusher), then in large diameter tumbling mills, followed by finish grinding in secondary (tumbling) ball mills. Here 5-13 cm diameter steel balls act as tumbling grinding media to complete the fine grinding of the ore. Although not shown in the simplified schematic (Fig. 2), "classifiers" act as size separators within the grinding circuit to remove finished size particulate, and recycle back to the grinding mills material that is still too coarse. Today most classifiers are hydrocyclones.

In terms of marine disposal systems, coarser grinds are generally desirable to accelerate particle sedimentation rates. Coarser grinds can also minimize the abundance of "slimes" (particles nominally < 10 μm in

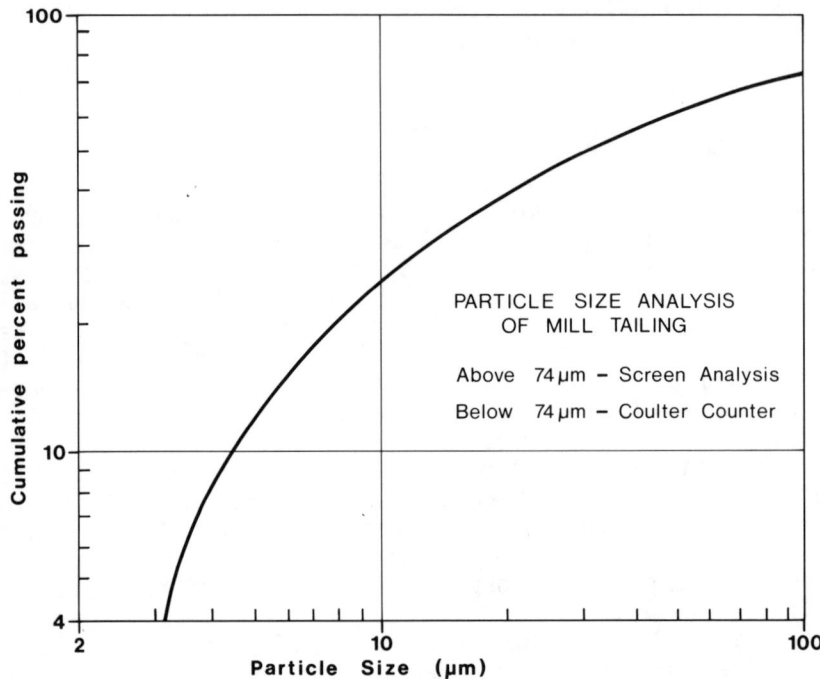

Figure 1. Typical particle size distribution of tailings from low-grade porphyry copper-molybdenum concentrator.

diameter), which individually have such low settling rates that they can move with very low water currents. These slime fractions can sometimes cause undesirable turbidity in the upper reaches of the water column. More will be said about these aspects later.

Once the valuable mineral has been liberated adequately, physical separations of these mineral constituents in an ore can be based on differences in properties such as: specific gravities, magnetic susceptibilities, electrical conductivities, particle sizes, particle shapes, colors and surface chemistries. The example flowsheet presented in Fig. 2 illustrates froth flotation as the primary mineral separation technique. This method depends on differing surface characteristics of the valuable and gangue mineral particles present in the finely ground ore slurry.

Process chemicals called "collectors" are added to the mineral slurry to make certain specific minerals

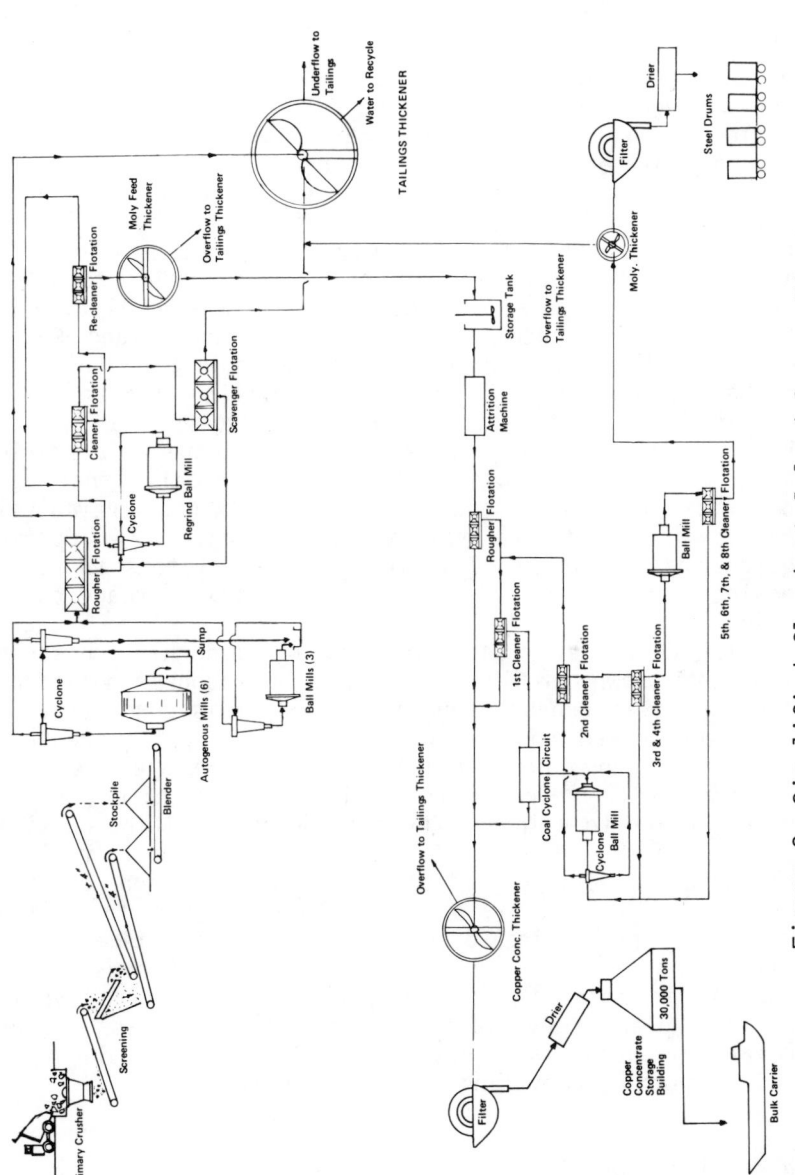

Figure 2. Simplified flowsheet Island Copper Mine.

hydrophobic or air-avid, while the rest remain hydrophilic or preferentially wetted by water. Air bubbles are dispersed into the flotation tank or cell. These air bubbles attach preferentially to the hydrophobic particles and buoy them to the surface of the pulp, where they enter a column of stable froth. This froth column holds them up until the froth overflows into a launder as a froth concentrate. Mineral particles that remain hydrophilic pass through the cell and exist as flotation tailings. The stable froth column is generated by the addition of alcohol-type reagents called "frothers".

Many other process chemicals can be added to enhance these froth flotation separation mechanisms. Most of these reagents are added in relatively minute amounts and most will report with the mineral concentrates rather than with the tailings. Table I presents a complete listing of typical milling chemicals as used by the Island Copper Mine (1). This table also presents a realistic evaluation of the eventual disposition of these milling chemicals.

The above discussion is not to suggest that milling chemicals present no concern to the prospect of marine mill tailings disposal. Certain chemicals that are sometimes used as flotation reagents can be relatively toxic while most are relatively non-toxic (2). Optimal design and operation of marine disposal systems therefore requires that certain of the more toxic reagents be avoided.

To continue with the example mineral processing flowsheet shown in Fig. 2, after a bulk copper-molybdenum concentrate is produced and thickened, this concentrate is next reprocessed to separate it into a copper concentrate and a molybdenum concentrate. Each concentrate must also be dewatered by thickening, filtering and often thermal drying, prior to shipment for sale or for smelting on site.

As shown in Fig. 2, most of the mill tailings originate directly from the primary flotation circuit. These are the materials under consideration for either on-land or submarine disposal methods. Sometimes these tailings are partially dewatered in thickeners before disposal in order to reclaim process water for recycling to the plant. Availability of fresh process water, operating costs, and environmental concerns all enter decisions about recycling of process waters.

Physical and Chemical Characteristics of Tailings

As a result of mineral processing operations, the concentrator produces one or several mineral concentrates plus mill tailings. The main characteristics of these mill

Table I. Milling chemicals used by Island Copper and their environmental considerations.

| Reagent type and name | Chemical composition | Amount added lb/ton ore | Approx. fractional distributions in mill products | | | |
| | | | Adsorbed on solids | | In liquid fraction of tailing | Volatilized |
			In mineral concentrates	In tailing		
Xanthate-collector	$C_5H_{11}OCSSK$	.005	>1/2	<1/4	<1/2	~0
Aerofroth 71R-frother	C_6-C_9 alcohol	.05	<1/4	<1/4	>1/2	<1/2
Dowfroth SA-1012 frother	Polypropylene glycol	.004	<1/4	<1/4	>1/2	~0
Lime-coagulant and pH regulator	CaO	1.3	<1/4	<1/4	>1/2	~0
Cyanide-depressant	NaCN	.04	>1/4	>1/4	~1/4	<1/4
Hydrosulphide-depressant	NaSH	.22	>1/2	<1/4	<1/4	<1/4

Table I, continued

Reagent type and name	Chemical composition	Amount added lb/ton ore	Approx. fractional distributions in mill products			
			Adsorbed on solids		In liquid fraction of tailing	Volatilized
			In mineral concentrates	In tailing		
Exfoam 635 froth suppressant		.013	< 1/4	< 1/4	> 1/2	~ 0
Aerodri 100- drying agent	Sulpho-succinic acid	.008	> 1/2	< 1/4	< 1/4	~ 0
Alchem 8863- flocculent	Polyacrylamide	.046	< 1/4	> 1/2	~ 0	~ 0
Alfloc 84046- flocculent	Polyacrylamide	.046	< 1/4	> 1/2	~ 0	~ 0

tailings that will make them either suitable or unsuitable for marine disposal, or most affect their behavior in the marine environment are:

·mineralogical (or chemical) composition;
·particle size distribution; and
·residual mill reagents.

Mineralogical Compositions of Mill Tailings

Table II presents a detailed mineralogical analysis of tailings solids derived from the low-grade porphyry copper-molybdenum operation of the Island Copper Mine near Port Hardy, B.C. This table also provides a comparison with the mineralogical composition of natural bottom sediments present in Rupert Inlet prior to start-up of the Island Copper operation. Close examination of Table II enables one to conclude that the tailings are very closely related, mineralogically, to the natural sediments in the inlet. The tailings are noticeably higher in copper (700 ppm compared to 44 ppm) and molybdenum (40 ppm compared to 2 ppm) since they result from an ore-body which contains economically viable concentrations of these two minerals. Since the concentrator has recovered 80-90% of the valuable mineral content present in the original ore feed, further significant reduction of the copper and molybdenum levels in these mill tailings is economically unrealistic.

Those tailings described in Table II are more or less typical of tailings to be derived from low-grade porphyry-type deposits. Most of the residual copper-iron and molybdenum contents would exist as small, discrete metal sulphide crystallites occluded within larger silicate particles. While some sulphide surface would be exposed to the seawater in a marine disposal system, most would be isolated within a silicate matrix.

Tailings derived from massive sulphide ore-bodies might, on the other hand, differ markedly from those described above. Table III presents data on the nature of the tailings discharged from the Greenex A/S Black Angel Mine which processes a high-grade, massive lead-zinc ore-body (3, 4). While the majority of the lead and zinc are extracted (at recoveries of 85-95%) from the ore, the tailings are still composed of about 1/4 metal sulphide minerals. Most of this residual metal sulphide is pyrite (FeS_2), which is a gangue mineral. In these tailings much more of the residual lead, zinc and iron sulphide minerals would be exposed to the seawater. Concerns for the chemical stability of these phases would naturally be much greater than for the sort of tailing solids described

Table II. Typical chemical and mineralogical compositions of tailing solids and natural sediments in Rupert Inlet, B.C.

| Element or Oxide, etc. | Content of Sediments | | Ratio A:B | Mineral Species | Tailing Content |
	A Tailing %	B Natural %			
SiO_2	62			Quartz	50-70%
Al_2O_3	14			Feldspar	2-20%
Ca,K,Na, & Mg Oxides	10			Biotite and Chlorite	5-10%
Fe Oxides	8			Magnetite	2-4%
Fe Sulphide	2-3	2-3	1:1	Pyrite	2-4%
CO_2	2	-	1:1	Calcite	~2.5%
Total	98-99	~99			

Element	ppm	ppm			
Cu	700	44	16:1	Chalcopyrite	0.2%
Mn	650	640	1:1	Mn Oxides	n.d.
Cr	140	125	1:1	In silicates	n.d.
Zn	80	88	1:1	Sphalerite	0.02%
Mo	40	2	20:1	Molybdenite	0.01%
Co	20	20	1:1	In silicates	n.d.
Ni	20	40	1:2	In silicates	n.d.
Pb	20	25	1:1	Galena	0.002%
As	5	5	1:1	Arsenopyrite	n.d.
Cd	3	2	3:2	In sphalerite	n.d.
Hg	0.03	0.06	1:2	Cinnabar	$> 4 \times 10^{-6}$%

n.d. = not determined

Table III. Typical chemical and mineralogical composition of tailing solids from the Black Angel Mill.

Element	Major Mineral Species	Metal Content, ppm			tons/day
		Ore	Concentrate	Tailings	
Pb	PbS	30,000-70,000	680,000-700,000	2,500	7
Zn	ZnS	150,000-200,000	580,000-590,000	5,000	14
Fe	FeS_2	200,000 FeS_2		~250,000	700
Cd	CdS			50	.14
	tremolite				
	marble			~750,000	
	talc				
tailing slurry					2,800

in Table II.

Chemical Stability of Tailing Solids

One of the most crucial questions in submarine disposal of mill tailings concerns whether their heavy metal contents will remain stable in the solid phase or dissolve and enter the seawater column. The deep ocean composition has been close to constant for at least 100 million years and the levels of dissolved heavy metals are extremely low. Since nature has provided a continuous input of dissolved heavy metals from erosion of land sources, the ocean's sediments are well recognized as a major sink for originally dissolved heavy metals (5). On the basis of thermodynamic calculations, pH and redox (reducing or oxidizing) conditions within the aqueous phase are expected to exert primary influence on which is the most stable phase, the solid metal sulphide or the dissolved metal ion. The basic environment (pH = 7.9-8.1) of the sea favors the solid phase being more stable than the heavy metal ions. Fig. 3 indicates the influence of pH on solid (metal hydroxide): ionic (dissolved) equilibria for a large variety of metal ions. Prevention of the formation of acidic conditions in the sea generally enhances long-term stability of heavy metal solids as compared to on-land systems in which acid generation due to sulphides is all too common.

Recognizing that the slightly basic pH is generally beneficial in preventing metal release from sediments (either natural or tailings), the redox condition in the sea in general is the principal controlling factor. In fact, careful experiments of Lee and Chen (6) have recently confirmed that the direction of metal migration either from the sediments to the seawater or from the seawater to the solid phase is regulated mainly by the redox condition in the overlying or interstitial seawater. The type of sediment that they studied as clay, silt or sand had a decidedly secondary effect. These researchers found that as the redox conditions became more oxygen deficient or more reducing, only Fe and Mn exhibited increased release to the water column. The amounts released agreed well with expected releases based on thermodynamic calculations.

Under oxidizing redox conditions the amounts of released Cu, Pb, Zn, Cd and Ni increased, but not nearly as much as expected based on thermodynamic calculations. Lee and Chen (6) believe that readsorption of released metal ions in the form of humic metal complexes back onto the solid surfaces probably accounted for this lack of

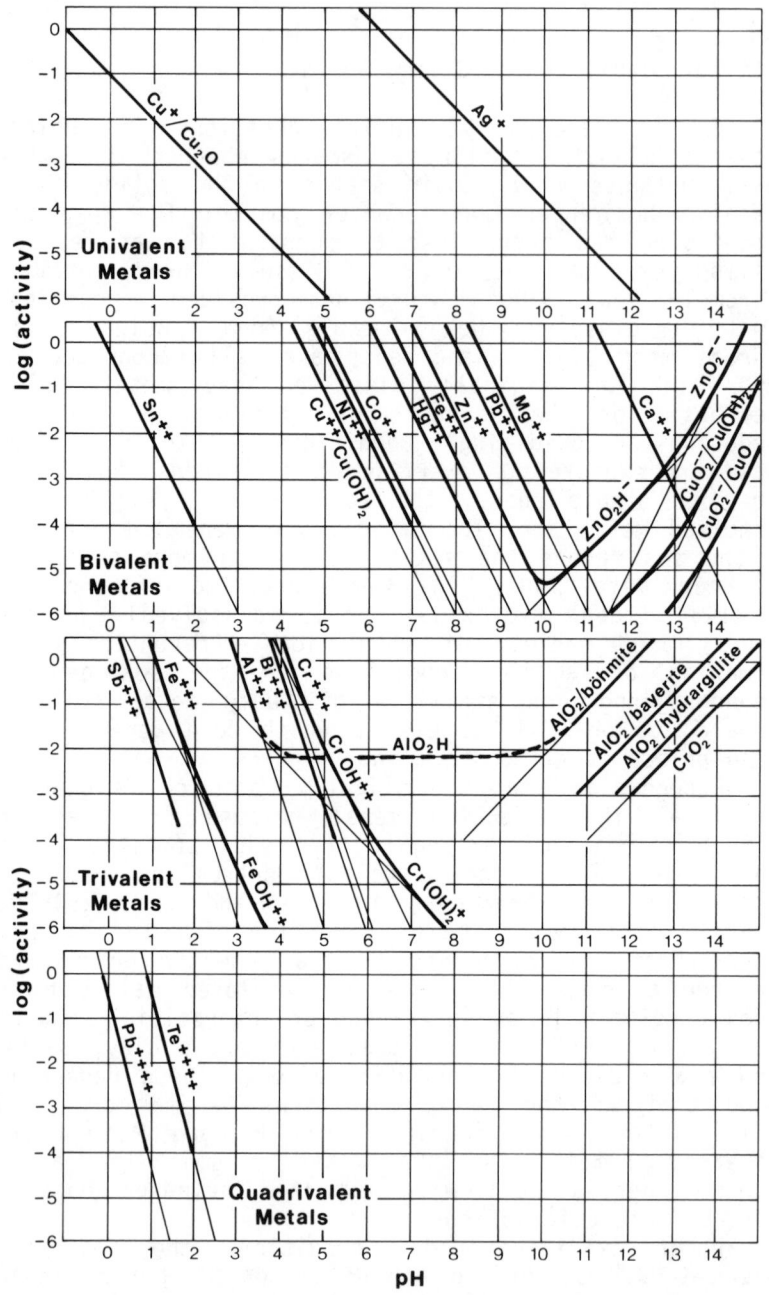

Figure 3. Influence of pH on equilibria in heterogeneous systems - solid-solution type (solubility of hydroxides). Modified from (7).

theoretical and experimental agreement.

Results of Lee and Chen (6) agree closely with over ten years of detailed marine monitoring data on the effects of discharging Island Copper mill tailings into Rupert Inlet, B.C. In this system, only Mn has shown a slight but statistically significant increase in dissolved concentration in the seawater column (1, 8).

Of the several mechanisms for migration of heavy metals between sediment solid and seawater, chemical transformation, adsorption-desorption, bio-oxidation, dissolution, precipitation, complexation and diffusion are well recognized (6). For tailings from massive sulphide ore-bodies an additional galvanic electrochemical mechanism might be important. Since sulphides are semiconductors their intimate electrical contacts in relatively massive sulphide tailings can accelerate the dissolution of the less noble mineral. Galena-pyrite contacts have been shown to accelerate the decomposition or dissolution of galena (9).

Realistic leaching tests should be conducted on samples characteristic of the expected tailings to determine experimentally whether leaching problems can be expected in the marine environment. Some process changes might be made or additives used to minimize leaching effects if they are expected (10, 11).

Physical Behavior of Tailings in Marine Environments

Sedimentation characteristics of tailings affect their disposition within the marine environment. Enhanced settling rates can minimize the areal extent of the spread of the tailings over the ocean floor or inlet bottom. Enhanced settling, with incorporation of slimes fractions within "floc" or "coagule" aggregates can also minimize the chance of creating turbid conditions in the upper euphotic zone in the marine water column. By ensuring that the tailing particulates remain below the euphotic zone, photosynthetic processes should not be adversely affected by the marine disposal system.

Fine particles settle in water at rates directly proportional to the square of the particle diameter. For this reason, individual slime particles (the fraction nominally less than 10 μm diameter, see Fig. 1) settle extremely slowly and could be carried upward into the euphotic zone by commonly encountered up-welling currents in inlet systems. One way to cause the fine slime particles to aggregate into clusters is by coagulation. Reduction of the charge carried within the ionic atmosphere (electrical double layer) surrounding each solid particle to near zero enables particles to aggregate by the weaker but ever present Van der Waals attractive forces. This is the process of coagulation.

Fortunately seawater acts as a natural coagulating medium for most minerals including both natural sediments and mill tailings. Fig. 4 shows how the surface charges (zeta potential) of both pure silica and mill tailings decrease towards zero in full strength seawater of $35^\circ/_{oo}$ salinity (12). Addition of flocculating agents such as high molecular weight polyacrylamides can further enhance sedimentation characteristics by adsorptive bridging of particles into aggregates called flocs (13, 14). Both coagulants, in the form of lime ($Ca(OH)_2$), and synthetic flocculents are often added to tailing thickeners to enhance both settling rates and supernatant clarity (1, 13, 14). Preparing flocs of sufficient strength to persist through a submarine outfall system might assist sedimentation behaviors of tailings in submarine disposal systems.

Figs. 5 and 6 illustrate how the sedimentation rates of mill tailings at Texada, B.C were enhanced by seawater additions (12). Fig. 5 illustrates that the coagulating action of the seawater markedly enhanced the clarity of the supernatant water column overlying the settling sediment (12).

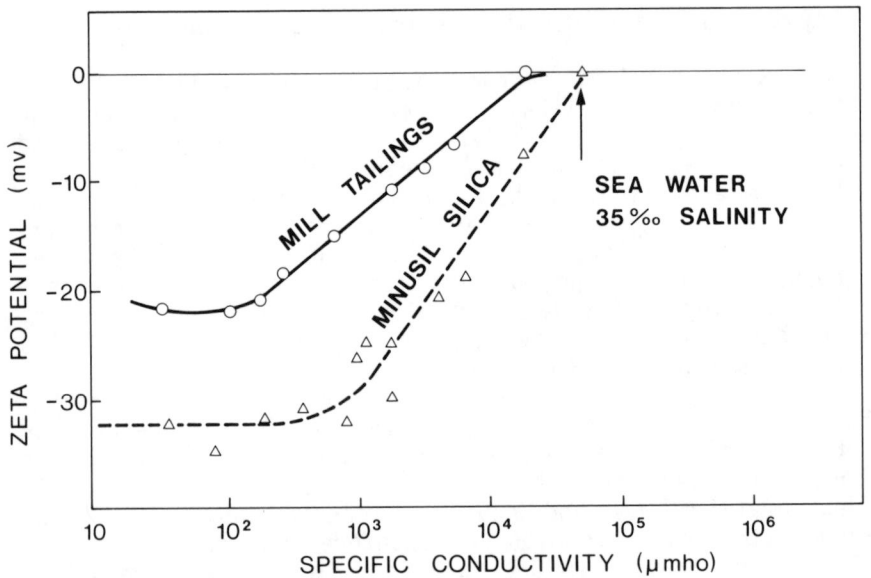

Figure 4. Effect of seawater addition on zeta potential.

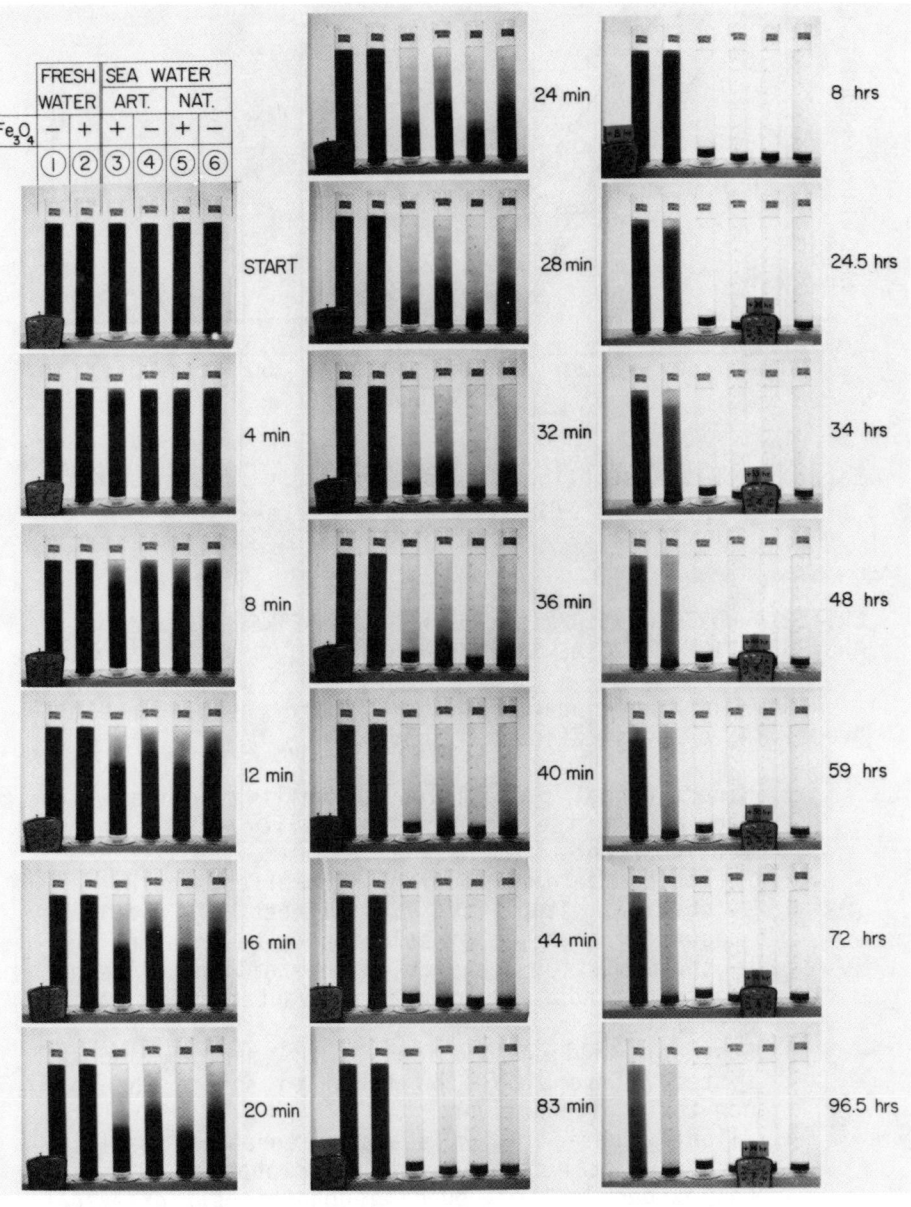

Figure 5. Settling of tailing (6%) solids in freshwater and in seawater.

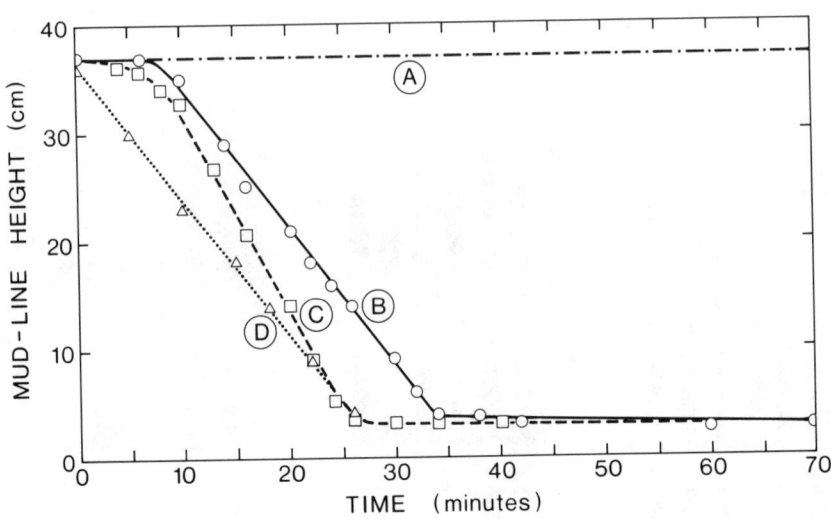

Figure 6. Batch settling curves of tailing (6%) solids in: Freshwater A; 1.8% seawater B; 5.4% seawater C; and 100% seawater D (magnetite retained).

CRITERIA WORTH CONSIDERING IN CHOOSING BETWEEN ON-LAND AND SUBMARINE DISPOSAL SYSTEMS

Some of the characteristics of both the tailings and "receiving systems" discussed above are:

1. Mineralogical composition of tailing solids which can affect the chemical stability of the heavy metal contents.
2. Particle size distribution which affects both:
   a. chemical stability via effects of available surface area for dissolution or adsorption; and
   b. sedimentation behavior of the tailing solids.
   Coarser grinds should offer advantages for marine disposal.
3. Residual milling chemicals, for marine disposal systems. Care should be taken to select the most non-toxic alternatives.
4. Slightly basic pH of seawater enhances stability of heavy metal compounds as solid phases.
5. Low oxygen levels (reducing conditions) in several marine systems favor stability of most heavy metals in the solid phase. Manganese and iron can be exceptions.

6. Natural coagulating action of seawater causes coagulation and enhanced sedimentation characteristics of most tailings.

Several of the above factors tend to favor marine disposal systems over on-land impoundments for mill tailings. Some other important criteria that need to be considered are:

7. Probability of earthquake acceleration loadings on dams or dykes being considered for on-land impoundments. Fig. 7 shows the epicenters of earthquakes along the west coast from Alaska to Washington state from 1899 until 1966. Severe earthquake acceleration forces can cause tailing dam failures and severe downstream damage. Earthquakes can also cause sudden slumping of submarine tailing deltas which can set up damaging wave actions.

8. Geographic and climatic conditions such as steep-walled canyons and/or high rainfall or snow loads can make on-land impoundment impractical. Steep-walled canyons or valleys can make dams too high to store safely the requisite volume of tailing solids over the life of a mine. High levels of precipitation into a tailing impoundment area can make a zero-liquid discharge system impossible. Detailed engineering evaluation of an on-land alternative to the submarine disposal system at the Island Copper Mine indicated the following: decant of supernatant waters from the tailings pond would almost match the current liquid component of the tailings in their direct marine disposal system. Since the on-land system would expose the Island Copper mill tailings to atmospheric, oxidizing conditions, the total input of dissolved heavy metals in the tailings pond decant could well be expected to exceed that emanating from their direct discharge systems.

9. Submarine disposal of mill tailings into a "contained" system need not alienate these materials from possible future reprocessing for recovery of other minerals, as is usually claimed (15). Modern submersible dredge pumps operate readily at depths as low as 600 ft. Such pumps could readily retrieve bottom tailing sediments for reprocessing if new economic conditions warranted.

10. Capital and direct operating costs of submarine mill tailings disposal systems are often

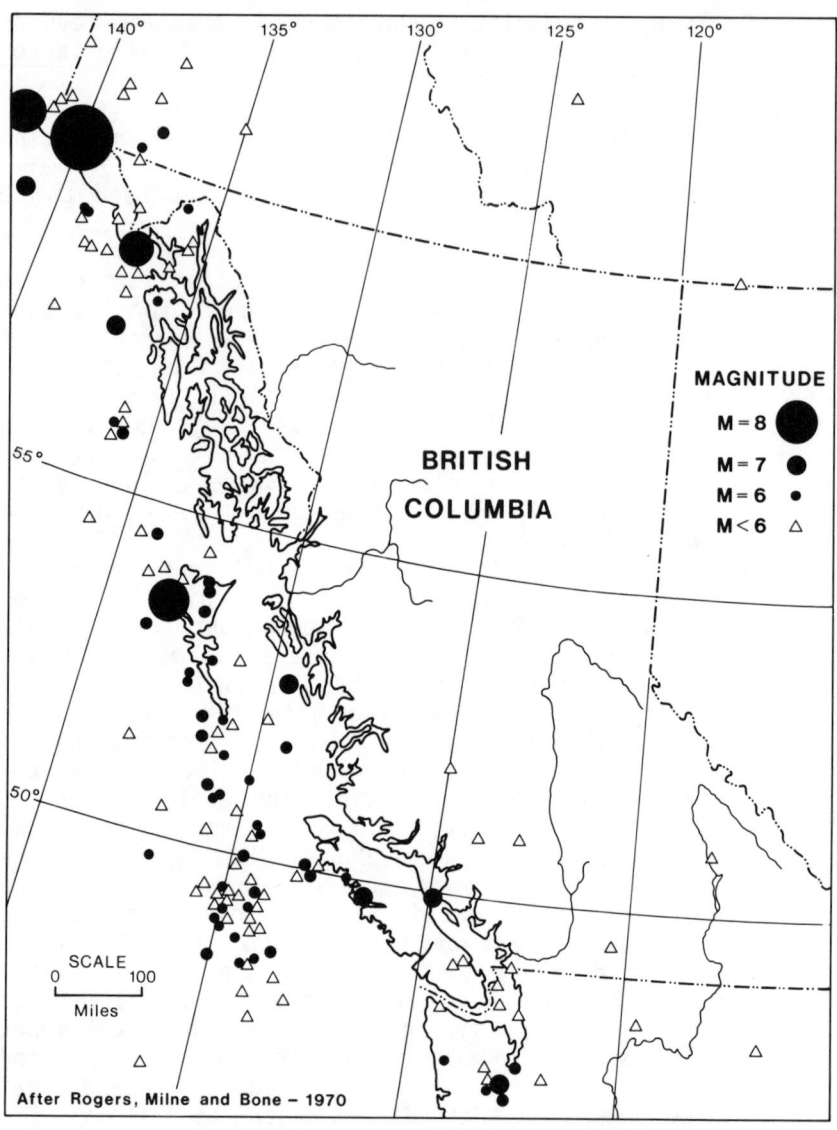

Figure 7. Epicenters of earthquakes in western Canada
        greater than magnitude 5 (1899-1966).

significantly less than on-land impoundments.
11. Aesthetic impacts of well designed and well
    operated submarine disposal systems can be minimal
    compared to some alternative on-land impoundment
    systems.
12. Problems in abandonment following termination of

the mining-mineral processing activity should normally be much less in marine disposal systems. On-land systems often require expensive reclamation or revegetation, at the very least.

CONCLUSIONS

Not all mill tailings nor all potential receiving sites are suitable for submarine disposal of the tailing wastes. More conventional on-land, closed water circuit impoundments are often preferred where practical. Submarine disposal can, however, offer a reasonable alternative at possibly lower overall environmental impact. This paper has attempted to outline some of the tailing and receiving site characteristics that should be considered in evaluating marine disposal prospects.

REFERENCES

1. Evans, J.B., D.V. Ellis, J. Leja, G.W. Poling and C.A. Pelletier. "Environmental Monitoring of Porphyry Copper Tailing Discharged into a Marine Environment", Proc. XIII International Mineral Processing Congress (Warsaw, Poland, 1979), pp. 650-690.
2. Hawley, J.R. "The Use Characteristics and Toxicity of Mine-mill Reagents in the Province of Ontario", Ontario Ministry of the Environment (1972).
3. Fish, Robert. "Mining in Arctic Lands - the Black Angel Experience", CMJ 8: 24-29 (1974).
4. Mikkelborg, E. "Looking over the Black Angel's Shoulder", CMJ 8: 30-36 (1974).
5. Horne, R.A. Marine Chemistry (NY: Wiley-Interscience, 1969).
6. Lee, C.S. and K.Y. Chen. "Migration of Trace Metals in Interfaces of Seawater and Polluted Surficial Sediments", Envir. Sci. and Tech. 11: 174-182 (1977).
7. Pourbaix, M.I.N. Thermodynamics of Dilute Aqueous Solutions (London: Edward Arnold, 1949).
8. Island Copper Mine. "1980 Annual Assessment Report. Volume I", Utah Mines Ltd. (1982).
9. Majima, H. and E. Peters. "Electrochemistry of Sulphide Dissolution in Hydrometallurgical Systems", Paper E-1, Proc. VIII International Mineral Processing Congress, vol. 2 (Leningrad, 1968), p. 5.
10. Kuit, W.J. "The Polaris Mine Tailings Disposal in the High Arctic", paper presented at the Marine Tailings Disposal Symposium, Ketchikan, Alaska, March 22-23, 1982.

11. Lee, G.F., J.M. Lopez and G.M. Mariani. "Leaching and Bioassay Studies on the Significance of Heavy Metals in Dredged Sediments", Proc. International Conference on Heavy Metals in the Environment (Toronto, 1975), pp. 731-764.
12. Poling, G.W. "Sedimentation of Mill Tailings in Freshwater and in Seawater", CIM Bulletin 66: 97-102 (1973).
13. Vreugde, M.J.A. and G.W. Poling. "Effects of Flocculents on Reclaim Water Quality for Flotation", CIM Bulletin 68: 54-59 (1975).
14. Rogers, D.W. and G.W. Poling. "Compositions and Performance Characteristics of Some Commercial Polyacrylamide Flocculents", CIM Bulletin 71: 152-158 (1978).
15. Down, C.G. and J. Stocks. "Environmental Problems of Tailings Disposal", Mining Magazine: 25-33 (July, 1977).

CHAPTER 3

DEEP CIRCULATIONS IN INLETS: INDIAN ARM AS A CASE STUDY

R.W. Burling
Department of Oceanography
University of British Columbia
Vancouver, B.C.
Canada V6T 1W5

INTRODUCTION

The increasing need to expand the extraction, processing and use of resources in and near the marine environment and the desire for long-term preservation of that environment, has lead to clearer recognition of the need for adequate and widely acceptable impact assessments. A basic requirement for assessment of the impact of any activity in any water body is an understanding of its circulations. These largely determine the physical, chemical, biological and geological character of oceans, seas and lakes and their margins by transporting and redistributing heat, dissolved salts (including nutrients), free-drifting animals and plants (plankton, including fish larvae), and all other suspended and dissolved materials including those introduced by man.

In this paper some aspects of circulations in inlets are described, with Indian Arm near Vancouver as a case study. The intent is to draw attention to circumstances which might lead to redistribution of already suspended material or to resuspension of sediments by stirring it from the bottom. The particular application in mind is that of dispersal and deposition of tailings in inlets. It is assumed that only "insignificant" amounts of tailings are expected to appear at depths less than a few tens of meters so currents at greater depths are those of concern. An attempt is made to quantify effects which deep currents such as those in Indian Arm might have on deposited or suspended tailings by stirring and uplifting them to shallow depths. In general such assessments can be made

only from appropriate observations extending over
sufficient time in each region of planned disposal.

It is also assumed that tailings are to be confined
behind a barrier, so inlets with a sufficiently deep basin
separated by a shallower region (a "sill") from more open
waters are those of concern. The practice of tailings
confinement is likely to remain desirable unless and until
the most difficult problem of impact assessment is
resolved, i.e., the long-term effects of deposited
tailings and their biological consequences.

These assumptions imply that neither "drowned river
valleys" (such as Chesapeake Bay) which are typically only
a few tens of meters deep and lack sills, nor "bar-built
harbors" (for example, Laguna Madre), which are even
shallower, are excluded from the discussion below. In both
cases any tidal action or occasional flooding would
preclude confinement. The type of estuary in the Pacific
Northwest region of North America in which there is
precedence of tailings disposal is the "glacier scoured"
fjord or inlet; most have deep basins and one or more
sills formed by glacial moraines. Also, the region is one
of high rainfall and high tidal range; their effects on
silled inlet circulations will be seen to be important. An
important point to notice is that while in the following
five sections the discussion is about changes over weeks
or longer, instantaneously the largest currents are
usually associated with tides; in many inlets these are
large enough to flood water several km up-inlet from the
mouth and back on the ebb. Obviously they are important.
We will mainly consider effects averaged over several
tides.

A very simple "classical" description of estuarine
circulations is presented first, followed by a description
of some salient features during one year of observations
in a shallow-silled inlet. Reference is made to the
problem of prediction following a period of observations.
Processes providing energy for and causing the
circulations, annual and long-term variations, and some
comparisons with other inlets are briefly described.

In keeping with the spirit of the workshop the aim of
this paper is to provide a simple but fairly comprehensive
description of circulations relevant to tailings (and
other effluent) disposal in inlets in such a way that
those not familiar with the topic may better fit their
concepts to a general understanding of this basic aspect
of the physical environment.

A much broader discussion of the many kinds of deep
circulations in inlets, and physical processes associated
with them, is given in an excellent review by Gade and
Edwards (1). This also includes many useful references.

ON CLASSICAL ESTUARINE CIRCULATIONS

Estuaries have been classified in several ways. A geographical classification is referred to briefly in the introduction. Here we are concerned only with the type of circulation which is controlled by the excess of freshwater input at the surface by precipitation and (river) runoff over the loss of freshwater by evaporation.

In the case of a freshwater deficit the estuary is called "negative"; the idealized circulation is like that sketched in Fig. 1A. Loss of evaporated freshwater increases the saltiness and thus density of surface water,

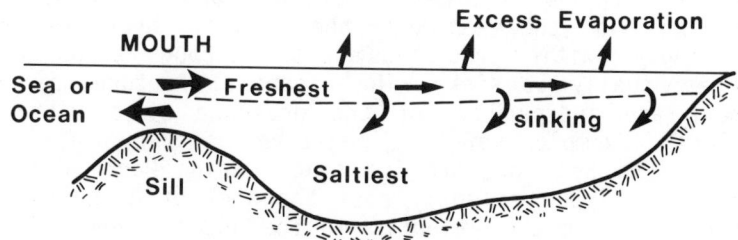

A. In a **NEGATIVE ESTUARY** evaporation exceeds precipitation and runoff.

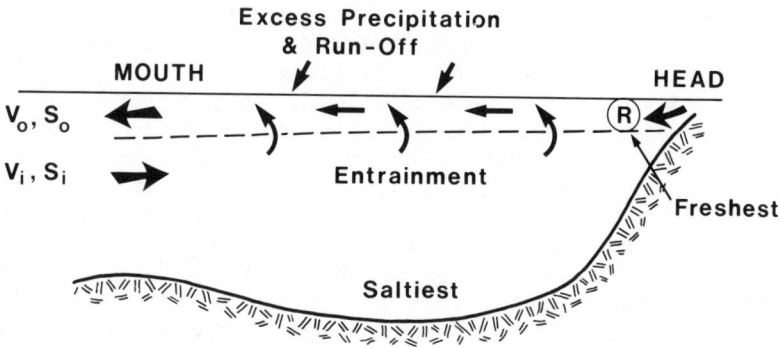

B. In a **POSITIVE ESTUARY** precipitation and runoff (R) exceed evaporation.

Figure 1. Classical, idealized circulations.

so it sinks (by convection) beneath lighter, less salty water nearby. The saltiest water is thus found beneath a fresher surface layer. If there were no connection to an outside sea the water volume would decrease and the saltiness increase. However, such estuaries, which are open to seawater at the mouth or entrance, do not empty over the years, so there must be an average net inflow from outside. Incoming seawater imports salt, so an outflow must export salt at the same rate. The inflow volume must balance both evaporation and outflow, so its volume must slightly exceed and its salt content be less than that of the outflow. The latter volume is clearly that of sinking water, so the outflow must be of salty water beneath a fresher incoming upper layer. In proceeding up the estuary the upper layer becomes increasingly saltier and thus denser through evaporation until eventually it sinks. The essential features in the argument are independent of the presence of a sill. An excellent example of a negative estuary is the Mediterranean Sea, despite its size. During World War II, submarines slipped silently past the Rock of Gibraltor by drifting in from the Atlantic in the upper layer and out again in the lower layer.

Fig. 1B illustrates a "positive" inlet or estuary in which on the average runoff and precipitation exceed evaporation, a circumstance typical of inlets in the Pacific Northwest. In these the long-term runoff greatly exceeds precipitation and in most a significant river input and the freshest water is at the head as indicated by R in Fig. 1B. This water is lighter than saltier water in the inlet so it spreads down the inlet near the surface, mixing with and incorporating saltier water beneath, a process called "entrainment". As these surface waters pass the mouth their salinity* S_0 is not much less than that of deeper water. Salt is thus exported in the surface layer. But the inlet's salt is not entirely removed, so there must be an inflow of salt water beneath the surface layer at the mouth. It is interesting to note that the outflow volume is usually much greater than river flow; the amplification factor is the ratio of incoming salinity to its excess above the slightly lower outgoing salinity at the mouth. This factor may be as high as 30 or even 100; it results from the constancy of volume and total salt in the inlet.

The average circulation just described is the

---

*Salinity is here defined as the ratio: mass of dissolved salts to mass of seawater in parts per thousand. Thus S = 25 $^o/_{oo}$ means 25 gm salt per 1 kg of seawater.

classical estuarine circulation. The rationale for the arguments is derived from observations of the kinds illustrated later in Fig. 3 in which the salinity diagrams all show that in Indian Arm the upper layer is the freshest, and that its salinity is greatest near the mouth. The "tongues" in the 11°C isotherm in August 1974, and the 12°C isotherm in August 1975, are strongly suggestive of an inflow beneath the surface. However, current meter observations, e.g. (2), show that actual profiles of currents with depth do not always conform with the simple flow pattern just described. There is often flow in several layers; they vary with the tides, are strongly influenced by winds (the surface flow may be reversed) and are affected by internal waves. These matters are referred to later.

Despite the short-term differences, the idealized flow is a useful concept which must be something like the average (or residual) flow over a fairly long time, say a few days or more. This becomes evident in the discussion of Indian Arm in the next section. It should be noted that the argument leading to the concept of a classical estuarine circulation is independent of the presence or absence of a sill. In the latter case the depth of the up-inlet flow beneath the surface layer may be determined by the "stratification" of water near the mouth and need not reach the bottom. Some differences between inlets are discussed later.

ON STRATIFICATION, DENSITY AND OXYGEN

In preparation for the following discussion it must be noted that in inlets the densest (heaviest) water is always (except very temporarily) at the bottom and the least dense (lightest) at the top. This configuration or "stratification" is called "stable". If the water is at rest the downwards weight of an arbitrary volume (chunk) of water is supported by the net pressure (which is upwards) exerted by the surrounding water. If some water were moved downwards, it would displace denser water and be subject to a greater upwards net pressure called "buoyancy", so it would bob back towards its initial position. Similarly, water moved upwards bobs back down; slowly moving water is similarly stable. If water flowing quietly over a sill (e.g. in the absence of tides) is denser than that at the same level inside an inlet it tends to flow down the sill slope beneath lighter water until it reaches a level at which the density equals its own, so its buoyancy is zero and it tends to spread out horizontally.

The density of water in Indian Arm deeper than about 30 m most often lies in the range 1018 to 1022 kg/m^3. Density is normally expressed as sigma-t, which for our purpose equals density minus 1000 kg/m^3 (the density of pure water at 4°C). The range above is thus 18 to 22 kg/m^3 in sigma-t units. The major control of density in inlets in temperate latitudes is by changes in salinity rather than temperature. An increase in salinity of 0.1 $^o/_{oo}$ produces the same increase in sigma-t, about 0.08 kg/m^3, as does a <u>drop</u> in temperature of roughly 0.6°C in inlet water with temperatures near 7° to 10°C. Temperature changes become more effective at higher temperatures and less effective in colder water. The seasonal range in salinity observed in Indian Arm is at least three times as effective as the temperature range in determining density (and sigma-t) near 30 m, and above four times at 200 m.

Dissolved oxygen is a valuable indicator of processes in inlets. It is measured in units of milliliters (mL) of oxygen dissolved in one liter of seawater. Near the surface it tends to come into equilibrium with the atmosphere's oxygen by exchange across the surface (equilibrium or saturation values at the surface are typically 6-7 mL/L in summer and up to 8 or 9 mL/L in winter). The other source of dissolved oxygen is photosynthetic production by plant life: the phytoplankton near the surface. Reduction of oxygen with depth is due to its removal from solution by bacteria to oxidize organic material, i.e., for the decay of plant and animal detrital material as it sinks towards the bottom.

Typically then, the highest oxygen values are near the surface where the sources are and the lowest, because of bacterial decay of detritus, near the bottom. However, this distribution is affected (as also are distributions of heat and salt) by bodily transport, called advection, with the circulation and by mixing, as seen in subsequent sections.

DISTRIBUTIONS OF PROPERTIES DURING ONE YEAR
IN INDIAN ARM

Indian Arm (see Fig. 2) is about 22 km long and averages 1.6 km in width. In this paper we are concerned with processes leading to the distributions of properties in vertical sections (Fig. 3), during several months selected from a series of roughly monthly observations extending over 18 months by L.W. Davidson (3). Fig. 3 shows the distributions of salinity, temperature and oxygen in vertical sections (vertical slices) roughly along the deepest trough between station

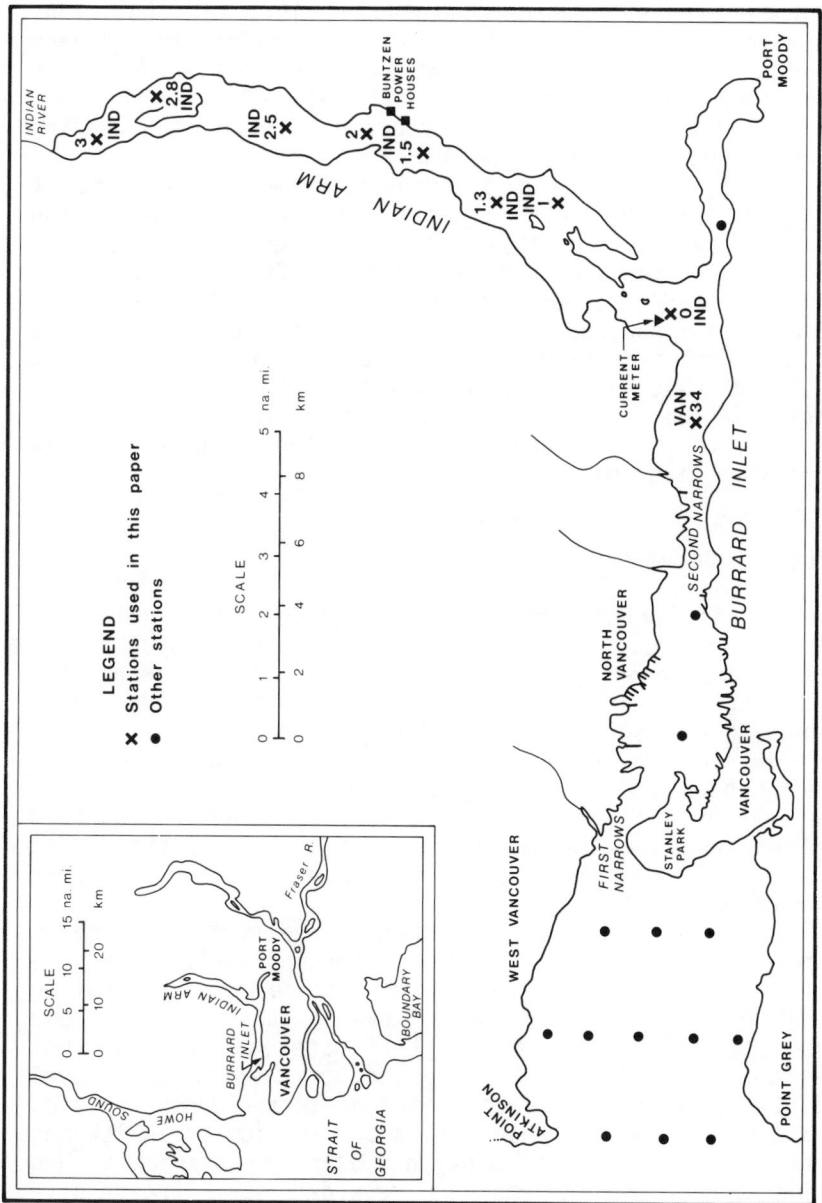

Figure 2. Map of Burrard Inlet and Indian Arm showing station locations.

VAN 34 just outside the sill at the mouth of Indian Arm to close to the head. Stations IND 0, 1, 1.5 and 2 (see Fig. 2) were used for all sections, and IND 2.5 and 3 in the Augusts of 1974 and 1975, but IND 1.3 and 2.8 in December 1975 and January, February and April 1974. The positions of these stations are marked with ticks at the top of each diagram. The vertical exaggeration is about 80; the slopes down from the 18 m deep sill, at the mouth near the left of the diagram are all quite gradual, nowhere exceeding about 5° and averaging less than 1.2°.

First, compare the distributions in the Augusts of 1974 and 1975 at depths greater than about 100 m. During the year salinity in this deep water increased by 0.3 to $0.5^{\circ}/_{\circ\circ}$, temperatures decreased by 0.4 to 0.5°C and oxygen content increased by about 2 mL/L. It will become evident from subsequent paragraphs that these changes are just the opposite from those to be described during eight months of that year, and indeed during all of some other years. The observed distributions of Fig. 3 will be used to infer those processes leading to the changes shown in successive diagrams.

August to December

It is noted that in the August 1974 diagrams saltiest, coldest and thus densest water is closest to the bottom, and salinity and thus density decrease towards the top, so the stratification is stable. Oxygen decreases downwards as described in the previous section.

The tongue in the 11°C isotherm in August 1974 has already been noted as suggestive of water entering over the sill beneath an outflowing surface layer as typified ideally by the classical estuarine circulation. The incoming tongue appears to be confined to depths shallower than about 80 meters.

By early December, just over three months later, at depths less than about 125 m (except very near the surface) the salinities had increased indicating increasing density at these levels; this is probably caused by inflow over the sill of water as dense as that down to 100 to 125 m so it spread out down to this level and of course must displace water at lesser depths upwards so it must leave the inlet. This combination of inflow and outflow is required to maintain a constant average volume inside the inlet and may be regarded as a progressive deepening of the incoming estuarine flow as its density increased progressively. This point will be discussed further, and confirmed, when describing Fig. 4. Ignoring for the moment tilts in the isotherms and isopleths of oxygen, the mean values of oxygen and temperature shallower than 125 m are readily explained in terms of the seasonal marches of values in incoming and inlet water as fall progressed; in particular the mid depth temperature maximum resulted from the entry of water warmer than $9^{\circ}C$ through November and surface cooling within the inlet.

At depths greater than 125 m salinities decreased by roughly 0.05 $^{o}/_{oo}$ and temperatures increased by up to $0.1^{\circ}C$; these both imply a small decrease in density at these depths. The inference is that these changes resulted from the mixing downwards from above 125 m of less saline, warmer water. Of course the higher salinity, colder deep water is simultaneously mixed upwards through 125 m. Oxygen content, however, decreased in the deep water during the fall (by up to 0.5 mL/L) due to decay of detritus, despite downwards mixing of higher values.

The slopes of isohalines down to the right in December are suggestive of inflow of salty, and so dense, water down the sill to mid depth; a similar feature in salt distributions obtains in some later diagrams. The isopleths of temperature and oxygen, however, slope in the opposite direction. Down to about 125 m these suggest transport of properties down the sloping bottom as water inflows and very slow transport along the inlet. The continued presence of these slopes to greater depths is suggestive that mixing is most intense in the vicinity of the sill slope, a point on which further evidence is presented later.

December to January

At the end of this seven week period water more saline (greater than 27.3 $^{o}/_{oo}$) than any in the inlet at the start appeared near the bottom. The continuation of the 27.3 $^{o}/_{oo}$ isohaline along the sill slope is only partially supported by the data; however, this value was observed at the depths indicated at the stations just outside and just inside the sill, at that where the isohaline crosses 115 m and at the two central stations

Figure 3 (pp. 94-99). Distributions of salinity, temperature and oxygen in vertical sections in Indian Arm, in selected months during one year. Note that the time elapsed between successive diagrams is varied. Data used were observed by Davidson (3).

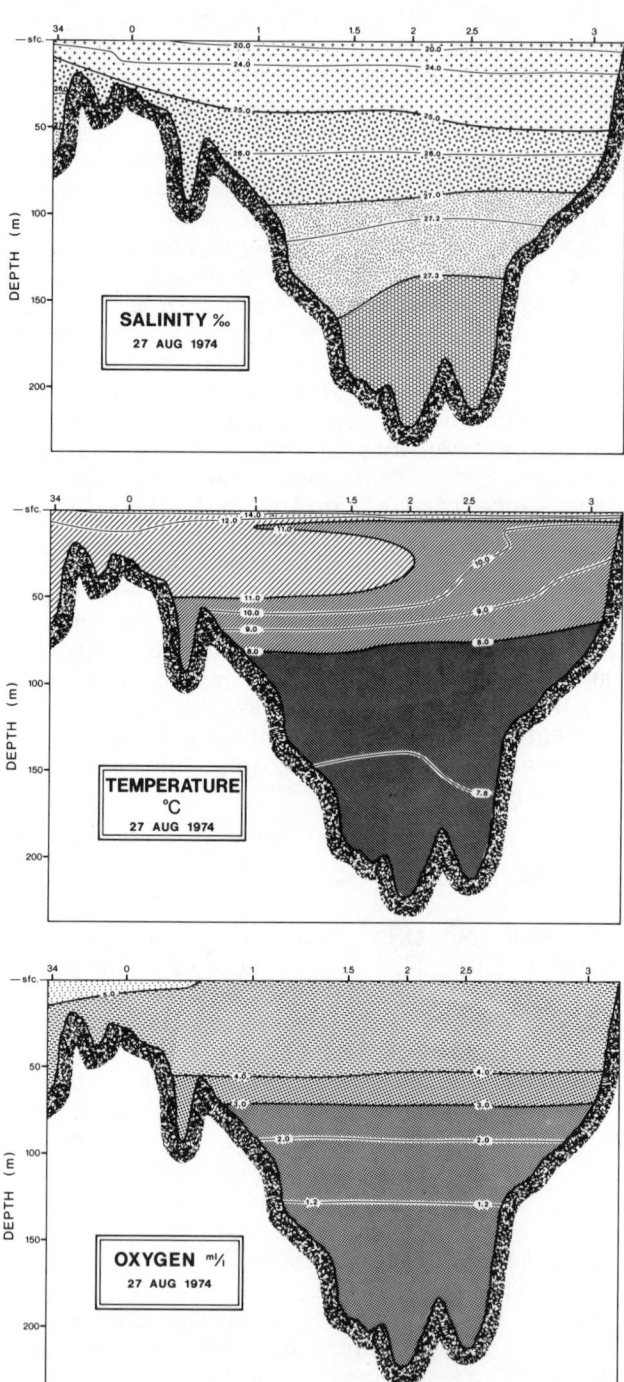

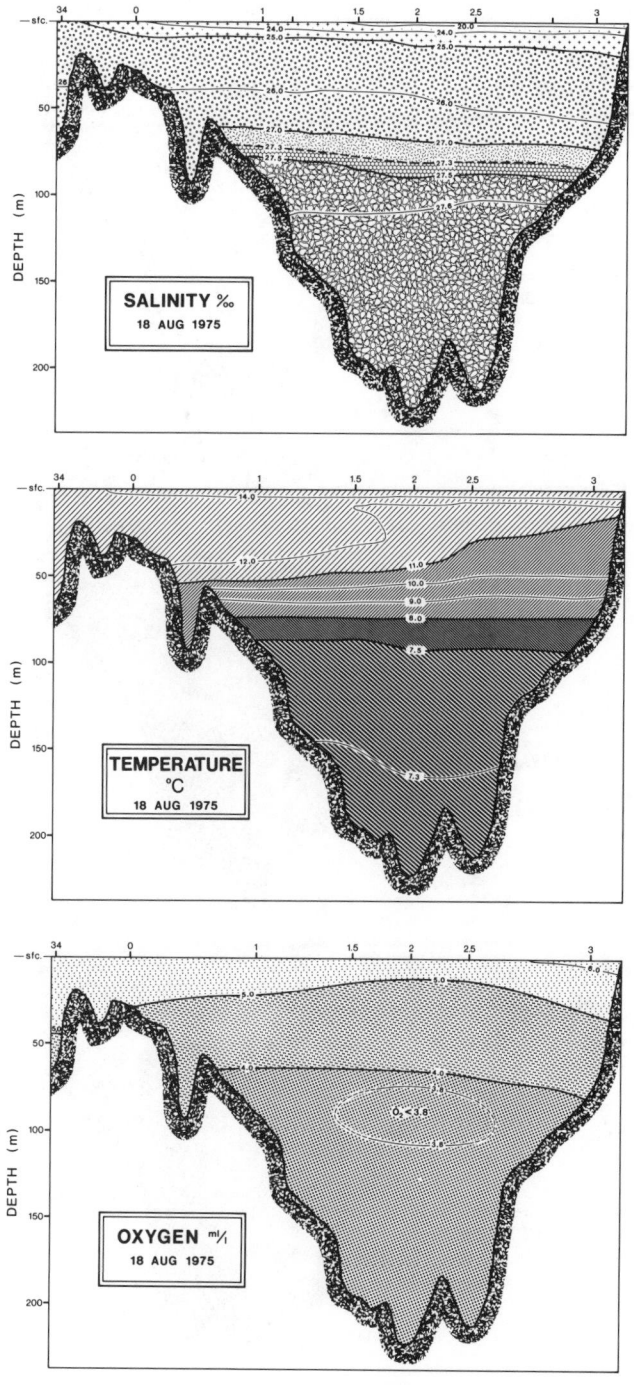

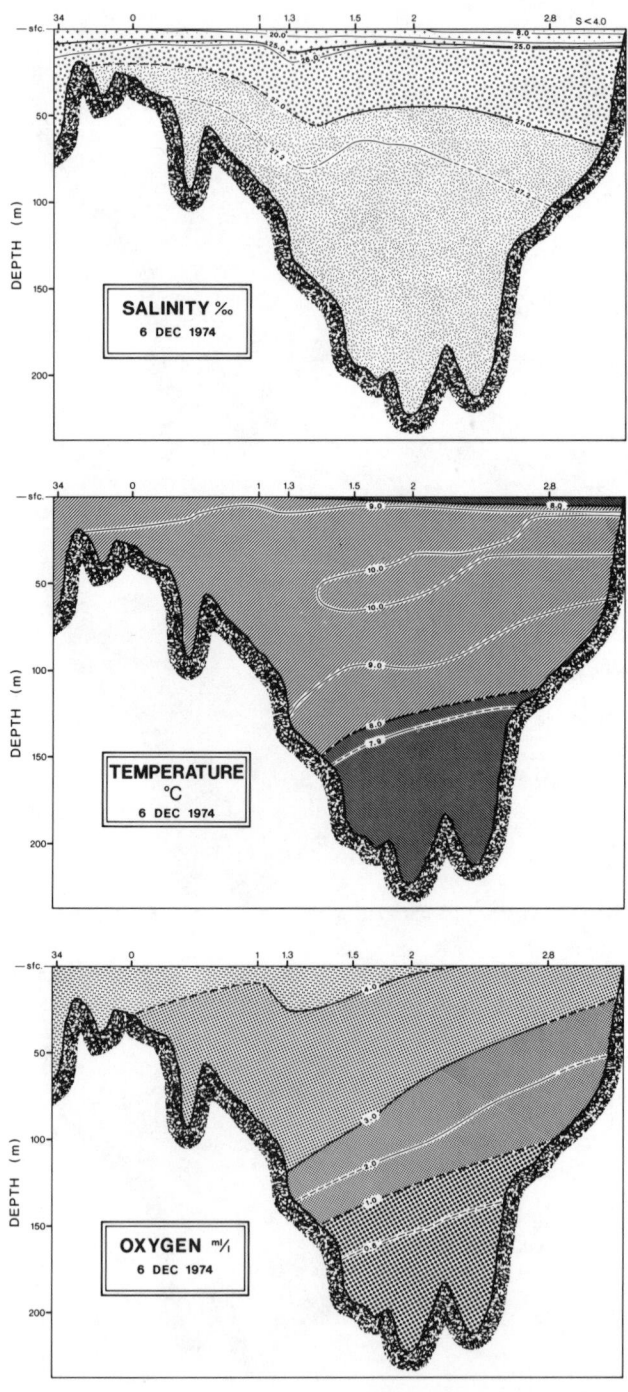

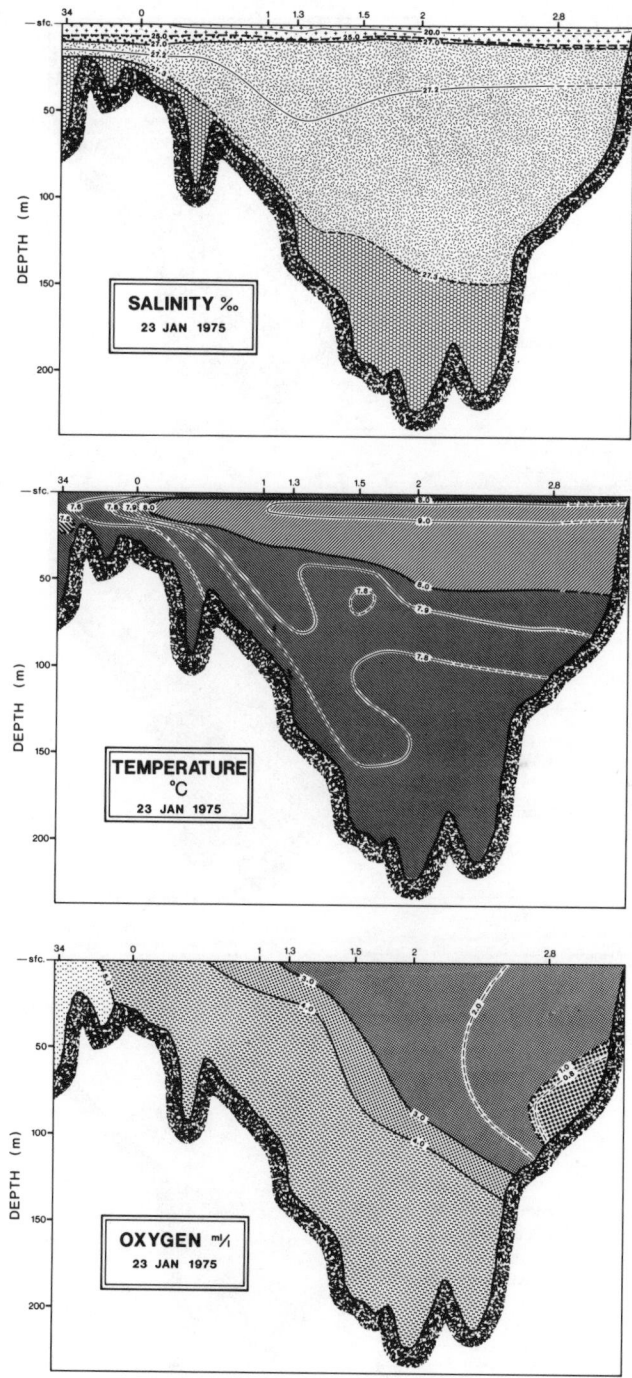

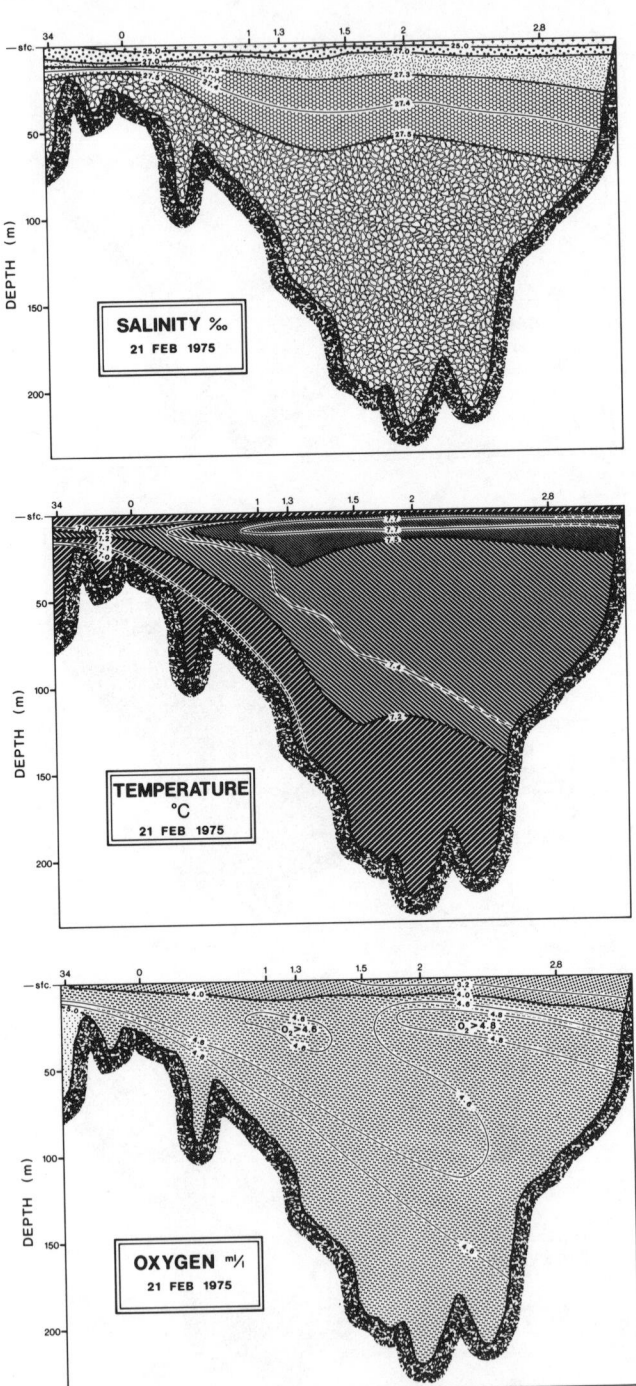

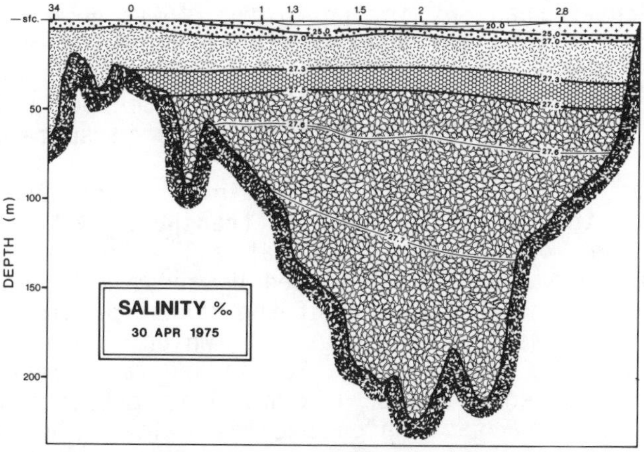
SALINITY ‰
30 APR 1975
DEPTH (m)
sfc.
50
100
150
200
34
0
1
1.3
1.5
2
2.8

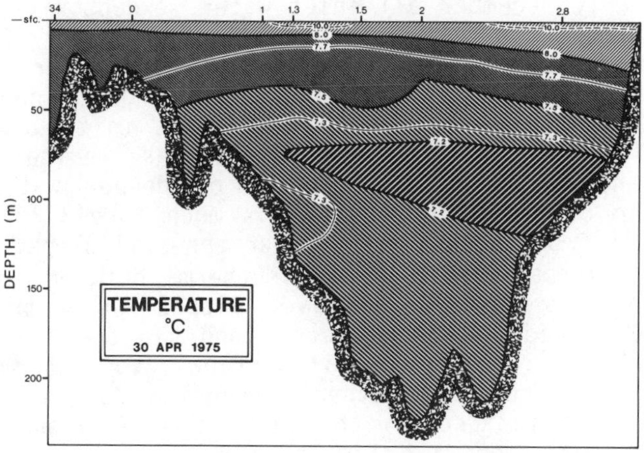
TEMPERATURE
°C
30 APR 1975
DEPTH (m)
sfc.
50
100
150
200
34
0
1
1.3
1.5
2
2.8

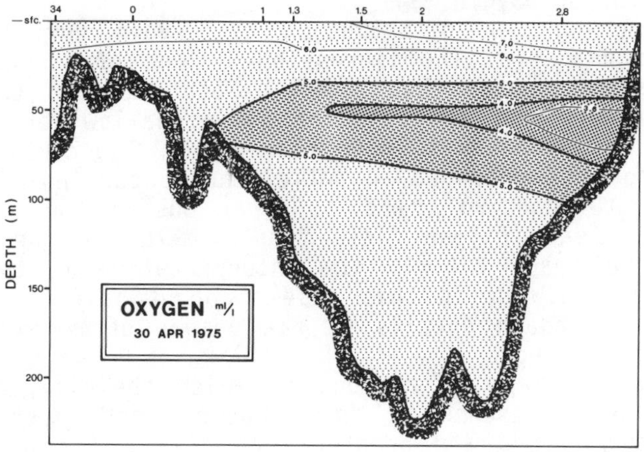
OXYGEN ml/l
30 APR 1975
DEPTH (m)
sfc.
50
100
150
200
34
0
1
1.3
1.5
2
2.8

over deepest water. The inference is clear that incoming water, denser than any in the inlet, flowed down the inlet slope to the greatest depths; in so doing it must have displaced water near the bottom, a process called "deep water renewal". Of course, as the new water pours in over the sill, an equal volume must pour out in surface water through the mouth.

From the distributions of salinity alone and on the assumption that advection (bodily transport) with the flow acted alone, one might infer that all water deeper than 130 to 150 m was simply displaced upwards. However, it is by now well-known that flows of the kinds (particularly in tidal cycles, see later discussion) which transport water over sills and down sill slopes are associated with vigorous turbulence and so mixing. This obscures precise interpretation, a fact which becomes evident in the next few paragraphs.

In early December all inlet water (except river water near the head) was warmer than $7.8^{o}C$; by January 22 water colder than $7.8^{o}C$ had entered. Again deep water renewal can be inferred, but through a greater range of depth, to all water beneath about 100 m. The loops in the 7.8^{o} and 7.9^{o} isotherms might be interpreted as indicative of upwards and mouthward displacement of deeper water near the center of the inlet, but it is much more likely that they result from intense mixing near the sill slope.

The reversal of slopes of isotherms between December and January is most informative. Slopes which trend up towards the head, as in December (and see the Augusts of 1974 and 1975), conform with inflows to intermediate depths, but slopes of isotherms trending down along the sill slope are in accord with the much more vigorous flow associated with renewal. Similar remarks are appropriate to isopleths of oxygen.

The change in distribution of oxygen during this period is dramatic. The impression is one of "turnover". High oxygen (greater than 4 mL/L) water appears to have poured in over the sill and the quite shallow water from near the head appears to have flowed out. The combination appears to have produced an anticlockwise turning of most isopleths through 90°, and to have pushed the deepest water with least oxygen (less than 1 mL/L) up the slope towards the head. Clearly the interpretation is that of deep water renewal, in this case of about half the inlet volume. Once again details of the volume of renewal are obscured by mixing.

If we overlap the regions in which the distribution indicates the presence of "new" water (for temperatures greater than $7.8^{o}C$ and greater than 4 mL/L for oxygen), the volume renewed would be about 60%. However, mixing reduces the contrast between "old" and "new" water so this is likely an underestimate. Observations made on January 3, 1975 at station IND 2 over the deepest basin showed that the renewal process had not started by that date. Davidson (3) assumed 55% renewal to have occurred during the interval 3 to 22 January. The consequent average rate of inflow (and outflow) was about 720 m^3/s, in not unreasonable agreement with an estimate of about 1200-1700 m^3/s made from data from a current meter moored near the bottom on the slope just inside the inlet. It is worth noting that the observed currents averaged over successive diurnal tidal periods during this time were all up-inlet and averaged about 9 cm/s. Indeed, during the 92 day measurement period between 5 December and 7 March, the diurnal averages were up-inlet on 77 days; they were largest between 3 and 20 January, and again $\sim$ 8 cm/s decreasing to $\sim$ 4 cm/s during February. Davidson's measurements certainly support the notion of inflow over the sill.

January to February

Without going into details, as in preceding paragraphs, it is clear that this was also a period of deep water renewal which was even more effective than that just discussed. On January 22 no water was as saline as 27.4 °/oo nor as cold as 7.6°C inside the inlet (except close to the sill), but by 21 February water more saline than 27.5 °/oo and colder than 7.5°C filled much of the inlet. Changes in the oxygen distribution support the renewal. From the temperatures more than 80% of the inlet's volume was renewed; Davidson's estimate from this, of the average rate of inflow during 26 days was 820 m^3/s, in good agreement with 670-1130 m^3/s estimated from his current meter observations.

February to April

Information observed on 26 March 1974, but not presented here, showed that little or no renewal had occurred following 21 February. However, between 26 March and 30 April there was substantial further renewal. New water with salinities higher than 27.6 °/oo and oxygen content in excess of 5 mL/L indicate a renewal of at least 60% of the volume during this spring month. The total volume of renewal from 3 January to 30 April was at least twice the inlet volume (of 2.3 x 10^9 m^3), which is several times that of Burrard Inlet and the bay east of Points

Atkinson and Grey. Clearly the ultimate source of renewal is the Strait of Georgia.

The high oxygen values shallower than about 30 m are almost certainly due to its production during the spring phytoplankton (plant) bloom, and the minimum near 50 m due to fast decay of the consequently abundant detritus. Temperatures mainly increased during April, except for deeper cold water displaced to 75 m, because the incoming water had warmed.

April to August

During this 11.5 week period the salinity of surface water down to about 10 m increased. This was simply due to higher runoff in April than in August. The down-inlet increase in salinity in the surface layer in both months indicates an average estuarine circulation with surface outflow over the sill as explained earlier, and in August the tongue in the $12^{0}C$ isotherm provides additional support for an inflow just beneath the surface layer. At all depths greater than 10 m the salinity decreased during the period. Down to perhaps 70 or 80 m this was probably due mainly to advection in over the sill, of water freshened by high runoff during May to July in rivers outside Indian Arm, by the inflow part of the estuarine circulation. At all depths greater than 80 to 100 m the decrease resulted from mixing downwards of fresher, and so lighter, water above.

Temperature increased during this period at all depths. At the surface this would result from summer heating but at greater depths the explanation is the same as for the salinity decrease. Down to about 80 m the warming was a result of estuarine circulation, and at deeper levels warmer temperatures were mixed down.

The change of oxygen content near the surface simply reflects the reduced photosynthesis and lower equilibrium values of warmer water in August than in April. The decrease of oxygen at levels beneath about 70 m was clearly due to the dominance of detrital decay.

Summary

The changes in deep water over the year, referred to in the early part of this section, can now be seen to have been achieved during the four months of deep water renewal in winter and spring. The net salinity increase and temperature decrease were somewhat reduced by downwards diffusion (i.e. turbulent mixing) of fresher, warmer water during the remaining eight months of late spring through the fall, and the net oxygen increase, by detrital decay.

The flows associated with the estuarine circulation and deep water renewal must be powered by some energy source or sources, and so too must the turbulence. The likely energy sources are discussed in a later section. Before doing this it is useful to discuss in the following two sections the circumstance controlling the type and degree of exchange of water with the region outside the inlet, and also to examine how often deep water renewal occurs there and to discuss briefly its prediction.

THIRTY-TWO MONTHS OF CHANGE IN INDIAN ARM

Fig. 4 shows the march of density at several depths, between March 1973 and October 1975, at Station IND 2 (see Fig. 2) over the deepest part of Indian Arm. In this plot, sigma-t (= density - 10^3 kg/m^3) is plotted (increasing downwards) against time. In the plot, each line represents a depth; an upwards trend with time represents decreasing density at that depth. The lines are for 10, 30, 50, 75, 100 and 200 m, the shallowest depths being represented by the uppermost line and the greatest depth by the bottom line.

The data used were observed on occasions marked by a tick below the top margin of the diagram (some gaps in lines represent missing data) and are available in University of British Columbia (UBC) data reports, numbers 35, 37 and 41. They include Davidson's (3) observations from May 1974 to October 1975.

A particularly useful aspect of Davidson's observations is that not only did they embrace the interesting year (Fig. 3) in which occurred both "normal" estuarine circulation with deep diffusion and an intense deep water renewal in Indian Arm, but they also included observations throughout Burrard Inlet (Fig. 2), an intermediate source (and sink) region for Indian Arm water. All source water from the Strait of Georgia passes through the two narrows where it is mixed and modified. Here, we discuss only the immediate source of inlet water, which is at station VAN 34 just outside the sill. Water entering Indian Arm must come from near sill depth (about 20 m) or possibly slightly deeper. To indicate its influence on Indian Arm, the trends of densities at 20 m (upper dashed line) and 30 m (lower dashed line) at VAN 34 are also shown in Fig. 4; water between these depths is represented by the shaded region.

Before discussing this diagram it should be noticed that density is chosen as the "dynamic" property

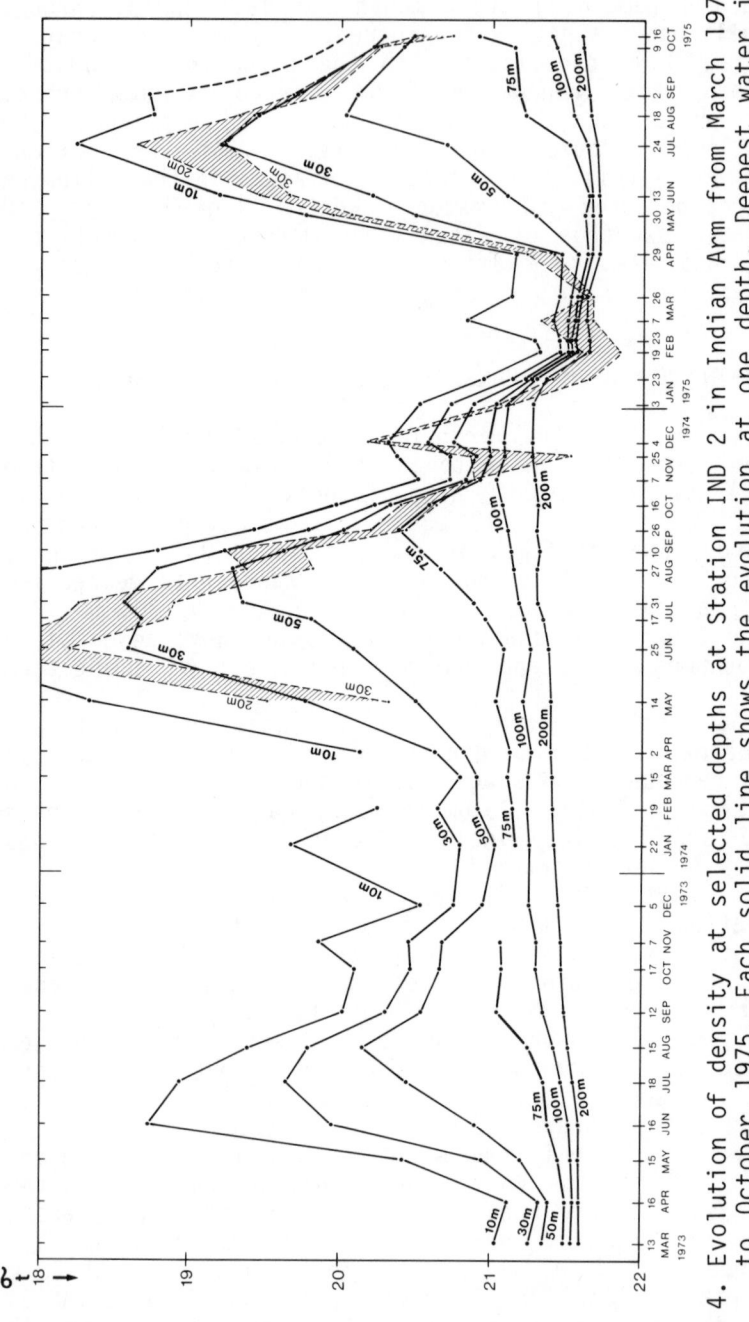

Figure 4. Evolution of density at selected depths at Station IND 2 in Indian Arm from March 1973 to October 1975. Each solid line shows the evolution at one depth. Deepest water is represented at the lowest part of the diagram and shallow water at the top. Density of "source" water from between 20 m (upper dashed line) and 30 m (lower dashed line) at Station VAN 34 is represented by the shaded region.

associated with buoyancy. A similar plot of trends of salinity at each depth (not shown here) is very similar to Fig. 4, as might be expected, except for minor differences in timing of some peaks and in the relative steepness of slopes of some lines. Thus we can relate the density changes to changes of salinity.

The most obvious characteristic of the density trends inside the inlet is the annual cycle, the change in density being greatest at 10 m and diminishing with depth; by 100 m and deeper an obvious annual cycle throughout the record is not apparent. This aspect is further pursued below and in the next section. Starting in the spring of each year (March to May) density decreases at depths shallower than about 50 m, an effect of snowmelt in the region. A decrease at 75 m starts somewhat later. Nearest the surface at 10 m, water is lightest and freshest in June or July, but at successively greater depths the peak in freshness occurs at successively later times until at 75 m it is freshest in September (in 1973 and 1974).

The reason for this sequence becomes apparent from inspection of the shaded region representing the immediate source water just outside the sill. As this entered as the estuarine inflow during the springs of 1974 and 1975 it clearly contributed directly to water in the center of Indian Arm down to 30 m or so, but the density decrease and freshening at 50 m and deeper is more likely to have been achieved by downwards mixing of fresher water from above. This mixing of light water down against upwards buoyancy requires energy (compare pushing down a football). It is notable that as the stability (proportional to the density gradient) increases during the spring, that is as the lines in the diagram move further apart, the energy needed increases. However, the effect of mixing does not decrease but becomes more noticeable and extends to greater depths as the season progresses. For example, the 100 m and 200 m lines slope upwards most steeply starting in about July of each year when the density contrast (gradient) between 30 and 200 m is greatest. This implies a source of energy for mixing which is great when the stratification is greatest. Possible energy sources are examined later.

To continue examining the effect of the source water, we notice that as its density starts to increase (from June 1974), where the shaded region has peaked and starts down, so too does the density (and saltiness) at 30 m, and successively each depth line peaks and turns down just where the shaded region reaches it. This shows that as source salinities increase following maximum snowmelt runoff, the estuarine inflow penetrates deeper and deeper inside the inlet to depths where the density equals that of the source. This series of events during fall 1974 was referred to in the previous section. At each stage shallower, lighter water is displaced upwards and largely replaced by denser water; this is evident by the slope down of all shallower lines. That there is mixing of some of the shallower and source water at these depths is clear because the lines remain vertically spaced but are closer; this means that while stability is maintained it is reduced successively to increasing depths.

The sequence through fall of 1974 continued until November when it reached 100 m, as shown by the small down-slope of that line. In fact, it reached down to 125 m but not 150 m or deeper. The very dense source water at 30 m in late November either did not enter the inlet or its density was reduced by mixing during inflow. Note that in about August 1975, the density at 50 m started to increase but source water was apparently lighter. This would seem to imply that source water at station VAN 34 denser than that shown in Fig. 4 was present in August between the observations. A continuously recording technique would help to resolve such uncertainties.

A sharp observed increase in rainfall and river flows in November and December 1974 undoubtedly caused the decrease in density of source water on December 4; this led to freshening above 75 m. Density of source water increased following the freshet and by January 3, 1975 had again resumed the density increase down to 100 m.

In January and February source water was denser than any water in the inlet. This led to the "deep water renewal" described previously. Fig. 4 shows that density decreased at all depths in this period.

A small freshet in local runoff in March 1975 freshened the source and inlet waters to 30 m and essentially stopped the renewal process. There was probably, however, some inflow, perhaps to 100 m (seen by small down-slopes in the 50 to 100 m lines) and mixing down to 200 m (a small up-slope) until late March. In April further deep water renewal is evident. Source water dense enough to have caused this was not observed; again it was probably missed by the time sequence of observations.

Two final features are to be noted on Fig. 4. One is the relative consistency in seasonal trends of density shallower than about 75 m during spring through fall of all three years. Differences between years in these trends are mainly of degree of freshening. The large decrease in density in 1974 was undoubtedly due to a larger than normal (about 20%) snowmelt from local rivers, including the Fraser River into the Strait of Georgia.

The second feature is the marked difference at all

levels between the density trends during the winters and early springs of 1973-74 and 1974-75. That the latter was a period of deep water renewal is explained above. That the former was not is by now obvious because the density at 200 m either decreased or was constant from March 1973 to January 1975, the slow changes being due to downwards diffusion of less dense water from above.

Summary and Discussion

To summarize this and earlier sections, it is emphasized that an estuarine type average circulation always occurs. It is driven by maintenance of a surface layer which is less dense than water below inside the inlet. Its direction (compare negative and positive estuaries) should normally be controlled by the surface freshwater budget. In Indian Arm, a shallow silled inlet in a region of quite high precipitation, the circulation is positive; the outflow is of surface water balanced by an only slightly smaller (or equal, if runoff were zero) inflow over the sill. The nature of this inflow and its effects on the inlet are determined entirely by properties of source water just outside and near the same depth as the sill, and in particular by the relation between the density of source water and the stratification inside the inlet. The latter controls the depth of penetration, and consequently of replacement. The case of deep water renewal is a particular and intense example of an estuarine circulation forced by sufficiently dense source water.

Clearly, conditions and processes outside the inlet control the density and stratification of the source water itself, and thus they control the intensity of the inlet's estuarine circulation. The question arises as to whether outside processes may entirely control an inlet's circulation. This question is partially addressed in the final section.

DEEP WATER RENEWAL IN INDIAN ARM AND ITS PREDICTION

The objective in this section is to attempt to determine the effectiveness of a series of observations for evaluation of the likelihood of deep water renewal, which is the most likely of the average flows to stir and transport deep bottom sediments, and to infer the types of information needed. In the previous section we saw the value of observing source water just outside the inlet. I believe this should be standard practice while observing changes inside.

Fig. 5 shows the annual variations and trends of temperature salinity and oxygen at 100 m and 200 m at station IND 2 from early 1968 to April 1972. Only data observed by members of the Institute, now the Department of Oceanography at UBC, have been plotted. Where data are sparse each observation is plotted as an ● from 200 m, and a o at 100 m, but a solid (200 m) or dashed line (100 m)is drawn where observations were more frequent than about six per year. Data for 1968 to 1974 were plotted in a rather similar fashion and discussed by Pickard (4). (In the present plot one oxygen value at each depth from early in 1969 has been omitted as probably in error.)

From the discussion of Fig. 4 it is evident that during periods of deep water renewal in Indian Arm the density must increase at 200 m; thus we might expect salinity to increase sharply (upwards in Fig. 5) because it is the major determinant of density. A simultaneous sharp increase in oxygen at 200 m is anticipated as oxygen depleted deep water is replaced by source water with high values typical of shallow depths (note values of O_2 near the sill in Fig. 3). During replacement temperature might change in either direction as it has a relatively small effect on density; even a casual glance at Fig. 5 shows that such changes do occur.

Between brief periods of renewal we expect salinity at 200 m to decrease slowly (i.e. the line should trend downwards) because of the diffusion of fresher water from above, and oxygen should slowly decrease (also a downwards trend) through its use for detrital decay. Again a glance at Fig. 5 shows that these events occur.

From salinity observations at 200 m alone, and based on the above expected changes, we can infer apparently with certainty that renewal occurred about the time of the New Years starting 1969, 1970, 1971, 1973, 1975 and 1979 (even though data in this last year are sparse). Note that oxygen changes at 200 m around each of these New Year's Days are consistent with renewal in 1969, 1970, and 1979 where data are sparse (taking into account an anticipated decay between mid 1969 and mid 1970); the 1971, 1973 and 1975 salinity events are clearly confirmed by the oxygen changes as deep renewals. Thus from observations of salinity alone extending over nine New Years (1969 to 1975, 1981 and 1982) when salinity observations were dense, we can infer five renewals and one other occasion (1979) with certainty when two sparse observations happened to straddle a large salinity change.

From the early period of more intense observations (1968 through 1975), from salinity alone we would infer that there were two New Years of non-renewal, 1972 and

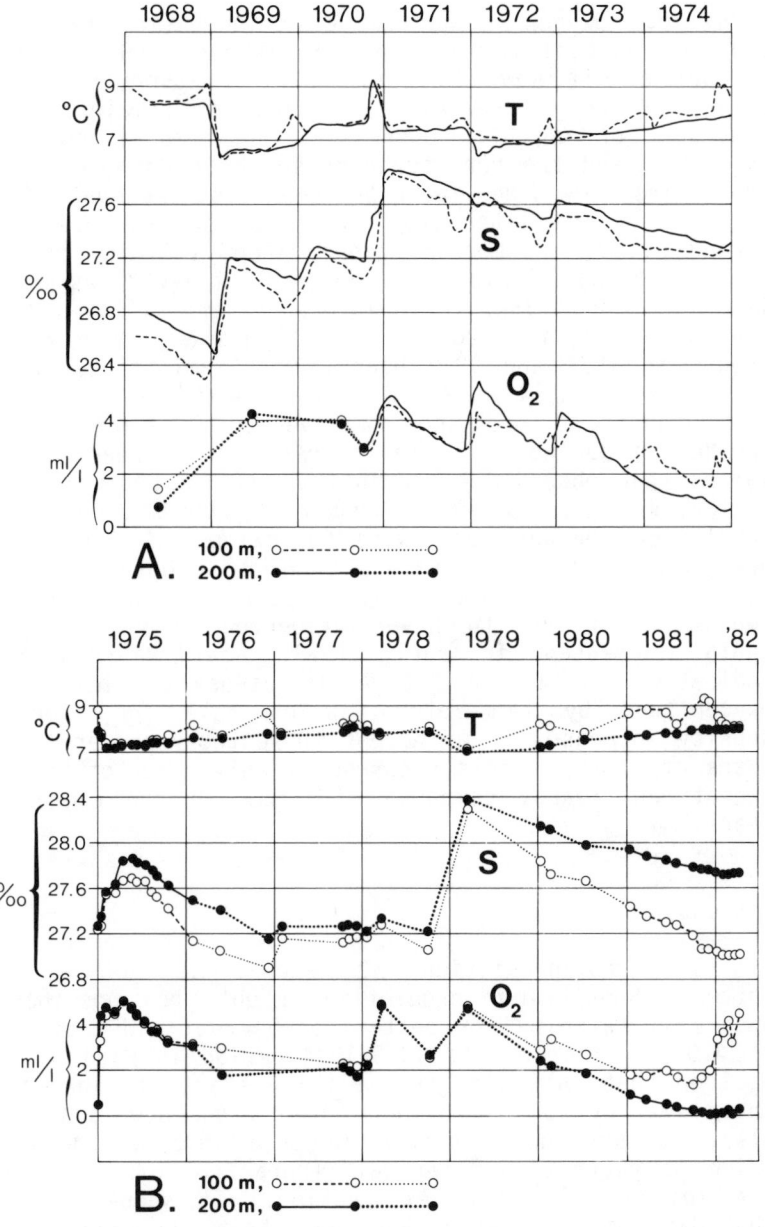

Figure 5. Long-term and seasonal changes in temperature, salinity and oxygen at 100 m (dashed line) and 200 m (full line) at Station IND 2. Figure 5A, for 1968 to 1974, uses data described earlier by Pickard (4). Figure 5B is for the period 1975 to April 1982. Dotted lines are drawn where data are sparse.

1974. The oxygen trend and also increasing temperature confirm non-renewal in 1974. But oxygen shows a definite renewal at the New Year in 1972. The reason, however, is clear. Although salinity at 200 m remained nearly constant, or slightly decreased, the temperature dropped by 1.2°C, and the density increased by 0.04 kg/m^3. During these first seven New Years there were six renewals.

During the later seven years of sparse observations through 1980, and more frequent since, salinity trends at 200 m **(dots)** indicate non-renewal at New Years in 1976, 1980, 1981 and 1982. These inferences are all supported by trends in oxygen at 200 m. As 1979 was a year of renewal, events in 1977 and 1978 bear close examination. Salinity increases slightly at 200 m over the 1977 New Year, and by the year's end had the same value. Assuming diffusive freshening during 1977, salinity must have reached higher values than observed early in the year. A renewal is thus indicated in early 1977. Oxygen at 200 m supports this conclusion, because otherwise following mid 1976 it would have decayed to lower values than those observed in late 1977. Also, the temperature appears to have remained constant and not to have influenced density. Salinity again increased in early 1978; again a renewal is indicated. It is this time unequivocally supported by oxygen and by temperature, which fell and aided the density increase. If we accept these two years and 1979 as years of renewal, this occurred at only three of the seven recent New Years, compared with six in the first seven years.

Discussion

Clearly, prediction of renewal is not a simple matter. Short-term prediction might be possible by observing salinity, or better density, at mid depth. From the 100 m line for the 1969-1974 period in Fig. 5 it can be seen that in Indian Arm salinity starts to increase at that level by up to two months prior to renewals at 200 m. Also, it does not increase in late 1973 nor late 1981 prior to years of non-renewal. (Data from other years are not suitable for this examination.) It should be noted that data observed at lesser depth than 100 m would not be suitable for this purpose. It is evident from Fig. 4 that seasonal influences extended in each of the three years to 75 m. Only one was a year of renewal; the others would be false alarms. Continuous observations of source water might give very short-term warning (about a week?); at least they would indicate a possibility of renewal.

Meteorological conditions over an extended region

near and outside an inlet system might possibly be used to predict changes in source water for some inlets, a matter referred to later.

Assessment of probability of renewal from observations may not be possible unless a very long series of suitable observations is available. For example, from Fig. 5 we can see that renewal followed every occasion on which the salinity at 200 m late in the year was less than 27.4 $^o/_{oo}$, in two years out of four (except 1974 and 1976) when it was between 27.4 and 27.66 $^o/_{oo}$, that is 50% of the time, but in no year (i.e. 1980, 1981, 1982) in which it was greater than 27.66

The salinity at 200 m in Indian Arm decreased to 26.66 $^o/_{oo}$ in April of 1982. We may reasonably expect it to fall during the rest of 1982, but judging from observed rates of decrease, not to as low as 27.4 $^o/_{oo}$. Does this mean that there is a 50% probability of renewal at the next New Year? Pickard (4) discusses an earlier series of Indian Arm data from 1956 through 1961 during which time the density at 200 m was less than 27.4 $^o/_{oo}$ late in each year so renewal following each seems probable. Although the earlier observations were less consistent, renewal is demonstrated by these data to be certain for two of the five New Years, very probable in one, probably in another, but a definite no in the fifth year. Even with the fourteen years of data from within Indian Arm since 1968, simple predictions based on them are not yet completely reliable.

SOURCES OF ENERGY FOR DEEP CHANGES

This section starts with an examination of the implications, about deep processes, of changes in the potential energy at depth in the inlet. This is done by analogy to a simple reservoir. Then a rough estimate of the annual energy change deeper than about 50 m in Indian Arm is made. Finally, processes known to act on the inlet are described. It will be seen that many produce major changes near the surface, but only two, gravitationally induced inflow of dense water (already seen to be important) and tides, generate processes in deep water which are persistently able to effect the deep changes. Changes deeper than about 50 m are of chief concern because of their potential to be associated with stirring of bottom deposits and transport of materials below 50 m.

The Annual Energy Cycle and Its Implications

We can compare the implications of deep inlet changes with those of filling and replacing the water in a simple reservoir from an inflow source level with or higher than its top. During initial filling water flows down a slope or falls from the source. In doing so gravity accelerates it and does work on it to convert the potential energy it had at the elevated source to kinetic energy of motion. This kinetic energy is eventually all lost to heat, via action of frictional forces, as the reservoir fills and motion stops. When the reservoir is filled to depth H the final potential energy referred to the bottom in a "unit" column of water (beneath unit area = 1 m^2 of surface) is that of its mass/m^2, $\rho_1 H$ at height $H/2$, which is 0.5 $\rho_1 g H^2$ where ρ_1 is the density (taken constant here) and g (= 10 m/s^2) the acceleration of gravity. The initial potential energy of the same mass when at the source, height H_0 above the bottom was $\rho_1 g H H_0$. (In this examination we neglect the exceedingly small effects of dissipative heating and pressure on water density.)

The loss of potential energy, initial minus final energy, can be written in several ways as:

$$\text{Loss of PE}/m^2 = \rho_1 g H H_0 - 0.5\,\rho_1 g H^2 \quad\quad (1a)$$

$$= 0.5\,\rho_1 g H^2 + \rho_1 g H d \quad\quad (1b)$$

$$= 0.5\,\rho_1 g\,(H_0^2 - d^2) \quad\quad (1c)$$

in which $d = H_0 - H$ is the distance between the levels of the surface and the source. This is the energy lost to heat, which from Equation 1b exceeds the potential energy of water stored referred to the bottom (the first term), by an amount (the second term) given by the ratio of source height d above the top of the unit column to column length H. Perhaps the most useful expression for the energy eventually lost to heat is that in Equation 1c, which shows that it depends on the difference between the squares of the vertical distances below the source (depths) to the bottom and to the top of the water column - a form which will be exploited further below.

It is worth noting that the result expressed by Equation 1 depends only on the geometry and not on choice of reference level for potential energy. Equation 1 also represents energy which must be supplied to empty the reservoir.

The role of turbulence, a chaotic random motion, is worth noting. Because dissipation requires large velocity shears (i.e. large velocity differences in small distances, at most a few mm) turbulence is the only significant mechanism which can effect it. In cases where the mean flow of water times its thickness (a number proportional to a "Reynolds Number" used by physicists) exceeds about 0.004 m^2/s the mean flow is unstable and breaks down into turbulence. The condition may be assumed always to be met at some stage; then kinetic energy of the mean flow becomes that of the turbulence. It is this kinetic energy of turbulence which is dissipated into heat (by molecular friction at small scales). Between start and finish gravity does all the work to produce a kinetic energy of the flow which in turn becomes kinetic energy of turbulence which dissipates to heat; the net amount of energy which passes through each phase, and ending up as heat, is (essentially) given by Equation 1.

The case just described is analogous to filling an empty inlet; but we are interested in energy changes associated with replacing water between two levels. Consider a dense inflow of density ρ_2 replacing water of lower density ρ_1 between the depths d deeper than the sill and a greater depth D (which may sometimes be the bottom) below the surface. Then, the potential energy lost for a unit column between these depths is

$$\text{Energy/m}^2 = 0.5 \, \Delta\rho g(D^2 - d^2) \qquad (2a)$$

where $\Delta\rho = \rho_2 - \rho_1$ (positive in this case). Equation 2a is exactly analogous to Equation 1c, and remarks made about work supplied by gravity, kinetic energies of mean flow and turbulence, and dissipation to heat apply exactly as before; of course the magnitude of the energy dissipated is very much smaller in the ratio $\Delta\rho/\rho$.

In an inlet $\Delta\rho$ varies with depth; in this case we consider $D - d = \Delta d$ as a small height (within which $\Delta\rho$ is essentially constant) and write $0.5 \, (D + d) = \bar{d}$, as its average depth. Then Equation 2a becomes, for the unit column of small height:

$$\text{Energy/m}^2 = \Delta\rho \, g\bar{d} \, \Delta d \qquad (2b).$$

By adding values over successive depth ranges Δd, we can calculate energy losses between any two levels. By doing this for data shown in Fig. 4 (for periods during which lines for each depth slope down to the right) the energy supplied by gravity between August 1974 and April 1975 to increase the density between 50 and 100 m depth ($\Delta\rho$ averages about 1.2×10^3 kg/m^3) is about 4×10^4 Joules/m^2 and between 100 and 200 m ($\Delta\rho \sim 0.5$ kg/m^3) is about 7×10^4 Joules/m^2.

It is readily seen that during an inflow of dense water, gravity is given the opportunity to do this work inside by circumstances external to Indian Arm, i.e. in Burrard Inlet and the Strait of Georgia. These processes raise the density and hence potential energy of source water near sill depth (~ 20 m) just external to the inlet; the work required to do this is all done outside Indian Arm. This potential energy is thus available to do work inside the inlet as the dense source water flows in down the sill under lighter water then spreads out displacing shallower water upwards; of course there is mixing between source and inlet water at each stage.

However, by reference to Fig. 4, during periods of "light" water inflow, e.g. following April 1974, for which lines for each depth slope up to the right, the source water (shaded in the figure) "floats" above denser water beneath (it remains shallower than 30 m), so in this case gravity is not able to do work at deeper levels. Proceeding as before, the energy lost in the exchange is again given by Equations (3), but this time $\Delta\rho$ is negative so there is an energy gain. Thus a process other than gravity must do the work needed to achieve the changes. Because the source water "floats" it cannot advect energy; the changes must thus be achieved by mixing.

From data shown in Fig. 4 the energy required during April to August 1974 was roughly 3×10^4 Joules/m^2 ($\Delta\rho \sim 0.8$ kg/m^3) for the unit column between 50 and 100 m. During June to October 1974 the energy required to reduce densities in the unit column between 100 and 200 m was also about 3×10^4 Joules/m^2 ($\Delta\rho \sim 0.2$ kg/m^3). That these two estimates are about the same, even though $\Delta\rho$ for the deeper colunm is only 1/4 that for the top, is not surprising, because both the length and mid-depth ratios are 2:1 (see Equation 2b). This reflects the fact that the forces achieving the change must do work against buoyancy (which is upwards) by mixing light water down (compare pushing down a football) and denser up. For deep unit columns the net work required is clearly greater than for shallower ones of the same length.

The kinds of processes which might be able to do the work needed to mix water deeper than 50 m are described in following paragraphs. The power which is needed to mix the unit columns in Indian Arm during spring to autumn of 1974, is easily calculated. In both the 50 to 100 m and 100 to 200 m columns, the mixing energy of 3×10^4 Joules/m^2 was introduced during about four months, roughly 10^7 seconds, or about 100 days. Thus, for each of these columns the power needed* is

---

*Physicists should note that a complete investigation of

$$\text{Power/m}^2 \sim 3 \times 10^{-3} \text{ Watts/m}^2 \sim 300 \text{ Joules/m}^2\text{/day}   (3).$$

Runoff, Evaporation and Heat Fluxes

One consequence of positive runoff minus evaporation, the positive estuarine circulation resulting from salt and volume balances, was discussed early in this chapter. Part of this process is of course a tendency to freshen the surface layer, that is to increase its density contrast with water beneath and thus to increase the stratification and stability near the surface. This limits severely the capability of runoff energy to become available at deeper levels. Because stratification within an inlet controls the nature of its responses to other processes, for example generation of internal waves which is briefly mentioned below, it is clearly important. Input of heat by solar radiation similarly increases the stability near the surface, by decreasing the density there.

Loss of heat by radiation from the surface, transfer of "sensible" heat across it to the atmosphere, and loss of freshwater and of latent heat by evaporation all

---

energy in any estuarine circulation involves a large energy input/m^2 of the form $\rho'g(D - d)H_2$ for each unit column, which is effective at shallow levels. This is energy required to raise water of density $\rho'(<\rho_2)$ by entrainment, through an effective distance H_2 between inflow and outflow levels. An additional relatively very small amount of energy is needed for mixing at shallow levels. The dominant energy required for cases discussed in Indian Arm is, for $\rho' \sim 10^3$ kg/m^3, (D - d) $\sim$200 m, $H_2 \sim$ 10 m, about 2×10^7 J/m^2 and the power is of order 2W/m^2. For the whole inlet (area $\sim 3 \times 10^7$ m^2) the power needed is about 6×10^7W. This power is easily provided by a difference between the rates of working by pressure at the mouth, on the Arm, of form pressure x velocity x area at mouth. For inflow at depth $H_3 + H_2$ and outflow at depth H_3, a rough approximation is $\rho'gH_2$ x 0.1 m/s x 10^4m$^2 \sim 10^8$ W, taking mean flows near 0.1 m/s and both areas as 10^4 m^2. This is clearly adequate to do the work needed for entrainment. It is noted that energy is advected upwards by an entrainment velocity w at rate per meter squared, ρgw; this requires values for w of the order 10^{-4} m/s. The shallow processes discussed here do not, however, affect our arguments about energy changes at deep levels, in particular about the mixing energy required to decrease densities at depth.

increase the density near the surface, thus decreasing the stability near the surface. When runoff is small as in winter in some inlets, these processes might lead to deeper convection. This can be expected to lead to density increases in winter down to at most moderate depths in such inlets, but not to generate directly, vigorous flows to stir deep bottom sediments. All of the processes discussed under the present heading are thus expected to have their greatest effect in helping to control stratification and its seasonal variations and thus the type of motions possible.

In freshwater lakes surface heat transfers exert a major control over the stratification and cold river runoff can penetrate to the bottom; however lakes are not discussed here.

Winds

The rate of wind energy input (power) per unit area of sea surface in the open ocean is given roughly by

$$W \sim (\text{wind stress}) \times (\text{surface water speed})$$

$$\sim (\rho_a C_D U_a^2) \times (U_a/30)$$

where ρ_a = 1.25 kg/m^3 is air density, U_a is wind speed about 10 m above the surface, and C_D is called a "drag coefficient" and has values given by Large and Pond (5) near 1.2 x 10^{-3} for U_a less than 12 m/s increasing linearly with U_a to about 4 x 10^{-3} at U_a = 32 m/s (a strong gale). The stress is not well-known in confined areas like inlets. However, the formula is close enough for our purposes. Note that as wind speed increases, the power of the stress increases as fast as or faster than its cube, an extremely steep rate of increase. Some rough values are:

U_a	Stress	Power/m^2	
8 m/s	10^{-1} N/m^2	0.026 W/m^2	$\sim$ 2 kJ/m^2/day
32 m/s	5 N/m^2	5 W/m^2	$\sim$ 470 kJ/m^2/day

Measurements in the ocean, e.g. see Niiler and Kraus (6), show that when there is a pycnocline, which is where the density changes quite rapidly (i.e. strong stratification and very stable) between the upper layer and denser water beneath (as in spring through fall), only about 0.1% of wind energy input at the surface is transferred across the pycnocline, and most of that would be dissipated within a few tens of meters. The high wind of 32 m/s (∿64 knots) would then transfer energy to below about 10 m at about 470 J/m²/d, which is comparable with that in Equation 3, needed for deep mixing in Indian Arm over four months. However, such strong winds blow only for a very few days, so they may contribute only a small fraction of energy needed, and probably not beyond about 50 m. Weaker winds are clearly ineffective.

In winter when stratification is often quite small then strong winds may be much more effective. For example, they may aid convective processes, but in general they mix down water of lesser density than that below, a tendency opposite to that of increasing density in deep water observed in Indian Arm. This effect might be important, however, in some other inlets.

The chief effects of winds in the present context of deep energy changes are probably twofold. Inside an inlet they may aid a positive estuarine circulation or they may impede it; such events were apparently observed by Pickard and Rogers (2) using current meters in Knight Inlet. In spring and summer when stability is large in Indian Arm these effects would be unimportant. But in aiding or impeding the estuarine circulation in late fall or winter, winds might be a deciding factor in the import of dense source water only marginally available. In other words, an up-inlet wind might delay or prevent a deep inflow or complete renewal. On the other hand, a down-inlet wind might trigger deep water renewal.

A second major effect of winds arises when they blow over the region outside an inlet, in which case they can affect the kind of source water available for inflow. The two most obvious ways in which this may happen are both associated with the stratification of water outside a silled inlet. In one a wind blowing towards a sill from outside tends to tilt the extreme surface slightly upwards towards the sill. A raised level here, of course, even if small, opposes a positive estuarine circulation. But more importantly, it tilts the external pycnocline downwards towards the sill with a slope very much in excess of the surface slope (perhaps 100 to 1000 times), thus depressing denser water to well below sill depth and making it unavailable as source water. If the wind blows away from the mouth outside the inlet the surface tilts up slightly away from the mouth. A consequent slight fall in surface level there could aid the estuarine circulation. Most importantly, in this case the pycnocline slopes down quite steeply away from the mouth and rises in its vicinity.

Higher density water beneath may thus become available as source water for deep water renewal.

A second way in which winds exterior to an inlet have a pronounced effect, may occur when its source water is in, near or has access to an ocean, or quite large open body of water. The effect results from the manner in which the ocean's upper layer responds to a wind stress because it is on a rotating earth. In the northern hemisphere the depth-averaged motion in a large and deep enough body of water is to the right of the wind (this is a direct result of "Coriolis force", which is to the right of motion in the northern hemisphere, but is opposite in the southern hemisphere). The motion in this wind driven layer, called the "Ekman layer", is essentially all above some particular depth which in mid and more polar latitudes is commonly 30 to 100 m and tends to be more confined to the upper layer as the stratification increases. When a wind blows with a coast on its right in the northern hemisphere, the lighter surface water is thus forced towards the shore and tends to accumulate there, forcing downwards the pycnocline separating it from denser water beneath; the latter thus becomes less accessible as dense source water for inflow over sills.

On the other hand, when the coast is to the left of the wind, the average Ekman (or surface) current is away from the shore and water tends to flow up from below to replace it. This forces up the pycnocline and denser water beneath so it is more available as dense water for deep water renewal. Off the west coast of Vancouver Island and the Panhandle the prevailing wind in summer time is from the northwest, so the coast is on its left; thus the up-welling situation prevails and dense water inflow results in some inlets. However, in winter prevailing winds are from south of west so down-welling is typical. In some silled inlets (e.g. Rupert-Holberg and Alberni Inlets) with access to the open North American west coast, the deep water renewal is in summer in accord with this process. This is directly contrary to the season of renewal in Indian Arm (and also Howe Sound and Princess Louisa), which is well removed from the open coast, and in which renewals are in winter.

The summer up-welling of nutrients off the west coast is also associated with high biological production and good fishing in summer in this region.

Besides the effects of winds described above, their most obvious and direct effect is to produce surface waves. These have circular water motions which are about 4% of the surface values at depths near about half their wavelength, and on breaking they produce turbulence. But waves in inlets are not very large so neither the currents nor turbulence are expected to be significant deeper than 50 m; indeed if winds were so great that they were so large, other wind driven motions would be much more significant.

Variations in wind stress produce very small variations in surface tilt and much larger varying tilts of isopycnals (lines of constant density). Winds thus generate internal waves in inlets. Very little is known about the nature and magnitude of wind induced internal waves. However, it is likely that deep currents associated with them are small and of brief duration compared with those generated by tides; extended measurements are needed to check on this.

Tides

The feature which leads to the most significant aspects of tides in inlets is that the entire volume, $A_o h$ called the "tidal prism" of height h between low and high tide over the inlet's area A_o, must flow in during the flood and out during the ebb. In the Pacific Northwest both flood and ebb extend over about 6 hours (about 2.2 x 10^4s). If the cross-sectional area of the mouth is a_o, then the average speed V_o of both flood and ebb at the mouth is given by

$$V_o = (A_o/a_o)(h/2.2 \times 10^4 s). \qquad (4).$$

The first bracket is the ratio of inlet to mouth areas, and the second is the average rate at which the surface rises. The maximum average speed is about 57% larger than V_o. In Indian Arm (h averages about 3 m and A_o and a_o are roughly 3.4 x $10^7 m^2$ and 2 x $10^4 m^2$), V_o is about 0.23 m/s, and the "run" of water through the mouth (= hA_o/a_o) is about 5 km. While the distance water and its properties move up and down inlet is rather less, this "swish" up and down is clearly important. Of more importance dynamically, however, is the tidal energy associated with these flows, and herein appears to lie a major difference of significance between inlets with small mouth areas (i.e. shallow sills and/or constricted mouth width) and those with large mouth areas (i.e. deep or no sills and only moderately constricted width).

In paragraphs following we first examine the amount of energy available in tides, then the distribution of tidal velocities and kinetic energy along an inlet. It will be seen that in inlets in which A_o/a_o is large, the kinetic energy of tidal flow at the sill is large; if also the depth inside is large compared with sill depth, this kinetic energy is highly concentrated in the vicinity of the sill. Obviously this is important because this is energy which may produce mixing. Finally, some observations are briefly described which indicate that tides can generate strong deep flow near the sill which produce large amounts of turbulent energy. It is the latter which can produce vigorous deep mixing. In other words, tides will appear to be the agency responsible for deep mixing in shallow-silled inlets in which tidal ranges are moderately large (or larger).

The potential energy added to a unit column in the inlet by raising the surface through height h between a low tide and the following high tide during the flood is $0.5 \rho gh^2$ when the energy is referred to low tide level and ρ is average surface density. Assuming h is the same throughout the inlet, the total energy added is this amount multiplied by A_0. This total is (slightly) less than the total energy imported through the mouth during the flood; the difference is the energy needed for mixing and dissipation which is, from Equation 3, for both unit columns in Indian Arm between 50 and 100 m and 100 and 200 m, 300 $J/m^2/d$. In Indian Arm the average tidal range h is about 3 m; so the amount imported during the two tides in one day is about $\rho gh^2 \sim 10^5$ $J/m^2/d$. The total fraction of tidal energy needed beneath 50 m is thus about $300/10^5$ equivalent to about 0.3% for each column. While this seems a low efficiency, so we might expect the energy extraction to be easily achieved, we still have to explain how surface tidal energy can be made available at the deep levels. To see how this may be possible we next examine tidal energy distribution in a simple, silled inlet.

For our simple inlet of constant width b we take the depth D as increasing linearly with distance x from the sill where it is D_0 (at x = 0), to a maximum D_m, at distance x = L/2 from the sill, where L is the inlet's length. The area A of the inlet from distance x to its head, is b(L - x). Thus, between the sill and deepest part of the inlet,

$$D = (D_m - D_o)(2x/L) + D_o \text{ and } A = A_o(1 - x/L). \quad (5)$$

The small increase in surface elevation Δh during a small time Δt is assumed to be the same everywhere along the inlet, an assumption which is reasonably true in real inlets except those which are shallow and long (a rough criterion is that the depth should exceed about $3L^2/10^8 m$).

The volume filled, during time Δt, between distance x and the head is $A \Delta h$, which is also given by the amount crossing area a = Db, which is $(V \Delta t)Db$, where V is the depth-averaged speed at distance x. Equating these values, rearranging and substituting for $b = A_0/L$, we find

$$V = (A/Db)(\Delta h/\Delta t) = (A/A_0)(L/D)\Delta h/\Delta t. \qquad (6)$$

This is similar to Equation 4, which can be found in the same way. Because both A/A_0 and L/D decrease linearly with distance from the sill, V decreases very rapidly.

The property of interest is the average kinetic energy per unit volume, for which $0.5\ \rho gV^2$ is a crude estimate, and which decreases very rapidly leaving the sill. To compare tidal flow energy at the sill in different inlets, either Equation 4 or 6 may be used to show that the kinetic energy is proportional to A_0^2/a_0^2. To examine the distribution of kinetic energy within an inlet it is convenient to represent V as a ratio R to its value at the sill; $V_0 = (L/D_0)(\Delta h/\Delta t)$ so $R = (A/A_0)(D_0/D)$. Then the ratio of kinetic energy at distance x to that at the sill, is given by

$$R^2 = (A^2/A_0^2)(D_0^2/D^2) = (A^2/A_0^2)(a_0^2/a^2). \qquad (7)$$

This can be evaluated using Equation 5. The second form involving a_0/a is applicable to any inlet; in the simple inlet $a_0 = D_0 b$ and $a = Db$. To illustrate the rapid decrease of this ratio with distance we will take $D_m/D_0 = 10$ which is similar to the ratio of maximum depth to sill depth in Indian Arm. The distance x is expressed as its fraction $2x/L$ of the distance $L/2$ to maximum depth. Relative depths, areas, velocities and kinetic energies (expressed in %) are:

$2x/L$	$D/D_0 = a/a_0$	A/A_0	V/V_0	$R^2 \times 100$
0	1	1	1	100%
0.1	1.9	0.95	0.5	25
0.2	2.8	0.9	0.32	10
0.3	3.7	0.85	0.23	5.3
0.4	4.6	0.8	0.17	3
0.6	6.4	0.7	0.11	1.2
1.0	10	0.5	0.05	0.25

The table clearly implies that the kinetic energy is strongly concentrated in the vicinity of the sill; in fact 90% of total energy is within the upper one third of the sill slope area, i.e, within one sixth of the inlet length. The concentration is within a smaller fraction of the inlet if the sill slope occupies a smaller fraction of its length. It should be noted that changes in R^2 depend mainly on D_0^2/D^2 (or in a real inlet on a_0^2/a^2) which changes 100-fold while A^2/A_0^2 changes by a factor of four.

The argument about average kinetic energy just given is correct only provided the velocity is evenly distributed over the total depth and breadth of the inlet. In fact, instead of being proportional to V^2 (the square of the mean speed), the average energy is increased by an amount proportional to the average of the square of the differences of the speed from V. For example, if the velocity has uniform value nV in one nth of the depth, and zero at other depths, it has the same average value V, but the average of its square over the depth is nV^2. Note that n always exceeds unity.

In subsequent paragraphs observations are described which imply that tidal flood velocities may be very strong just above the bottom down the slope from the sill. This further implies, from the foregoing discussion, that large amounts of tidal kinetic energy are available for deep mixing near the slope, and also that the value of R^2 is maintained at a higher value than indicated by Equation 7 in this vicinity.

Starting about five years ago a remarkable series of observations by Farmer and Smith has been made in Knight Inlet (7, 8, 9). The instrument used was an echo-sounder which records sound back-scattered from material different from water. Included are particles which tend to accumulate in stable interfaces which are regions of rapid vertical density change. Stable interfaces in B.C. inlets normally occur less than 10 m from the surface. Not only does the salinity change rapidly across them but commonly the temperature (warmer above) and often the currents do also, i.e. the region is often one of current "shear".

Farmer and Smith's echo-sounder traces showed that even though the interface may be fairly steady elsewhere in Knight Inlet, in the vicinity of the sill the behavior of suspended material associated with it is often truly remarkable and varied. For a given stratification inside the inlet the type of behavior is primarily dependent on the strength of flow over the sill and thus on the stage of the tide. Only two aspects are touched on here since the primary concern is with effects which might extend to deeper than 100 m. (See also pp. 185-192).

1. In strong tidal flows, besides the traces indicating some contortions of the interface, some material splits from it and produces an echo-trace which separates from the interface near the crest of the 60 m sill. It then plunges down the steep slope on the lee side of the sill, approaching the sloping bottom as the depth increases. At some depth, the greatest observed in

Knight Inlet being over 120 m, the trace turns upwards away from the sill, becoming very jagged and diffuse further downstream. The interpretation is that a strong current shear exists at the interface and the current below it plunges down the sill slope carrying some material from the interface down with it. The narrowing space between the trace and the lee slope indicates an accelerating flow down this slope; this is a supercritical or shooting flow, described further in the next section. The important feature in the present context is that at the depth where the trace turns away from the slope a "hydraulic jump" must exist which produces intense turbulence and consequent mixing. The mixing may extend from the interface at least down to the jump. The phenomenon just described developed in the winter time when the difference in density between the surface and deeper layers is least, so the interface was weakest. Such flows during part of the tidal flood would appear to be relevant to renewal of deep water, or to intermediate depths in Indian Arm described earlier in this chapter.

2. In summer, when stratification near the surface is greatest, the trace again splits and plunges down the sill slope in moderate to strong flows. But at this time large disturbances to 50 m high are seen on the plunging trace and both these and a jump near 100 m or so in Knight Inlet may produce mixing at least down to this depth. In weak flows, the flow separates from the sill slope near sill level as a jet and "billow" disturbances form downstream. As flow increases it still separates but "lee waves", a certain type of internal waves, form downstream. Neither of these shallower phenomena appear to cause mixing to great depths; however as the flow increases further the point of separation moves increasingly down the slope and the plunging unstable flow produces resultant deep mixing.

Farmer and Smith postulated that factors controlling the type of flow, including depths of jumps and mixing, are shapes and sizes of sills, flow velocities and the strength and distribution of density stratification near the sill. Their observations support dependence on the latter two features. However the effect, for example, of the factor of three between sill depths in Knight Inlet (~ 60 m) and Indian Arm (~ 20 m) is unknown; also effects associated with the density of inflowing source water relative to stratification inside inlets are unknown.

Despite such uncertainties, these observations provide us with a plausible mechanism to help explain deep mixing and renewal phenomena described in Indian Arm, and will be incorporated into the discussion of the following section. More extensive discussion involving wide ranges of inlet types and circulations is included in the review by Gade and Edwards (1).

DISCUSSION AND COMPARISONS WITH SOME OTHER INLETS

Summaries and discussion of some features appear at the end of several preceding sections; of particular importance is the implication that prediction of deep water renewal in Indian Arm with reasonable probability is not yet possible, despite having a long series of observations. Two matters not addressed above are the possibility of deep scouring of bottom deposits, and of transport of deep suspended material to shallow levels (here taken to be shallower than 50 m). Scouring may be expected if velocities about 1 m above the bottom exceed about 0.2 m/s (or somewhat higher).*

Davidson's (3) observations showed that some tidal up-inlet currents peaked above 20 cm/s but below 30 cm/s during December 1974 and the period of dense water inflow of January and February 1975, near the sill of Indian Arm. These "might" cause scouring much further down the slope particularly if the dense water flow accelerates down it. From the Farmer and Smith observations, discussed (as type 1) in the preceding subsection, it is inferred that a "shooting" or "supercritical" flow develops during part of the flood, which on being slowed down sufficiently does so at a "hydraulic jump" with the flow energy being largely transferred to kinetic energy of turbulence. A "shooting flow" of water beneath air can be observed in a kitchen sink as a thin fast flow dispersing outwards along the sink's bottom from a water jet directed onto it; at some

---

*Scouring or erosion velocities for compacted sediments are a matter of much as yet unpublished controversy. Komar (10) discusses an equation giving erosion velocities in the range 0.05 to 0.2 m/s for uniform particle sizes from less than 10^{-6} m to about 10^{-4} m appropriate to tailings. However, sediment compaction, which depends on size distribution, requires that erosion speeds be much higher. Moreover, past work has been on velocities sufficient only to generate bed-loads in which particles mainly roll or jump (saltate) along the bottom. Scouring and uplift sufficient that particles enter a flow (as stream-load) and be available to upwards transport demands a more energetic flow. More work is needed to settle these points.

point the flow "jumps" to a thicker much slower flow which is clearly turbulent. Flow rates within the shooting flow exceed the speed of waves in such shallow water so no information about downstream conditions can be transmitted upstream. (Compare also tidal bores).

Shooting flows on sill slopes are slowed down perhaps slightly by friction but probably mainly by one of two other forces. One is by increasing upwards "buoyancy" as the flow penetrates deeper than the level at which the density of inlet water is the same as that in the flow. Such "overshoot" probably occurs in flows observed by Farmer and Smith (7, 9) in which the jump and consequent turbulence through a large depth range were evident. It was noted earlier that mixing down to the bottom was most effective in Indian Arm between about June through the fall. By reference to Fig. 4 we see that this is the period when incoming water (shaded) is successively as dense as that at successively deeper levels in the Arm. The implication seems to be that the incoming flow penetrates beyond the level of its own density, then jumps and causes mixing down to deepest levels. A further implication is that to avoid the possibility of scouring and mixing upwards of tailings, their deposition on a sill slope is preferably to be avoided.

The second manner in which a shooting flow down a sill slope must be slowed, occurs when its density excess carries it right to the deepest part; it is then slowed by the combination of friction and gravity as it starts to rise from the deepest part. Again the flow must jump and cause mixing. In this case, if the inlet is one of tailings deposition, tailings would likely reach these deeper parts of a slope down from the head. Two questions then are:

1. Will velocities be great enough to cause scouring?
2. If so, are the upwards velocities associated with deep water displacement upwards during renewal great enough to transport tailings to shallow depths?

A partial answer to the first question (see also footnote on p. 124) can be found by examining the ratio (called a Froude Number):

$$F = U/(gh\Delta\rho/\rho)^{1/2} \tag{8}$$

where U is the flow speed, $\Delta\rho$ is the excess of its density ρ over that of overlying fluid and h is its thickness. The denominator is the speed of internal waves on the flow surface. This representation is fairly crude;

however in a shooting flow, F exceeds unity (roughly), and the flow jumps where the ratio is about unity. A shooting flow "entrains" surrounding fluid which tends to slow it down, decrease $\Delta\rho$ but increase h. Assuming that the dense inflow during deep renewals is a shooting flow and reaches deepest levels, then jumps near there (where $F \sim 1$), then $\Delta\rho \sim \rho U^2/gh$ just before the jump. If U (which is decreasing) were still just sufficient to cause scouring, say U = 0.2 m/sec, then $\Delta\rho$ would be about 0.4 kg/m^3 assuming h is about 10 m. Reference to Fig. 4 shows that during January and February 1975, some source water was denser than deep water by about this amount; also the density at all levels was increased by a similar amount during these months. The argument is however, crude; h is not well-known and the interval between observations precludes precise analysis, so the present approach should be regarded as suggestive only. Clearly only current measurements can be definitively useful in supplying the necessary information about scouring during renewals.

The second question about upwards velocities applies, not only to scoured tailings mixed up from the bottom, but also to tailings already in suspension in an effluent plume. If we assume that all water in Indian Arm can be renewed in about 30 days, as suggested by the data of Fig. 3, then water from the bottom at 200 m must move to the surface in this time. This implies an average upwards velocity for this water of about 7 m/d. A second estimate results from noting that the entire volume ($\sim$ 2.3 x 10^{9}m^3) must flow up into the surface layer (area $\sim$ 3 x 10^{7}m^2) in the same time; the average velocity implied is about 4 m/d. It is noted that these may be low estimates as renewal may take less than 30 days because of intervals between observations. However, they are in the same ball park as the earlier rough estimate of 10^{-4} m/s = 10 m/d (see p. 115) which was based on an energy argument. Now particles fall through still water at a rate which is proportional to the square of their diameter and to the excess of their density over that of the inlet; if their specific gravity is 2.7 (typical of many tailings) the fall speed of particles 10 x 10^{-6} m (10 µm) in diameter is 5.6 m/day. Particles of roughly this size would be held in suspension. Those which are larger would fall, but smaller particles would be carried up by the vertical velocities associated with deep renewals. Once again the argument is fairly crude; the upwards velocities may exceed the average substantially at some places and times. They also depend on inlet geometry and occurence and mode of deep water renewal. The fraction of suspended tailings swept upwards depends on the tailings size distribution.

Circulations and their variations in inlets differ

greatly. Three obviously different cases which are relevant to our central problem of tailings disposal are discussed in the next few paragraphs. The importance of the ratio a_0^2/a^2 (of cross sections at sill and at a point in the inlet) appearing in Equation 7 in determining relative changes in tidal energy along the sill slope, and of A_0^2/a^2 (Equation 4 or 6) in determining the tidal energy at the sill, was discussed earlier. Because in most silled inlets D_0/D varies much more than the effective width ratio, and because it is more available from published data than a_0/a, the depth ratio is chosen for comparisons between inlets.

In inlets with no sill or a deep sill the ratio $(D_0/D)^2$ in Equation 7 varies only slightly, so the concentration of tidal kinetic energy in the sill vicinity must be much smaller than in shallow-silled deep inlets. In such inlets the major controls on the deep circulations may be expected to be associated with winds and with changes in the stratification just outside the mouth, which were discussed in the subsection on winds. These wind effects would be similar to results discussed by Klinck et al (11), of a numerical model of a tideless inlet with no sill, constant depth, open to a stratified ocean and driven by winds over the ocean. The inlet circulations predicted by the model conform with some observations in several inlets in Norway, in Alberni Inlet on the west coast of Vancouver Island and also in the Strait of Juan de Fuca. Providing the inlet depth is great enough (100 m or more?) deep currents strong enough to stir the bottom and/or to transport deep suspended material to shallow depths are probably not as likely as in shallow-silled inlets; nevertheless some extreme winds might cause them. Also unknown are deep effects of large tides in deep-silled inlets of broad extent, but with narrow mouths (so A_0/a_0 is large in Equation 4).

An inlet differing markedly in changes of properties at different depths is Saanich Inlet. The following is a direct quote from Pickard (4) with some insertions in parentheses.

Saanich Inlet communicates with the Strait of Georgia across a sill of depth 57 m. The series (of observations) extends from late 1965 to 1973, but there is only one closely spaced series of just over one year, in 1971, the remaining data being rather sparse (3 to 5 per year). The most notable feature is that while there is a marked annual variation of all properties at 100 m (by $1.0^\circ C$, $0.4\,^\circ/_{\circ\circ}$ and 0.6 mL/L), there is little variation at 200 m (greatest depth 227 m). All properties show a progressive increase in range (roughly doubling) of annual variation.

At 200 m in Saanich Inlet, oxygen was almost always essentially zero, indicating a long residence time for deep water. During three falls of the seven years a small rise to less than 1 mL/L lasting three or four months occurred simultaneously with a decrease at 100 m; this suggests brief mixing at 200 m. An estuarine circulation clearly extended to 100 m but did not reach the bottom. However, the increasingly large annual range in salinity implies that source water might become dense enough to cause renewal to the bottom some time after 1973 (this point has not been checked). So even in this very quiet inlet an eventual overturn cannot be ruled out.

It is worth noting that while the ratio D_m/D_o (maximum to sill depths) is about 11 in Indian Arm, it is about 4 in Saanich Inlet, so $(D_m/D_o)^2$ is less by the factor 7.6 and confinement of tidal energy to sill vicinity must be less. Also A_o^2/a_o^2 in the former exceeds that in the latter by a factor exceeding 16. Because the range of tidal heights is similar, tides must be a much less potent factor in producing large flows down the sill in Saanich Inlet than in Indian Arm.

An inlet which differs markedly from those discussed earlier is the Rupert-Holberg Inlet system on Vancouver Island. Drinkwater and Osborn (12) show that in 1971-72 the temperatures at 9 m, 30 m and 167 m depth over the deepest part were all roughly equal but their seasonal change was large, about 4 to 5°C. Average salinities increased from top to bottom (by about 1.5 $^o/_{oo}$, indicating stable stratification) but the seasonal change at each depth was much larger (about 4 to 5 $^o/_{oo}$). These show that the seasonal density change is about 4 to 5 kg/m^3 at each depth. This is in vivid contrast to changes in Indian Arm (see Fig. 4), in which annual changes at 10 m, 30 m and 100 to 200 m, are roughly 3, 2 and less than 0.5 kg/m^3, respectively; also at the deepest levels (see Fig. 5) these are not always seasonal. Oxygen variations at all levels in Rupert and Holberg Inlets are about 2.5 mL/L; again this is in sharp contrast to Indian Arm (see Figs. 3 and 5). The energy required to change the density in a water column occupying the deepest 100 m in the Rupert-Holberg system is at least 10 times larger than for corresponding columns in Indian Arm (despite the fact that the latter are deeper).

The differences causing these contrasts between the Vancouver Island inlets and Indian Arm inlets are no doubt geographical. The junction of Rupert and Holberg Inlets is

near their deepest basin: Rupert Inlet extends about 10 km
ENE and Holberg Inlet about 35 km WNW. Their common sill
region, Quatsino Narrows, averages about 25 m in depth, is
narrow (less than 0.5 km), long (about 3 km) and is on the
south side of the junction far removed from the two heads
of the system. The source of water which enters through
the Narrows is Quatsino Sound which connects with the
Pacific Ocean; an interesting consequence of this is that
deep water renewal occurs in summer and depends on
offshore winds for reasons explained in the subsection on
winds.

Stucchi (13) describes the energetic inflow from
Quatsino Narrows as a highly turbulent and high speed jet
which separates from the sill. When the jet is buoyant he
characterizes it by a Froude Number identical to Equation
8 in which U is initial discharge speed and h is the depth
of the discharge channel. Stucchi then shows, using data
observed in Rupert-Holberg Inlet near the sill, that
mixing generated by the energetic buoyant (i.e. it tends
to float) tidal jet extended to the bottom when the Froude
Number exceeded seven. This result conforms with
laboratory flow experiments and mathematical models, e.g.
by Shirazi and Davis (14). When the observed tidal jet was
denser than inlet water it spread rapidly to all depths
and was largest at the bottom at 150 m. This thickening
was associated with rapid entrainment (in which ambient
water joins the jet) at all levels. The results are
consistent with experiments by Ellison and Turner (15).

These results show that in Rupert-Holberg Inlet the
inflow emerges as an energetic jet which causes the
intense mixing needed to produce the observed coincident
changes at all times to all depths. This contrasts with
the tidal sill flows described earlier as those associated
with mixing and renewal in Indian Arm. The Froude Number
(Equation 8) near the sill, however, must be much less
than that in Rupert-Holberg Inlet. In the latter tidal
flows in the narrows are typically 1 m/s (and up to 3 m/s)
whereas at the Indian Arm sill they rarely reached 0.3 m/s
during a deep water renewal; this is roughly consistent
with comparative values of the ratio A_o/a_o in Equation 4.
Because Quatsino Narrows is long, intense turbulence is
produced, and surface water from inside entering the
narrows on the ebb and source water on the flood tend to
be mixed. This reduces the density contrast ($\Delta\rho$ in the
denominator of Equation 8) and helps to maintain high
values of the Froude Number (Stucchi's estimates ranged
between 2 and 10).

From Fig. 4, the source water for Indian Arm was
buoyant (less dense) relative to inlet water near sill
depth between about April and late June. During these

months density decreased noticeably down to 50 m but not significantly below. If we accept that the downwards mixing of light water to 50 to 75 m was that appropriate to a jet, the Shirazi and Davis model implies a Froude Number as high as 2 or 3 for sill flow in Indian Arm. However, whether this flow should be described as a jet or a sill flow producing billows or internal waves, described in the preceding section as a (type 2) flow separating near sill level, could only be determined from suitable observations. In any case this may be merely a matter of semantics; either description may be suitable for the buoyant inflow which clearly separates and is part of an estuarine circulation beneath an outflow driven by snowmelt during this season.

An outstanding difference between the inlets is that the average slope down from the sill in Indian Arm is about $1.2°$, while that in Rupert-Holberg Inlet is $7°$ or $8°$C. This seems to imply that flow down the sill is more likely in the former inlet, but separation more likely in the latter. However, Stucchi's observations less than 1 km from the narrows showed that during a moderately dense inflow the velocity was maximum at 145 m near the bottom (about 1.5 m/s), decreasing only very slowly with decrease in depth. So the flow must have remained intense along the bottom in Rupert-Holberg Inlet, but was not a thin shooting flow of the types discussed earlier as a concept for deep sill flows in Indian Arm. In the section on changes of properties during one year in Indian Arm, the distributions in some months (in Fig. 3) were seen to suggest flows hugging the sill slope. The reason for these differences in flow in the two inlets are not readily apparent. An explanation may lie in applying data from suitable observations to, for example, the Ellison and Turner entrainment model (15) which is not described here. If such an approach does not prove appropriate, it will be because of the special nature of sill flows and the internal waves and instabilities generated; in that event many observations in varied circumstances of the kind started by Farmer and Smith (7, 9) and supported by suitable observations of currents and properties will be needed.

Summary

In this final section I have attempted to show how information from Indian Arm might be used to infer possibilities for scouring and uplift of sediments in that inlet, and to show that circulations differ greatly in other inlets. Observations which are site-specific for any application will be required.

The explosion of information during the past decade or so, about inlet circulations and our understanding of them, is apparent on reading Fjord Oceanography (16), the volume in which several of the references in this chapter appear. Of particular note for further reading and additional references is the chapter by Gade and Edwards (1). This recent enhancement of knowledge implies that impact assessments involving inlet circulations, should rapidly become much more definitive than has been possible up to the present.

REFERENCES

1. Gade, H.G. and A. Edwards. "Deep-water Renewal in Fjords" in Fjord Oceanography, H.J. Freeland, D.M. Farmer and C.D. Levings, Eds. (NY: Plenum Press, 1980), pp. 453-489.
2. Pickard, G.L. and K. Rodgers. "Current Measurements in Knight Inlet, British Columbia", J. Fish. Res. Bd. Can. 16(5): 635-678 (1959).
3. Davidson, L.W. "On the Physical Oceanography of Burrard Inlet and Indian Arm", M.Sc. Thesis, University of British Columbia (1979).
4. Pickard, G.L. "Annual and Longer Term Variations of Deep Water Properties in the Coastal Water of Southern British Columbia", J. Fish. Bd. Can. 32: 1561-1587 (1975).
5. Large, W.G. and S. Pond. "Open Ocean Momentum Flux Measurements in Moderate to Strong Winds", J. Phys. Oceanogr. 11(3): 324-336 (1981).
6. Niiler, P.P. and E.B. Kraus. "One-dimensional Models of the Upper Ocean" in Modelling and Prediction of the Upper Layers of the Ocean, E.B. Kraus, Ed. (NY: Pergamon Press, 1977), pp. 143-172.
7. Farmer, D.M. and J.D. Smith. "Tidal Interaction of Stratified Flow with a Sill in Knight Inlet", Deep Sea Res. 27A: 239-254 (1980).
8. Smith, J.D. and D.M. Farmer. "Mixing Induced by Internal Hydraulic Disturbances in the Vicinity of Sills" in Fjord Oceanography, H.J. Freeland, D.M. Farmer and C.D. Levings, Eds. (NY: Plenum Press, 1980), pp. 251-257.
9. Farmer, D.M. and J.D. Smith. "Generation of Lee Waves over the Sill in Knight Inlet" in Fjord Oceanography, J.F. Freeland, D.M. Farmer and C.D. Levings, Eds. (NY: Plenum Press, 1980), pp. 259-269.
10. Komar, P.D. Beach Processes and Sedimentation (Englewood Cliffs, NJ: Prentice Hall, 1976).

11. Klinck, J.M., J.J. O'Brien and H. Svendsen. "A Simple Model of Fjord and Coastal Circulation Interaction", J. Phys. Oceanogr. 11: 1612-1626 (1982).
12. Drinkwater, K.F. and T.R. Osborn. "The Role of Tidal Mixing in Rupert and Holberg Inlets, Vancouver Island", Limnol. Oceanogr. 20(4): 518-529 (1975).
13. Stucchi, D.J. "The Tidal Jet in Rupert-Holberg Inlet" in Fjord Oceanography, H.J. Freeland, D.M. Farmer and C.D. Levings, Eds. (NY: Plenum Press, 1980), pp. 491-498.
14. Shirazi, M.A. and L.R. Davis. "Workbook of Thermal Plume Prediction, Vol. 2, Surface Discharge", Pacific Northwest Environmental Research Laboratory Report EPA-R2-72-005b, Corvallis, Ore. (1974).
15. Ellison, T.H. and J.S. Turner. "Turbulent Entrainment in Stratified Flows", J. Fluid Mech. 6: 423-448 (1959).
16. Freeland, H.J., D.M. Farmer and D.D. Levings (Eds.). Fjord Oceanography. (NY: Plenum Press, 1980).

CHAPTER 4

A NOTE ON TRACE METAL BIOACCUMULATION AND BIOMAGNIFICATION
IN NEAR-SHORE MARINE FOOD WEBS

D.R. Young*
    Dames and Moore
    1100 Glendon Ave.
    Suite 1000
    Los Angeles, Calif. 90024
    U.S.A.

D.V. Ellis
    Department of Biology
    University of Victoria
    Victoria, B.C.
    Canada V8W 2Y2

During the past several decades there has been persistent concern amongst those involved with the management of the sea, that the metal (mercury) contamination catastrophes at Minamata and Nigata (1) would not only recur, but were the forerunners of widespread and possibly irreversible problems akin to DDT residues causing population declines of hawks and falcons through development of thin and fragile egg shells (2). There has been an enormous amount of investigative effort given to understanding the chemistry, physiology and consequent species population dynamics of the problem,

---

*D.R. Young was unable to be present at the symposium to present an intended paper "Passage of Wastewater Metals through a Near-shore Marine Food Web". Accordingly, since trace metal contamination of the surrounding ecosystem is an important potential impact at metal producing sites, a note on the topic has been written for these proceedings. The function of the note is to introduce readers to certain concepts and site-specific results relevant to mining operations and marine tailings disposal.

and we refer readers to such summary reviews as Goldberg (3) and Phillips (4).

Mines as the medium for activating potentially toxic metals from their geological sinks are an obvious source of concern. Nevertheless, there is little information available to show whether the concern is needed or misplaced.

In recent years, some long-term field survey results have appeared which may lead to our understanding the circumstances under which mines may cause marine contamination at tolerable or intolerable levels.

At Island Copper Mine levels of copper, molybdenum, lead, zinc, cadmium, arsenic and mercury have been determined for the water column, sediments (affected by tailings, and not so affected) and organisms, annually or more frequently since pre-discharge in 1970. Species monitored include marine algae and grasses, zooplankton, fish and benthos, specifically mussels, edible crabs and shrimps, and a variety of epifaunal and infaunal species.

The following quotation reviews findings to 1980, i.e. ten years of assessments:

> Chemical analyses of rockweed and eelgrass tissue metal content have shown increases in those stations closest to the mine's dock and waste rock dump. The trend of increasing copper levels in rockweed has now ended and there was no trend evident in eelgrass metal levels. No significant difference could be determined for metal levels in zooplankton (including shrimp) between areas. Metal levels in Dungeness crab tissue have remained relatively low, not differing substantially from last year. The only bivalve species showing a consistent pattern of elevated metal tissue concentration in relation to mine activities has been the blue mussel. Since one collection site is the actual concentrate loading wharf, it is probable that the increase was a result of loading operations and not due to mill discharge. Further, while a spatial pattern exists, there is no temporal pattern of regular increase and present mean levels are lower than those of some preceding years suggesting better materials handling practices now than in the past. There were no consistent spatial patterns of metal concentrations in the fish species collected. Temporally, only zinc showed an increasing trend in some species; however, this increase was not related consistently with area and it is

unlikely that it is associated with mine operations. (5)

Extensive investigations of trace metals in water, sediments and biological tissues have also been made since 1971 around southern California municipal wastewater outfalls by the Southern California Coastal Water Research Project (6, 7). The highest levels of contamination occurred in the discharge zone of the Los Angeles County Joint Water Pollution Control Plant (JWPCP). For example, the following surficial sediment contamination factors (median outfall/median baseline) were obtained for eight target metals measured during 1975 in a 45 square km area of the outfall monitoring zone: cadmium - 36, chromium - 12, copper - 20, lead - 17, mercury - 23, nickel - 5.4, silver - 27, and zinc - 7.7 (8).

This input resulted in distinct contamination of benthic invertebrates collected during 1975-77 from the zone (9). Median values for silver in muscle tissue of abalone, scallops, and lobsters were three to five times above those obtained for corresponding control samples. Similarly, cadmium concentrations in the muscle tissue of the JWPCP scallops and lobsters were two to three times those measured in the controls, and corresponding copper and mercury values for the scallops were elevated approximately twofold above the background levels. Further, median muscle tissue concentrations of nickel in abalone, scallops, sea urchins, and crab were two to six times background levels, and the median values for chromium in abalone and scallop muscle were ten times background levels.

This "excess" bioaccumulation of metals did not appear to be passed up the food web; median muscle tissue concentrations measured in five popular benthic feeding sport fish from the area generally were indistinguishable from control values.

Subsequent studies of this and several other marine ecosystems of southern California were conducted using the Cs/K ratio, shown by Young (10) and Isaacs (11) to be a useful indicator of the biomagnification potential, or "structure" of a food web. Despite the structure observed in most of the ecosystems investigated (12, 13, 14), a qualitative evaluation of the data revealed no indication of a systematic increase in tissue concentration with trophic level for most of the target trace elements. In contrast, ten of the elements (silver, arsenic, cadmium, chromium, copper, iron, manganese, nickel, lead, and zinc) were found to decrease exponentially with a numerical assignment of trophic level in one or more ecosystems, and no exponential increases were observed (15). "Excess" (outfall zone minus control zone) concentrations of chromium in the JWPCP outfall seafood survey organisms also decreased exponentially with the trophic level assignment value (16). These results strongly support the argument that most trace elements of present concern in pollution studies are not biomagnified in marine food webs.

The one trace element for which tissue concentrations did increase with trophic level assignment was mercury, measured as total mercury or organic mercury. The data indicate that, in most cases, a large majority of the total mercury measured in organisms from the higher trophic levels was in an organic form. The fact that organic forms of mercury have relatively long biological half-lives appears to be the cause of these trophic level increases. However, no such increases of "excess" mercury were observed around the JWPCP wastewater outfalls (16), a predominant coastal point source of anthropogenic (largely inorganic) mercury (17). These results suggest that mercury concentrations can increase naturally through marine food webs due to the dominant organic form(s). No evidence was found that the (inorganic) mercury wastes released via this major wastewater system were biomagnified following discharge to the coastal marine ecosystem.

In summary, two decade-long investigations of point source trace metal injections to the marine environment have not been able to detect biomagnification of the type that can potentially lead to Minamata type catastrophes.

It may even be that where biomagnification is physiologically possible, it will actually occur only within a short, direct food chain, i.e. much of each trophic level in the chain is eaten by only one or a few species of less abundant predators, and each predator within the food chain is heavily dependent on the contaminated food source. This is what appears to have happened at Minamata and Nigata where the local populations of fishermen subsisted on the contaminated fish and shellfish stocks.

We suggest that future assessments of marine contamination from mining activities include determination of:

- ·the existence of (or potential for) excess bioaccumulation (test zone minus control zone) of associated trace metals;
- ·the degree of "structure" of the food web that could lead to biomagnification of excess metals; and
- ·the relationship between excess concentrations of the metals in a given tissue (e.g. muscle, liver) and the estimated trophic level position of the

target organisms.

Such an approach is necessary to test adequately (e.g., disprove) the hypothesis that, relative to control zone values, mining related trace metals are biomagnified in marine food webs.

We also ask that more long-term time-series investigations, i.e. those of ten years or more, of marine mine-metal bioaccumulation and biomagnification, be published in the professional literature. Bioaccumulation obviously occurs near coastal mine sites but it is not necessarily a serious environmental problem, unlike metal biomagnification in the sea, which on very rare occasions (not tailings induced) has been catastrophic.

REFERENCES

1. Förstner, U. and G.T. Wittmann. Metal Pollution in the Aquatic Environment (Berlin: Springer-Verlag, 1979).
2. Hickey, J. Peregrine Falcon Populations (Madison, WI: University of Wisconsin Press, 1969).
3. Goldberg, E.D. The International Mussel Watch (Washington, D.C.: National Academy of Sciences, 1980).
4. Phillips, D.J.H. Quantitative Aquatic Biological Indicators (London: Applied Science Publishers Ltd., 1980).
5. Island Copper Mine. Utah Mines Ltd. "1980 Annual Environmental Assessment Report", vol. 1 (1982).
6. Young, D.R., D.J. McDermott, T.C. Heesen and T.K. Jan. "Pollutant Inputs and Distributions off Southern California", in Marine Chemistry in the Coastal Environment, T.M. Church, Ed. (Washington, D.C.: American Chemical Society, 1975), pp. 424-439.
7. Young, D.R., T.K. Jan and T.C. Heesen. "Cycling of Trace Metal and Chlorinated Hydrocarbon Wastes in the Southern California Bight", in Estuarine Interactions, M.L. Wiley, Ed. (NY: Academic Press, 1978), pp. 481-496.
8. Hershelman, G.P., H.A. Schafer, T.K. Jan and D.R. Young. "Metals in Marine Sediments near a Large California Municipal Outfall", Mar. Poll. Bull. 12 (4): 131-134 (1981).
9. Young, D.R., M.D. Moore, T.K. Jan and R.P. Eganhouse. "Metals in Seafood Organisms near a Large California Municipal Outfall", Mar. Poll. Bull. 12 (4): 134-138 (1981).
10. Young, D.R. "The Distribution of Cesium, Rubidium, and Potassium in the Quasi-Marine Ecosystem of the Salton Sea", Ph.D. Thesis, Univ. Calif. San Diego (1970).

11. Isaacs, J.D. "Unstructured Marine Food Webs and 'Pollutant Analogues'", Fish. Bull., U.S. 70: 1053-1059 (1972).
12. Young, D., A. Mearns, T. Jan, T. Heesen, M. Moore, R. Eganhouse, G.P. Hershelman and R. Gossett. "Trophic Structure and Pollutant Concentrations in Marine Ecosystems of Southern California", Calif. Coop. Oc. Fish. Invest. Rept., vol. XXI: 197-206 (1980).
13. Mearns, A.J. and D.R. Young. "Trophic Structure and Pollutant Flow in a Harbor Ecosystem", in Biennial Report for the Years 1979-80, W. Bascom, Ed. (Long Beach, CA: Southern California Coastal Water Research Project, 1980), pp. 287-308.
14. Mearns, A.J., D.R. Young, R.J. Olson and H.A. Schafer. "Trophic Structure and the Cesium-Potassium Ratio in Pelagic Ecosystems", Calif. Coop. Oc. Fish. Invest. Rept., vol. XXII: 99-110 (1981).
15. Young, D.R. and A.J. Mearns. Unpublished data.
16. Young, D.R., A.J. Mearns, T.K. Jan and R.P. Eganhouse. "The Cesium-Potassium Ratio and Trace Metal Biomagnification in Two Contaminated Marine Food Webs", Proc. Oceans 81, IEEE Washington, D.C. (1981).
17. Eganhouse, R.P., D.R. Young and J.N. Johnson. "Geochemistry of Mercury in Palos Verdes Sediments", Environ. Sci. & Tech. 12: 1151-1157 (1978).

Note: At the time of going to press, a preprint of a detailed review of this topic as it related to the mining industry was seen by the editor. It consists of responses by M.J. Waldichuk to a set of questions to be discussed by panelists and participants at a seminar "Submarine and Lake Disposal of Mill Tailings" to be held on October 22, 1982 during the XIV International Mineral Processing Congress in Toronto, Canada. It is intended that panelists' responses will be published in the proceedings of the congress.

CHAPTER 5

THE EFFECTS OF SUBMARINE CHANNELS ON MINE TAILING DISPOSAL IN RUPERT INLET, B.C.

A.E. Hay
   Department of Physics and Newfoundland
      Institute for Cold Ocean Science
   Memorial University of Newfoundland
   St. John's, Nfld.
   Canada A1B 3X7

The evolution of submarine channel systems in the Rupert Inlet mine tailing deposit, as determined from time-series of seismic profiling, bathymetric and side-scan sonar surveys, is presented. Three distinct morphological phases occurred: a meandering channel regime, which was destroyed during the subsequent apron regime, and a final rechannelized phase. A single, well-defined leveed channel extended to an axial distance of 5 km from the outfall during the meandering channel phase. Channel depth and width were about 10 and 100 m near the outfall, decreasing down-channel except for an initial increase probably due to an internal hydraulic jump in the discharge plume. The channel consisted of a left-hooking upper reach with an average axial slope of 2.2°, a middle reach (1° slope) with pronounced meanders ($700 - 1100$ m wavelength), and a straight lower reach (0.5° slope). Levee slopes in the upper reach were about 25°. The tailing apron overlaid the upper reach, and was scarred by slumps and incipient channels on its flanks. The new channel developed on the western flank of the apron. The morphological character of these systems, together with grain-size distributions and copper concentrations in the surficial sediments, the distribution of turbidites in the sediment column, and acoustic sounding profiles of the discharge plume suggest that the leveed channels result in more rapid transport of tailings to greater depths and distances from the outfall, and increased frequency of turbidity current generation by slope failure.

INTRODUCTION

Tailings disposal by Island Copper Mine (ICM) into Rupert Inlet began in October, 1971. A study of the accumulation and dispersal of the tailings was conducted during the period from August, 1976 through September, 1979. The purpose of the project was to study the formation of leveed submarine channels in the tailing deposit by turbidity currents associated with the discharge. These currents are of two types: (a) continuous flow driven by the discharge, and referred to as the "discharge plume"; and (b) surge-type flows generated by slumping of previously deposited tailings. Both flow down-slope due to the gravitational acceleration acting on the excess density provided by solids in suspension. These currents have long been thought to be responsible for the formation of the submarine canyons and leveed channels in the deep sea (1, 2), and for the deposition of turbidites (3).

In this article an attempt is made to focus on the practical problem of marine tailings disposal. Three questions are addressed:

1. How do submarine channels form as a result of tailing discharge?
2. How do submarine channels affect the dispersal of mine tailings when these tailings are discharged as a slurry from a submerged outfall?
3. Are submarine channels beneficial or not? That is, does the presence of a channel adversely affect the dispersal of tailings from either the engineering or the environmental viewpoint?

Previous studies of tailing dispersal have been conducted in Rupert Inlet by Johnson (4), who was concerned with the effects of the inlet circulation pattern as reflected in the distribution of sediment types, and in Lake Superior by Normark and Dickson (5, 6), who mapped submarine channels and detected turbidity currents on the Western Reserve fan. Carstens and Tesaker (7) and Tesaker (8) have studied the formation of a submarine channel by an erosive tailing discharge plume in Ranafjord.

The tailings are discharged into Rupert Inlet from a submerged pipe 1.07 m in diameter at a depth of 49 m (outfall, Fig. 1 and Frontispiece) as a slurry of solids, freshwater and seawater in the ratio 1:4:5 parts by volume

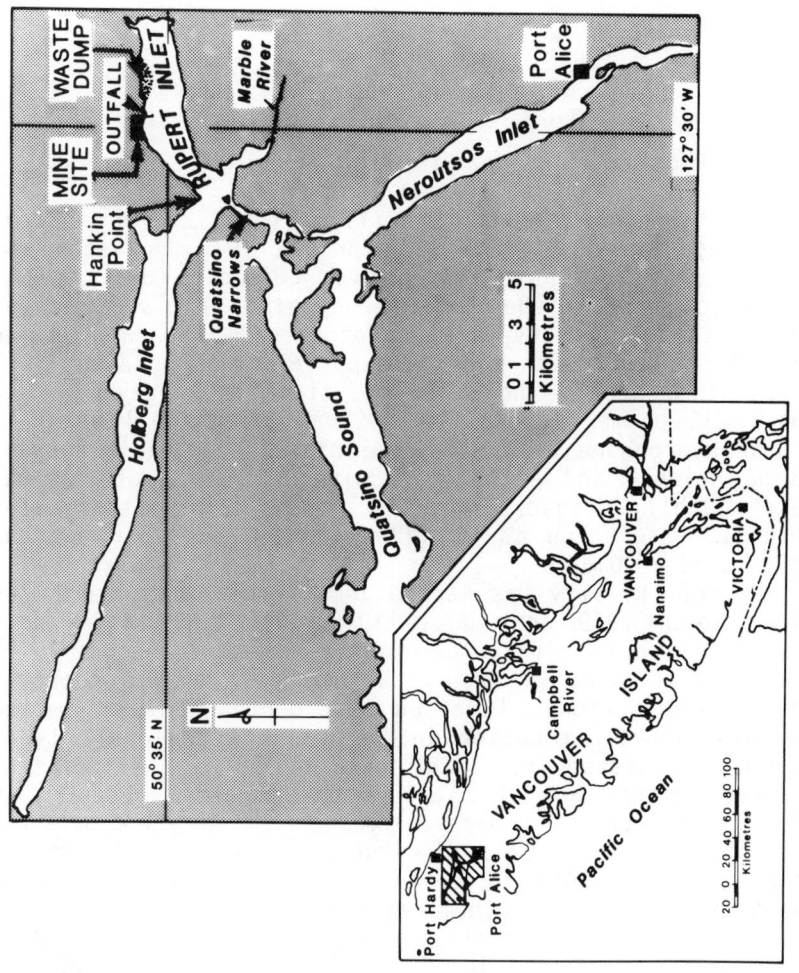

Figure 1. Location map showing Rupert Inlet and the mine site.

respectively. The solids discharge rate is about 380 kg/s. The tailing particles have a median grain diameter before discharge of 0.03 mm and 65-75% are smaller than 0.074 mm. Low grade rock is dumped into the inlet at a rate approximately three to four times that of the tailings (waste dump, Fig. 1). Further details concerning the ore mineralogy and processing may be found in Evans and Poling (9) and Poling (10).

METHODS

Seismic Profiling and Bathymetric Surveys

Continuous seismic profiling (CSP) surveys were conducted annually from 1971 through 1975 and again in 1977 by the University of British Columbia (UBC) to map the thickness of the tailing deposit. Ship's positions during the early surveys were determined by radar, and in the 1977 survey with microwave positioning equipment (Del Norte Trisponder). The seismic system included an EG & G Model 236 boomer (approximately omni-directional beam), a linear hydrophone array (10^0 calculated beam width between -3 dB points) and a custom receiver (0.4 - 2 kHz nominal band width). Because of ringing in the transmitted pulse, vertical resolution is at times limited to 1.5 m but can be as good as 0.3 m.

The bathymetry was mapped in November 1976, September 1978, March 1979 and August 1979. Ship's positions were obtained with the Trisponder system except in August 1979, when they were made with manned, radio-reporting theodolite stations. Echo-sounding depths were corrected to datum by using cosine interpolation between the tidal extrema predicted for Coal Harbour (11). The relative accuracy within a given survey, based upon the depths at line intersections, was estimated to be 0.2 m. The absolute accuracy, once the survey is corrected for the mean speed of sound in the water column, is probably of the order of 0.3 m at a nominal depth of 100 m.

In addition, a side-scan survey was conducted in June 1977, and a bathymetric survey in the vicinity of the outfall in December 1977. Radar positioning was used in both surveys.

Sediment Sampling and Analysis

Surficial sediments were sampled with a Shipek grab in December 1977, February 1979 and August 1979. Two

100-150 cm^3 subsamples were taken from each grab by the repeated use of a small piston corer made from a 50 cm^3 disposable syringe. One subsample was stored at ambient temperature and used for size analysis; the other was frozen and used for Fe and Cu determinations.

The sediment column was sampled with a Kullenberg gravity corer (6.2 cm core diameter) in November 1976 and in March 1979 with a Benthos Boomerang gravity corer (6.6 cm core diameter) modified for winch coring. The cores were X-rayed and then split before sampling for grain size and metal analysis.

Size analyses of surficial sediments were made with 30-40 g (dry weight) of the homogenized subsample after desalination with distilled water. This material was split into coarse-grained (sand) and fine-grained (mud) fractions by wet-sieving through a 0.0625 mm sieve. The sand fraction was dried and sieved at 0.5 phi intervals on a Ro-tap following Folk (12) and Carver (13). The mud fraction was dispersed in a 5 g/L aqueous solution of sodium hexametaphosphate and analyzed on Micromeritics Model 5000 and 5000D Sedigraphs. Because of the smaller quantities of material involved, the samples from the cores were not split but were analyzed in toto on the Sedigraph. The size analysis procedures are described in detail elsewhere (14).

Metal concentrations were determined on a Tectron AA4 Atomic Absorption Spectrophotometer (oxy-acetylene flame) at 324.8 nm for Cu and 386.0 nm for Fe, after digesting approximately 0.5 g of dry, unwashed sediment in a 4:1 aqueous solution of concentrated nitric and perchloric acids at approximately 200°C. After evaporation of the nitric acid, the digest was washed through a 0.005 mm nominal pore size Gelman vinyl membrane filter and the filtrate made up to 100 mL with deionized distilled water for absorption determinations. Standard solutions were prepared from known quantities of the metal or a metal salt. A blank digest was run with every set of 12 samples. The precision of the sample Cu concentrations was 2.7% $\pm$ 0.1%, based upon duplicate digests of 10 samples. Systematic errors due to interference by chemical species present in the digests but not in the standards were determined to be less than 3% by measuring the change in absorbance induced by spiking a sample with the standard solution.

Acoustic Backscatter

The acoustic sounders used in this program were manufactured by Ross Laboratories, Inc. and included the 42.5, 107 and 200 kHz Model 200 transceivers and 5° x 10°

elliptical-beam transducers on the CSS Vector, and a 192 kHz custom portable model with a 2.3^0 circular beam pattern which was used from a 5 m launch. The envelope-detected signal was recorded on a Hewlett-Packard Model 3960 FM instrumentation tape recorder and the recorded signal digitized at an effective rate of 20 kHz on the UBC Department of Oceanography PDP-12 minicomputer with a 12 bit A-D converter. The digitized data was processed on the UBC Computing Centre Amdahl 470 V/S Model II. The recorded acoustic backscatter was compared to concentrations obtained from bottle samples and to the amplitude of the echo from a standard target. The details are available elsewhere (14).

RESULTS

Seismic Profiling

The first definitive evidence of the existence of a submarine channel in Rupert Inlet was obtained in the November 1974 CSP survey (Fig. 2). The channel was present in each of the two subsequent surveys in 1975 (Fig. 3) and 1977 (Fig. 4). These three figures illustrate the general thickening of the tailing deposit and its westward growth with time, and the development of the waste dump and deposition at its base. The mine-derived material in this area is readily distinguished from tailings because of the lack of horizontal coherence in the sub-bottom echoes and the hummocky nature of its surface, due to the highly disturbed nature of the deposit and the presence of rocky material.

The channel starts close to the outfall, runs southerly down-slope and across the inlet, afterwards turning to follow the axial trend of the inlet toward Hankin Point. The flanking levees are readily distinguished in the cross-inlet portion (upper reach) as zones of heavy deposition, as is the outer levee immediately to the south of the first bend. In the 1977 survey (Fig. 4), the channel exhibited a series of pronounced meanders below the upper reach, and then straightened further to the west. The positioning during the 1977 survey was both more accurate and more frequent than had been the case in the earlier surveys. It is possible, therefore, that the meanders had also been present earlier.

There are several areas of more rapid accumulation to the west of CSP 11. These are the sites of large topographic holes in the pre-mine bathymetry (Fig. 5).

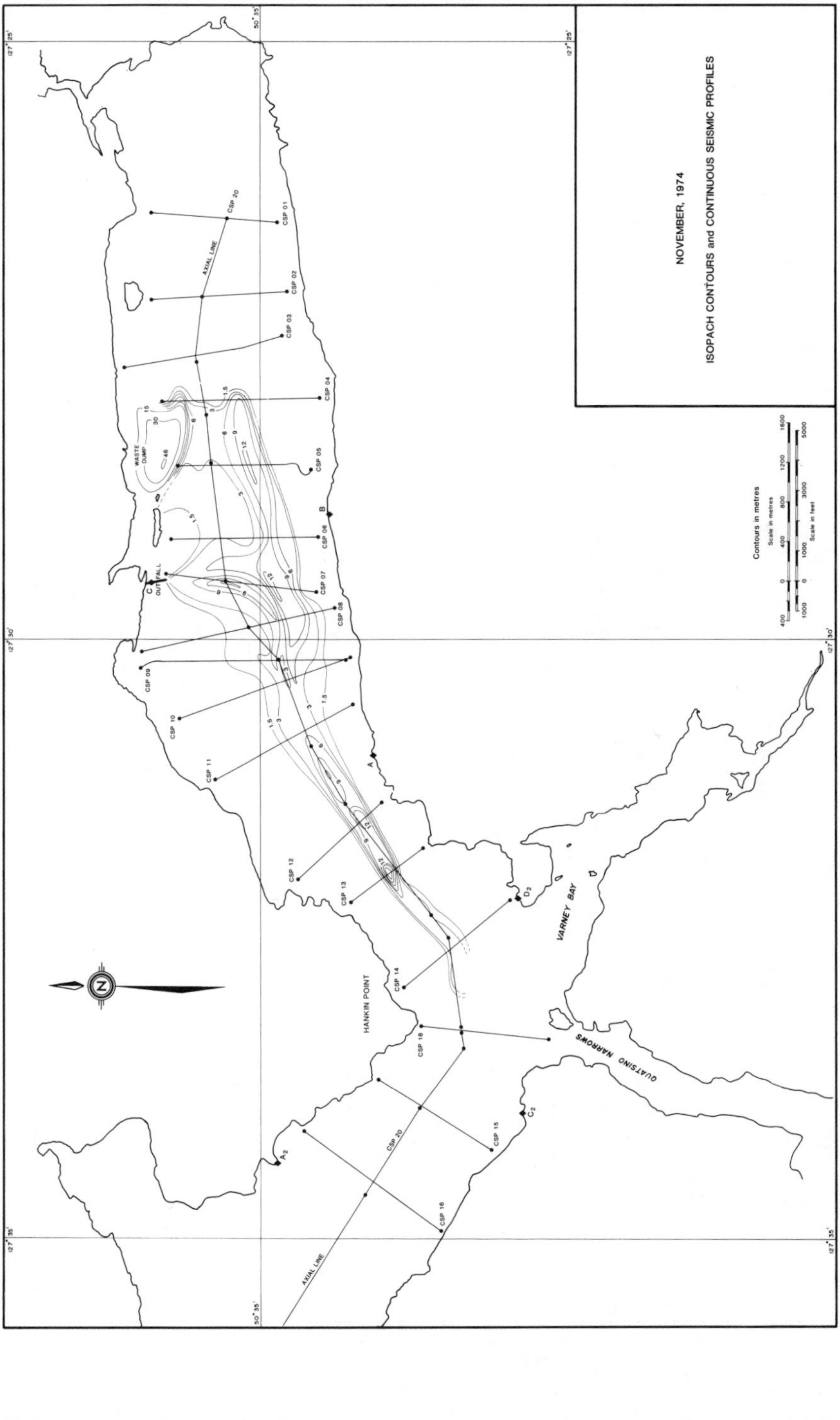

Figure 2. Tailing thickness in meters, 29 November 1974. Heavy line is the channel axis.

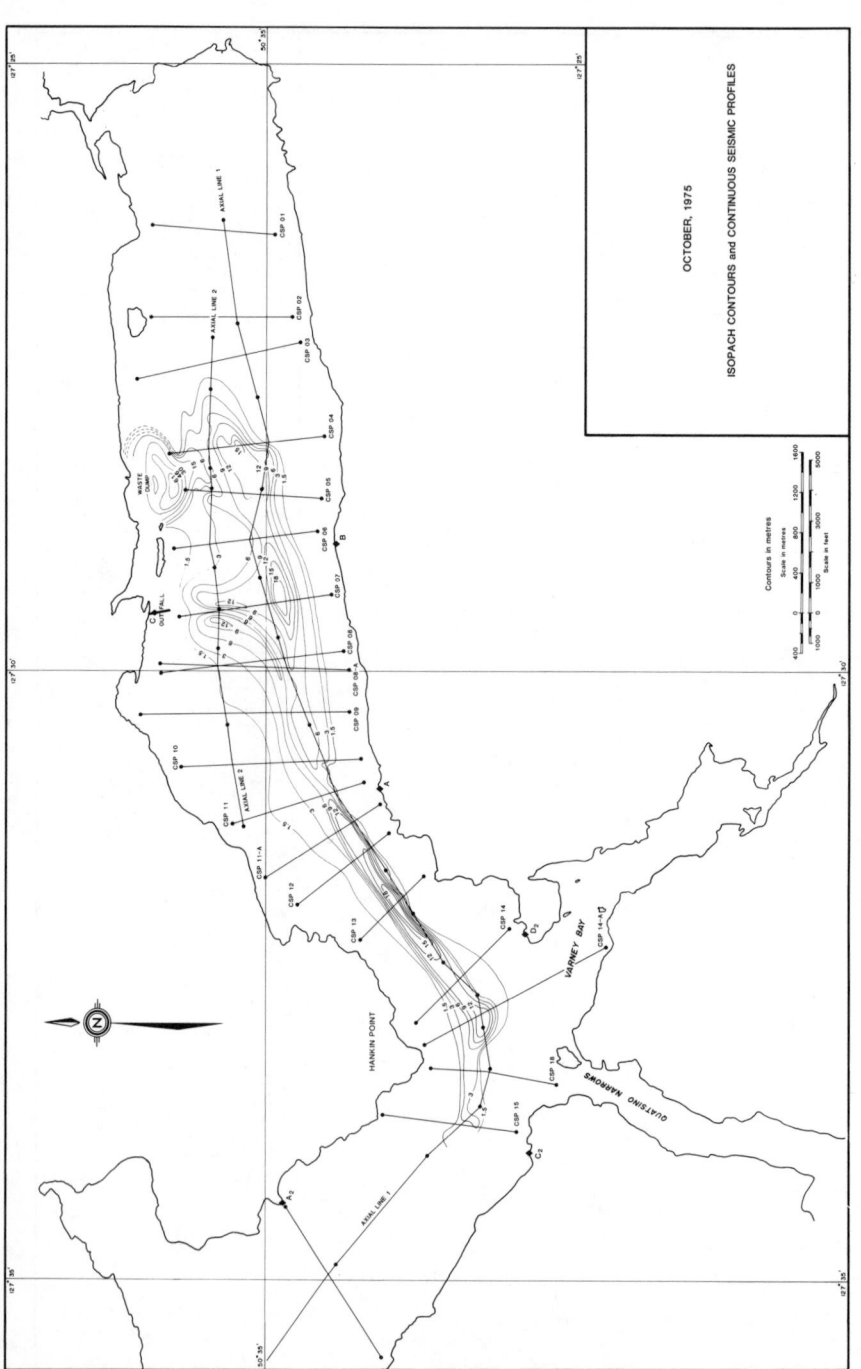

Figure 3. Tailing thickness in meters, 21 October 1975.

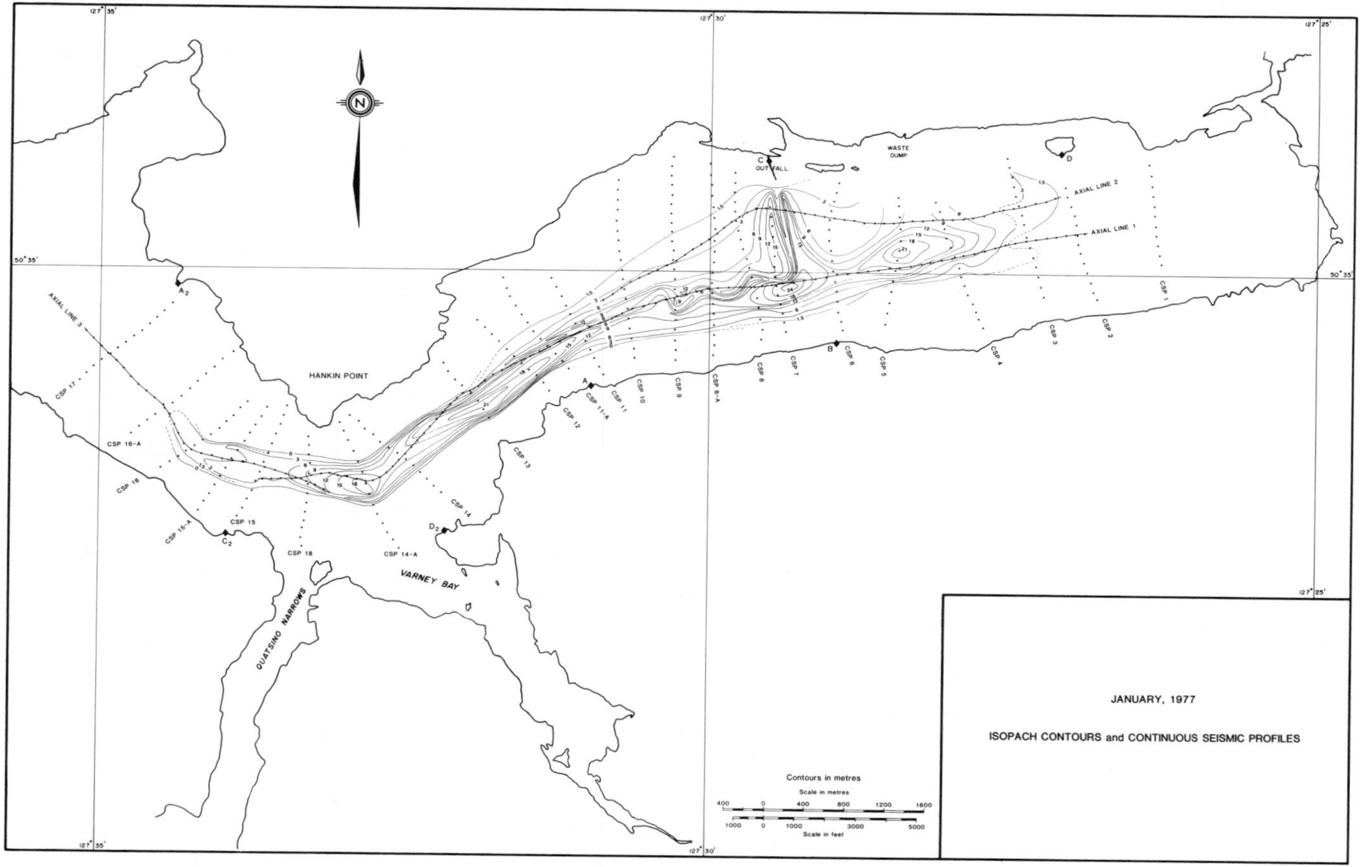

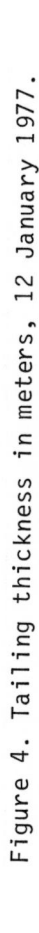

Figure 4. Tailing thickness in meters, 12 January 1977.

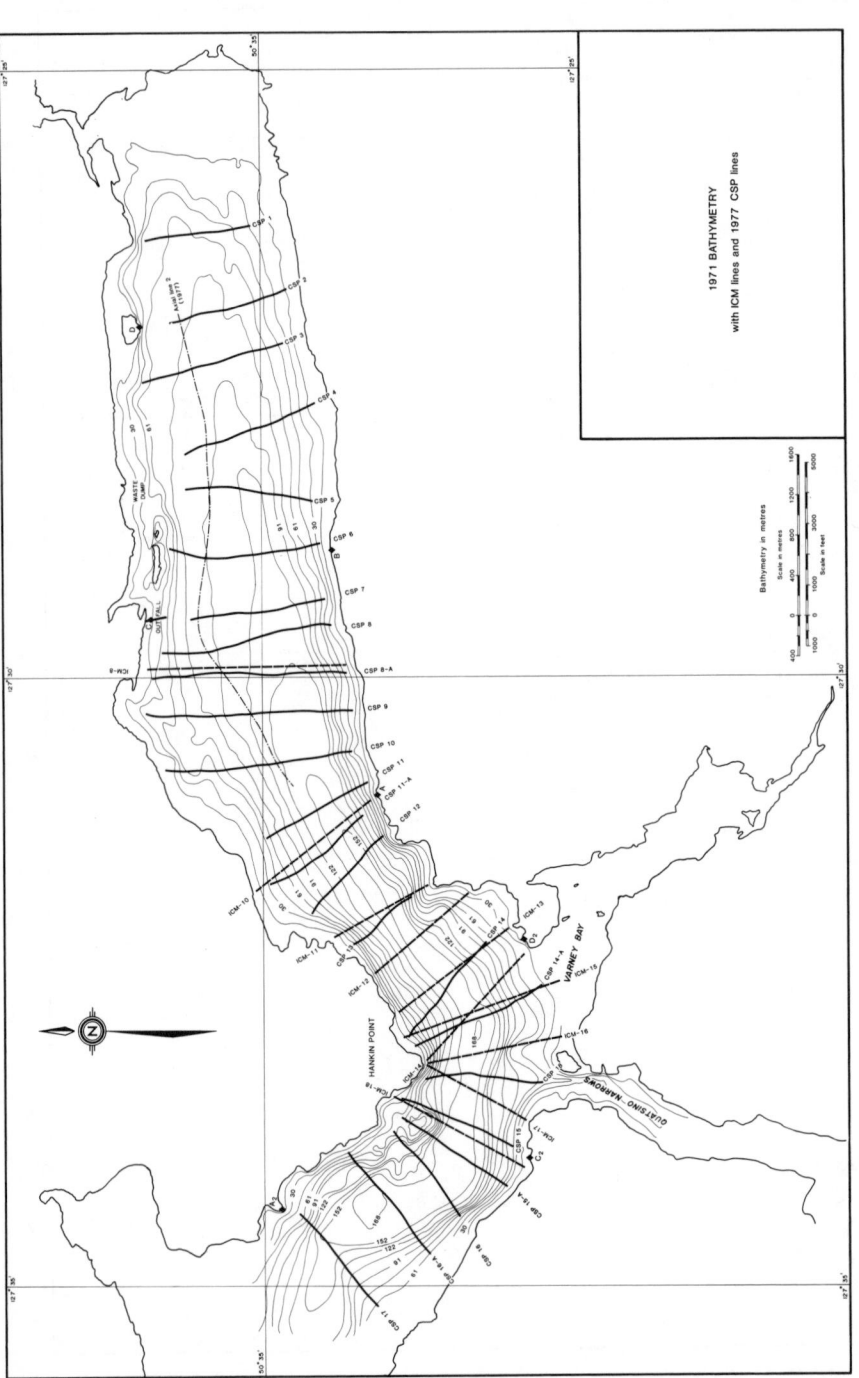

Figure 5. Pre-mine bathymetry in Rupert Inlet. Locations of seismic reflection lines (CSP) and the

Note that in the 1974, 1975 and 1977 surveys (Figs. 2-4), an axial line was run across the upper reach at a reasonable distance from the outfall. The use of such lines transverse to the outfall and approximately parallel to the pre-mine isobaths in the near-outfall zone is necessary if the initial stages of formation are to be studied. It is the absence of such lines in earlier surveys which makes the year of initial channelization uncertain.

Bathymetric Surveys

The soundings used to construct the bathymetric maps were uncorrected for the difference between the actual sounding speed in the water column and that for which the echo-sounder was calibrated (1463 m/s). A correction equation to allow true depth (z') to be obtained from the apparent depth (z) is given with each map.

The bathymetry was accurately surveyed for the first time subsequent to the beginning of tailing disposal in November 1976 (Fig. 6), about two months previous to the 1977 CSP survey discussed previously (Fig. 4). The density of survey lines was much greater than in the CSP surveys, with a typical interline spacing of 50-100 m. The upper reach exhibited a left-hook and the meanders in the middle reach were well-defined and similar in form to those in the 1977 CSP survey (Fig. 4). The lower reach was relatively straight and disappeared before reaching the Hankin Point area. The channel axes in plan view for these two surveys and a side-scan survey in June 1977 are plotted in Fig. 7. The channel shape is, within the margins of error due to different positioning methods and density of coverage, essentially the same in the three surveys. Deep-sea submarine channels also have a tendency to hook to the left in the Northern Hemisphere (2) and some develop meanders (15). The meanders in the Rupert Inlet system are geometrically similar to those in subaerial rivers (14).

The axial profile of the channel is shown in Fig. 8. The points on this figure are taken directly from the original sounding records and the depths are corrected to datum using the true sounding speed determined from temperature-salinity profiles. The number of points is an indication of the density of survey lines. The upper, middle and lower reaches can be distinguished on the basis of slope. In the upper reach, the slope decreased from 9.5-12° near the outfall to 1.9° at the entrance to the first bend; in the lower reach the slope was nearly constant at 0.91°, although it decreased between bends 5

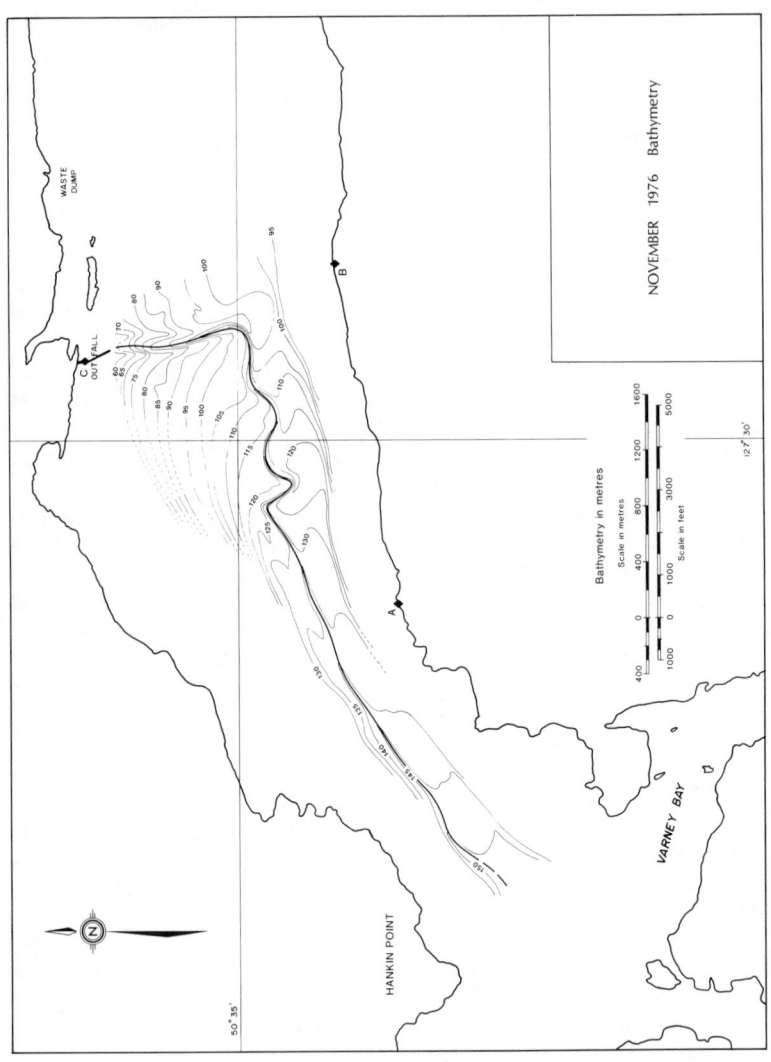

Figure 6. Bathymetry in November 1976. To obtain "true" depths (z), z' =
[(z-3.7)/1.031] c_s'/c_s where c_s' = 1483.4 m/s is the sounding speed to 100 m depth. The
heavy line indicates the channel axis.

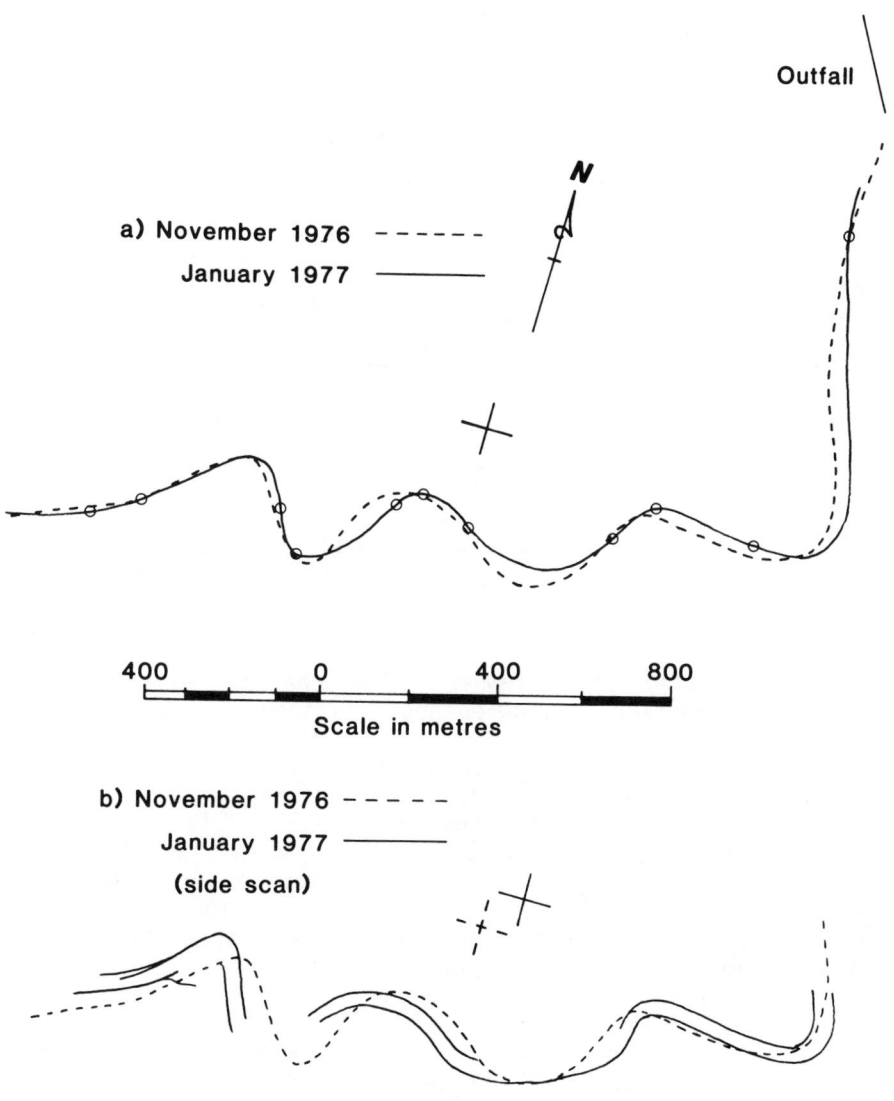

Figure 7. (a) Channel axes in January 1977 (solid line) and November 1976 (dashed line).
(b) Channel axes in June 1977 (solid lines) and November 1976 (dashed line). Note offset of side-scan profile due to ship to tow-fish separation.

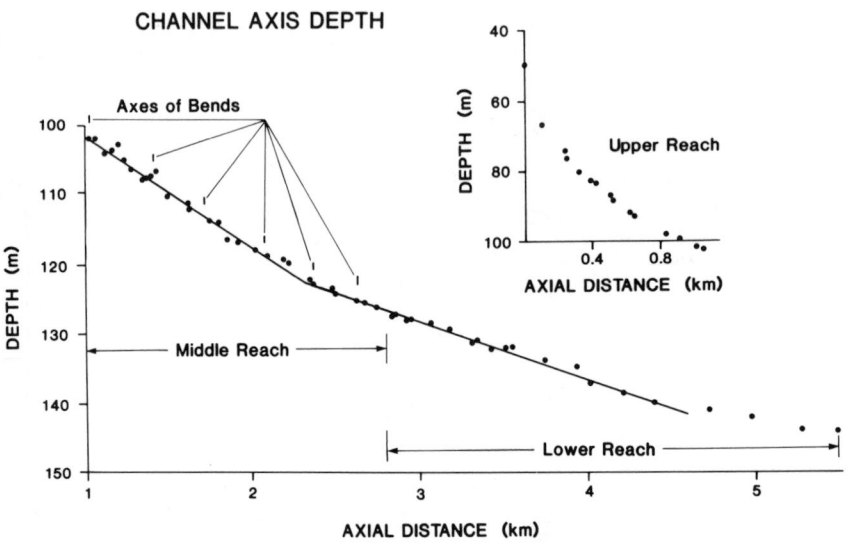

Figure 8. Depth of November 1976 channel axis as a
         function of long-channel distance.

and 6; and in the lower reach the slope was again constant
at 0.47° up to 4.5 km from the outfall.
     Cross-channel profiles, looking down-channel, are
shown in Fig. 9. Fig. 10 shows the profile locations. In
the upper reach the channel cross-section increased
initially between profiles 1 and 2, and then decreased
downstream in all successive profiles. This increase in
area may reflect the occurrence of a hydraulic jump
between profiles 1 and 2, since the flow is probably
supercritical when discharged onto the high initial slopes
and the downstream decrease in slope provides the
necessary deceleration of the flow. The profiles are also
noticeably asymmetric, the right levee being higher than
the left in the left-hooking section of the upper reach
(see Fig. 6). This is also reflected in greater
thicknesses of the tailing deposit under the right levee
(Fig. 4). The same is true of the outer levee at the axes
of bends (profiles 8-14 in Fig. 9). These features are
important to the model of channel development presented
later.
     Except for a survey of the near-outfall area in

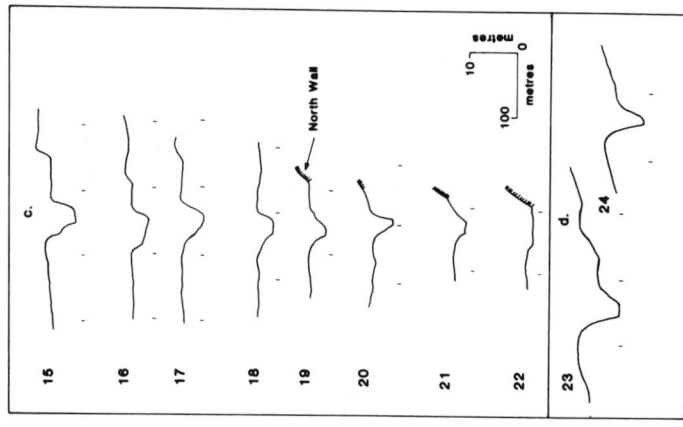

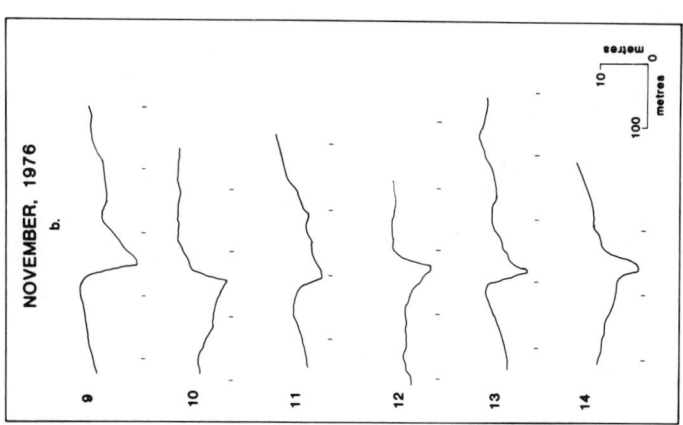

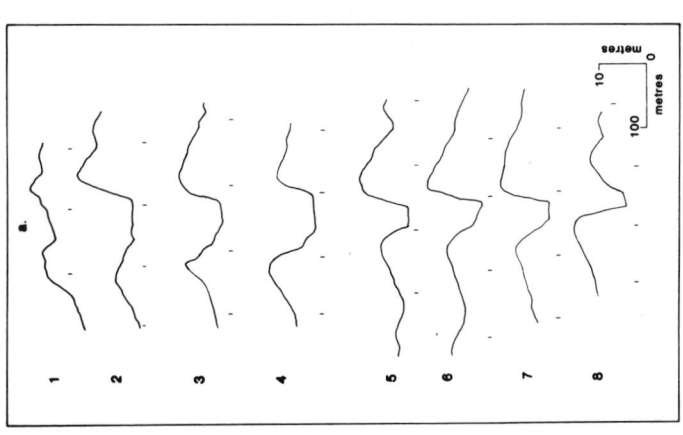

Figure 9. Cross-channel profiles, looking down-channel, in November 1976. (a) Upper reach; (b) middle reach; (c) lower reach; and (d) meander reach at channel crossovers. See Fig. 10 for locations.

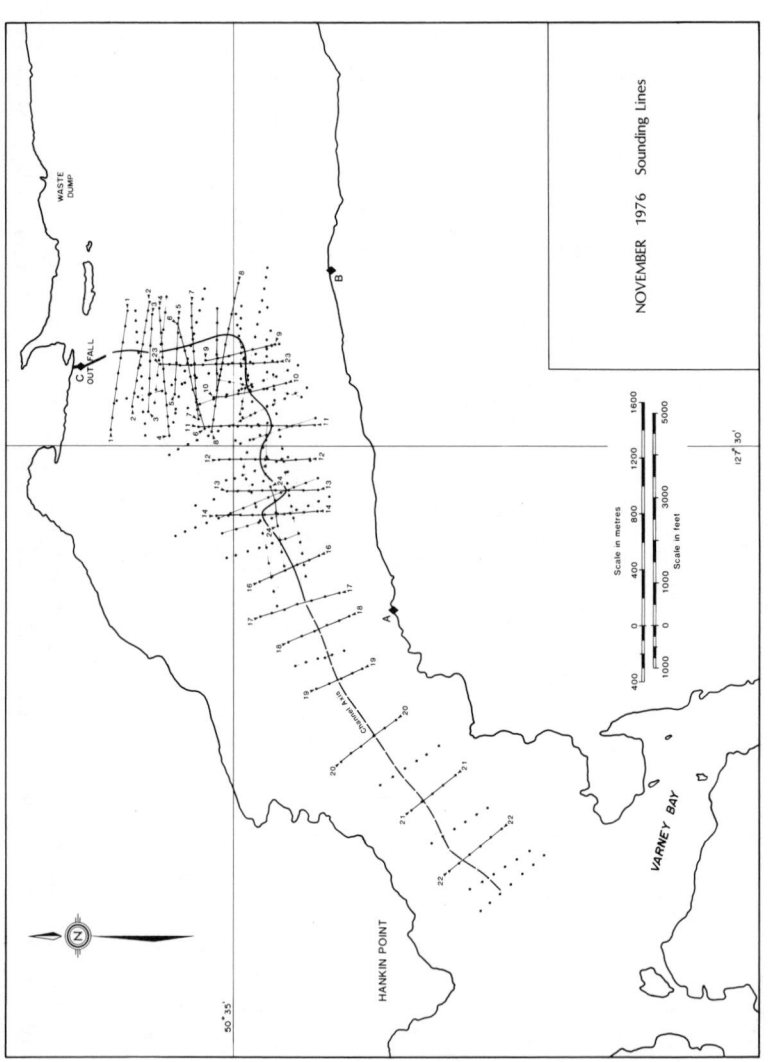

Figure 10. Sounding transects for profiles in Fig. 9.

December 1977 (Fig. 11), no further survey was conducted until September 1978, with the exception of the monthly echo-sounding lines run by ICM since March 1975. The September 1978 survey (Fig. 12) shows that the channel system had been completely obliterated. The superceding system will be referred to as the "apron regime", because of the apron of tailings overlying the upper reach. The same system was present in February, 1979 (Fig. 13). In both surveys the apron flanks, particularly the west flank, exhibited features which are interpreted as being either slump scars or incipient channels. Neither the time of the transition, nor its duration, from the channelized regime to the apron regime are well-known. Similarly, the cause of the transition is a matter of speculation. The ICM sounding profile of the channel in the middle reach was observed to change in April, 1978 after having been stable over the previous two years (16). The difference in depth, corrected for sound speed, between the November 1976 and September 1978 surveys is shown in Fig. 14. The absence of areas of negative accumulation at the locations of the former levee crests suggests that the levees did not collapse, but rather that the upper reach was filled in with material from the discharge. The volume of the upper reach of the channel had been 5.9×10^5 m^3, which would take a minimum of 25 days to fill at a discharge rate of 380 kg/s assuming a bulk density of 1.4 g/cm^3 for the deposit. It is possible, therefore, that the cause of the transition occurred several months previous to the change observed in the ICM profiles of the middle reach. ICM did not run sounding lines across the upper reach.

A final survey was conducted in August 1979, by which time a second channel system had developed on the west flank of the apron (Fig. 15). It is clear that this channel formed in one of the slump scars or incipient channels surveyed earlier, and down the steepest slope accessible to it. The bathymetry of the Hankin Point area shows a scour hole which is maintained by strong bottom currents generated at flood tide during periods of deep-water exchange (17, 18, 19). Leading into the scour hole is an extremely steep-sided channel which is not physically connected with the channel emanating from the outfall. The channel has no levees which, together with the steepness of its wall, indicates that it is an erosional feature.

Surficial Sediments

A study of Cu levels in the tailing deposit measured by ICM indicated a trend to decreasing concentration with

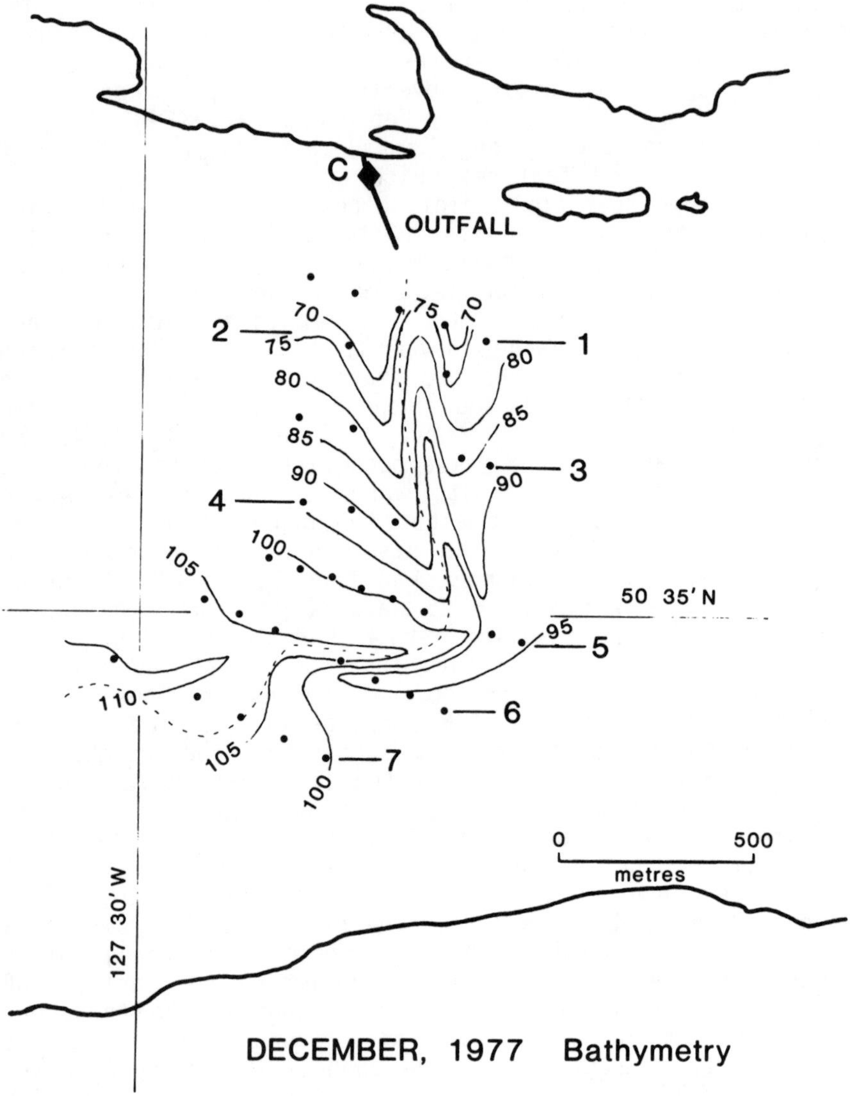

Figure 11. Bathymetry in December 1977. Axis of November 1976 channel is also shown (dashed line).

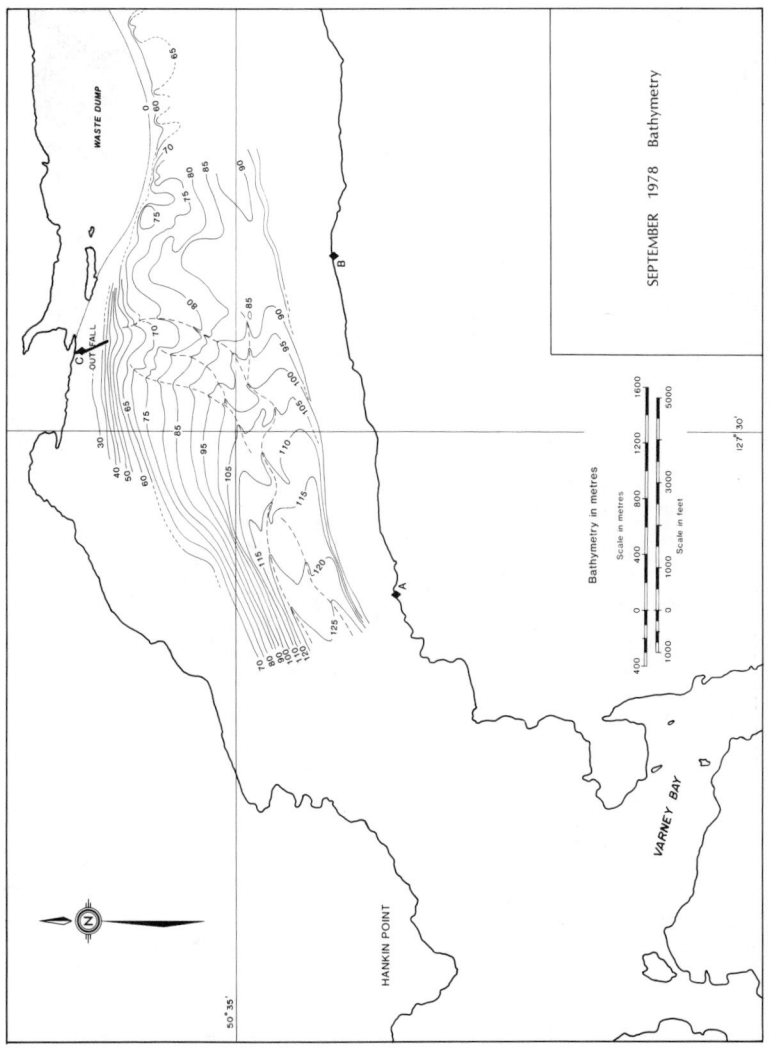

Figure 12. Bathymetry, September 1978. $z' = zc_s'/c_s$, where $c_s' = 1492.8$ m/s.

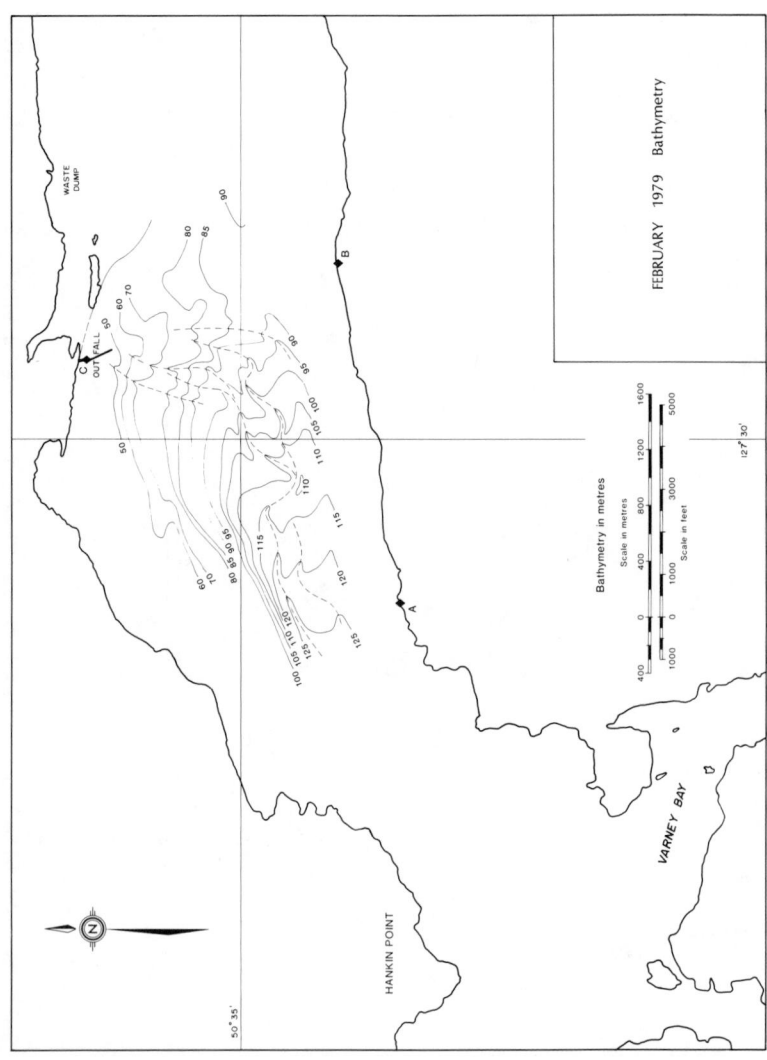

Figure 13. Bathymetry, February 1979. $z' = zc'_s/c_s$, where $c'_s = 1473.2$ m/s.

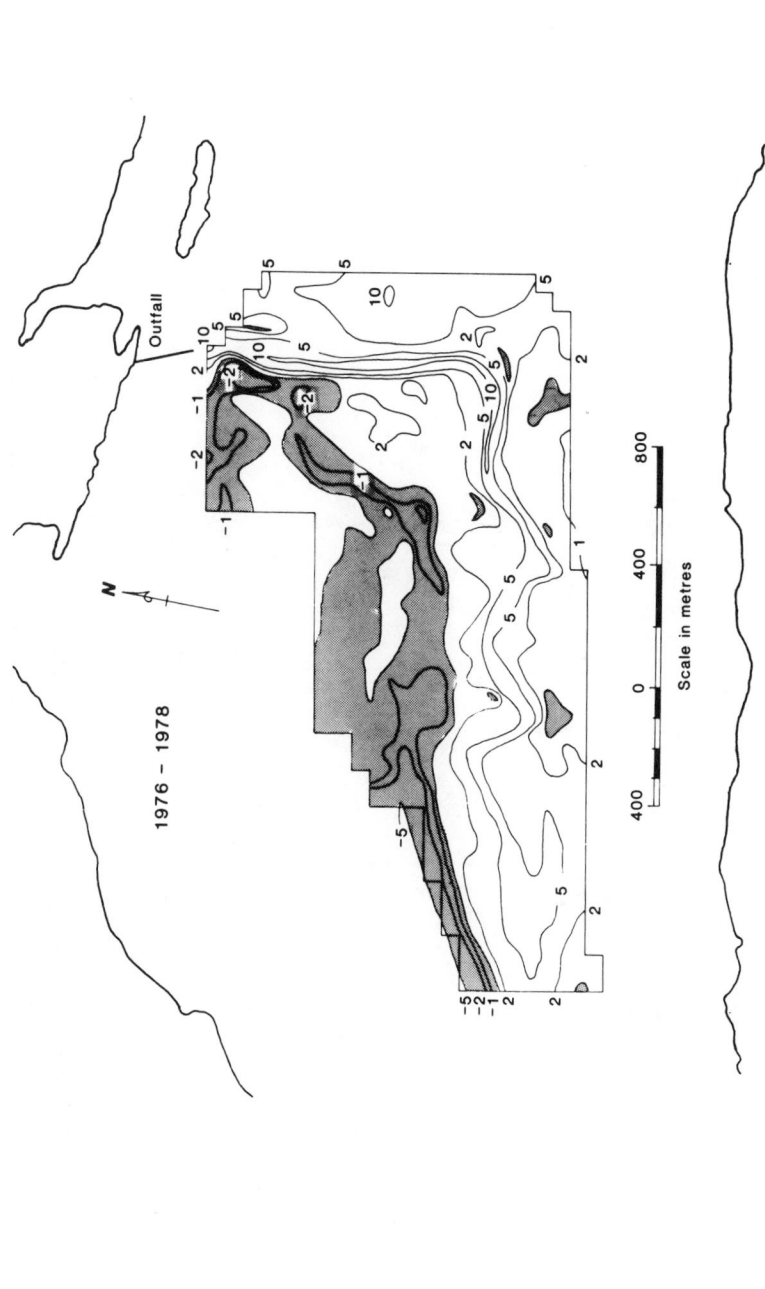

Figure 14. Depth difference map, November 1976 – September 1978. Positive values indicate net deposition. Negative values are shaded.

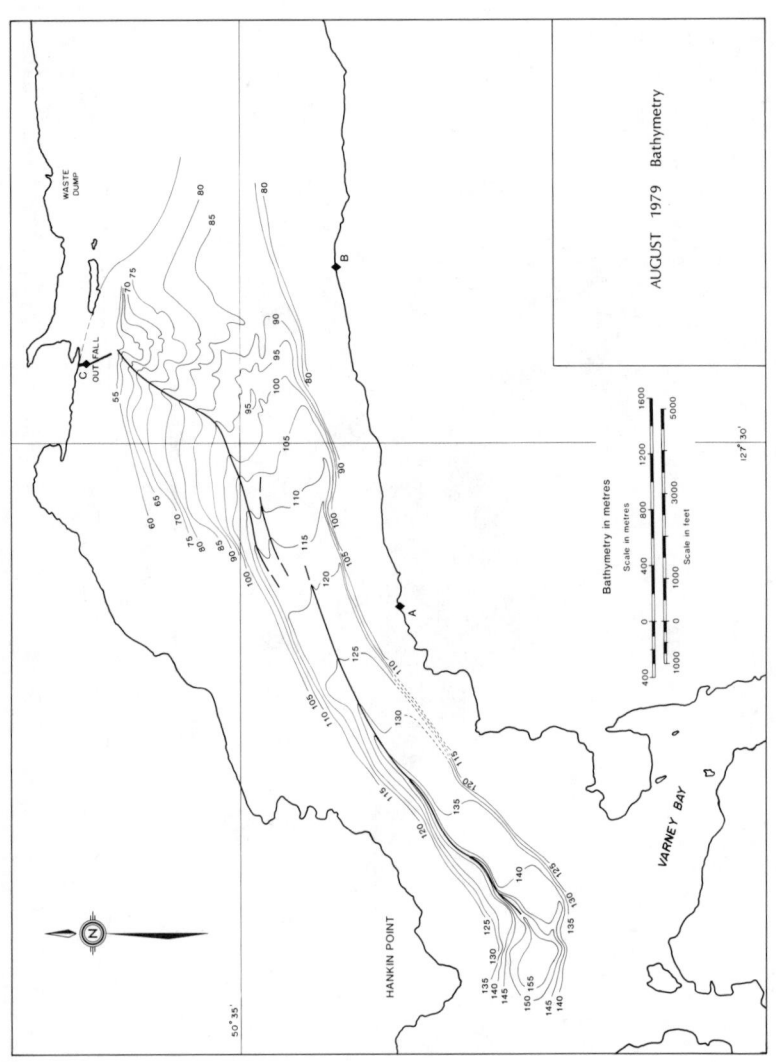

Figure 15. Bathymetry, August 1979.

distance from the outfall (20). This suggested that Cu might be used as a tracer for deposits, or more specifically turbidites, laid down by mass movement of material out of the outfall zone.

The only surficial sediment samples from the meandering channel regime were taken in December, 1977. The size analyses (by sieve and hydrometer) and Cu concentration determinations for these samples indicated a decrease from 921 to 495 ppm Cu with distance along the upper reach and some evidence of a lateral decrease in Cu concentration (14). Although the highest Cu levels were associated with the coarsest deposits, too few samples were taken to properly define the trends.

The sediment sampling stations in March 1979 during the apron regime are shown in Fig. 16. Coarser-grained, Cu-rich sediments were found on the apron flanks, the coarsest being on the west flank, separated by fine-grained, copper-poor mud along the apron crest (Figs. 17, 18 and 19). A coarse-grained, Cu-rich deposit was found at the base of the west flank. These observations are consistent with the interpretation suggested by the bathymetry: that the discharge plume is deflected down either flank, but principally down the west flank.

Surficial sediments were also sampled during the rechannelized regime at the stations shown in Fig. 20. The distributions of sand and Cu concentration are shown in Figs. 21 and 22. The highest Cu levels and largest sand fractions are found along the axis of the channel and decrease with lateral distance from it. There is no discernable axial trend.

Fig. 23 shows the strong correlation between the quantity of sand-sized material and the Cu concentration in the samples from these two surveys. The average Cu concentration in mud (samples containing no sand) from the tailing deposit is about 300 ppm, and represents the average background level found in the deposited mine waste. The lowest levels (60 ppm average) were found in gravity cores of pre-mine sediments.

The approximately linear relationship between the size of the sand fraction and the Cu concentration (Fig. 23) suggests that Cu behaves conservatively in the surficial tailings, being determined by the relative amounts of Cu-rich sand and Cu-poor mud. The scatter in these data is greater than the probable error in the measurements ($\pm$ 3% in Cu; 3% in percent sand), probably reflecting variations in ore-types.

It is believed that the increase in Cu concentration with larger amounts of sand is due to the reduced extraction efficiency of chalcopyrite by the froth-floation process with increasing grain diameter.

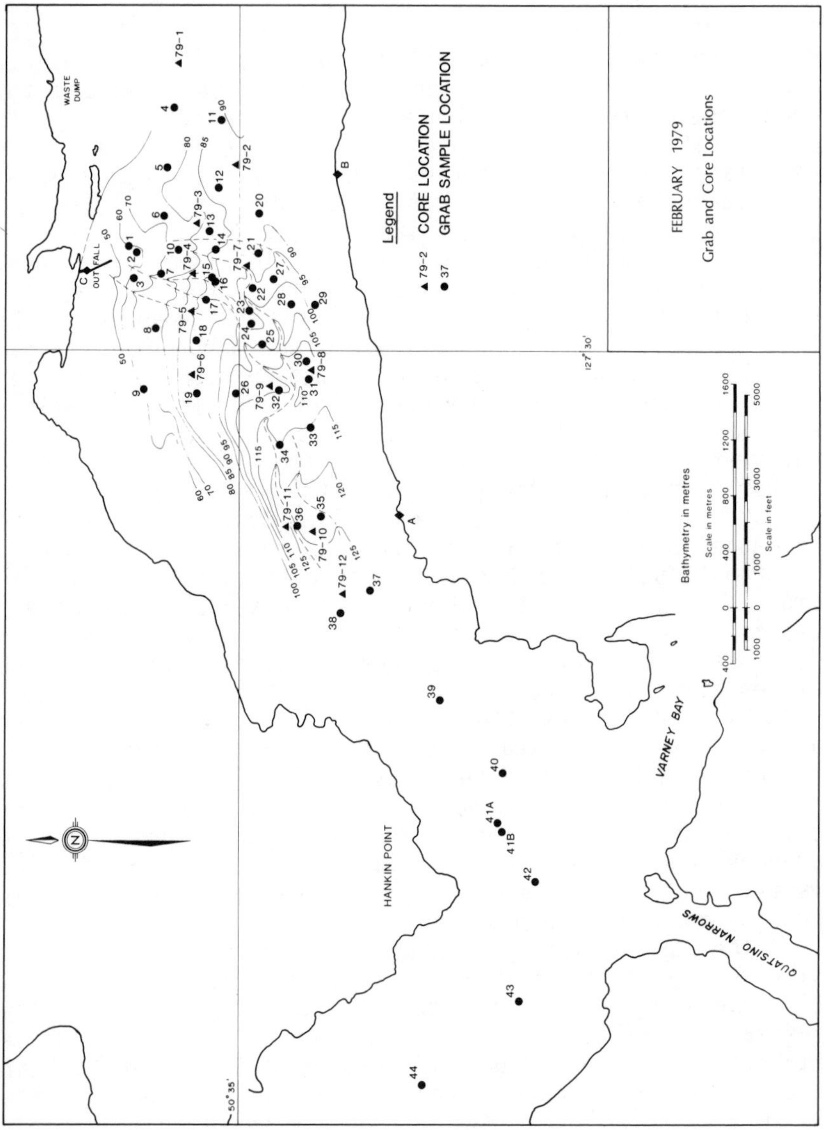

Figure 16. Core and grab sample locations, February 1979.

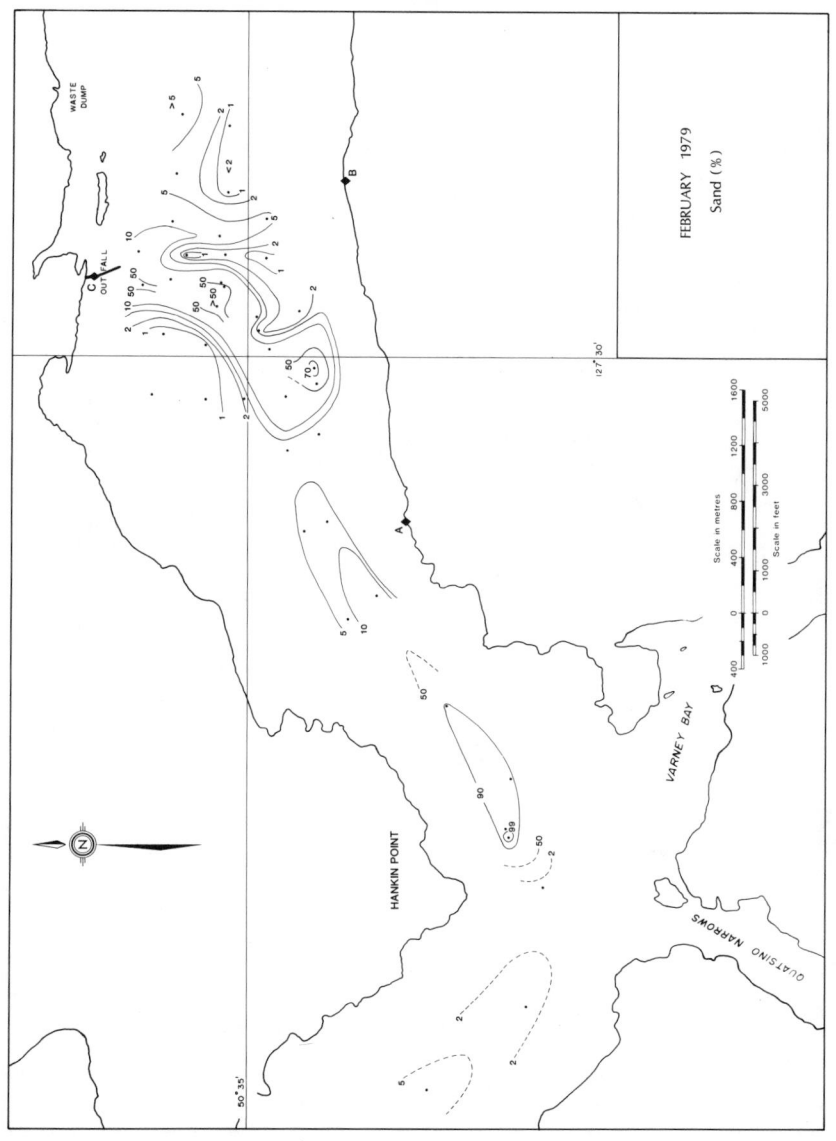

Figure 17. Sand content (%) of surficial sediments in February 1979.

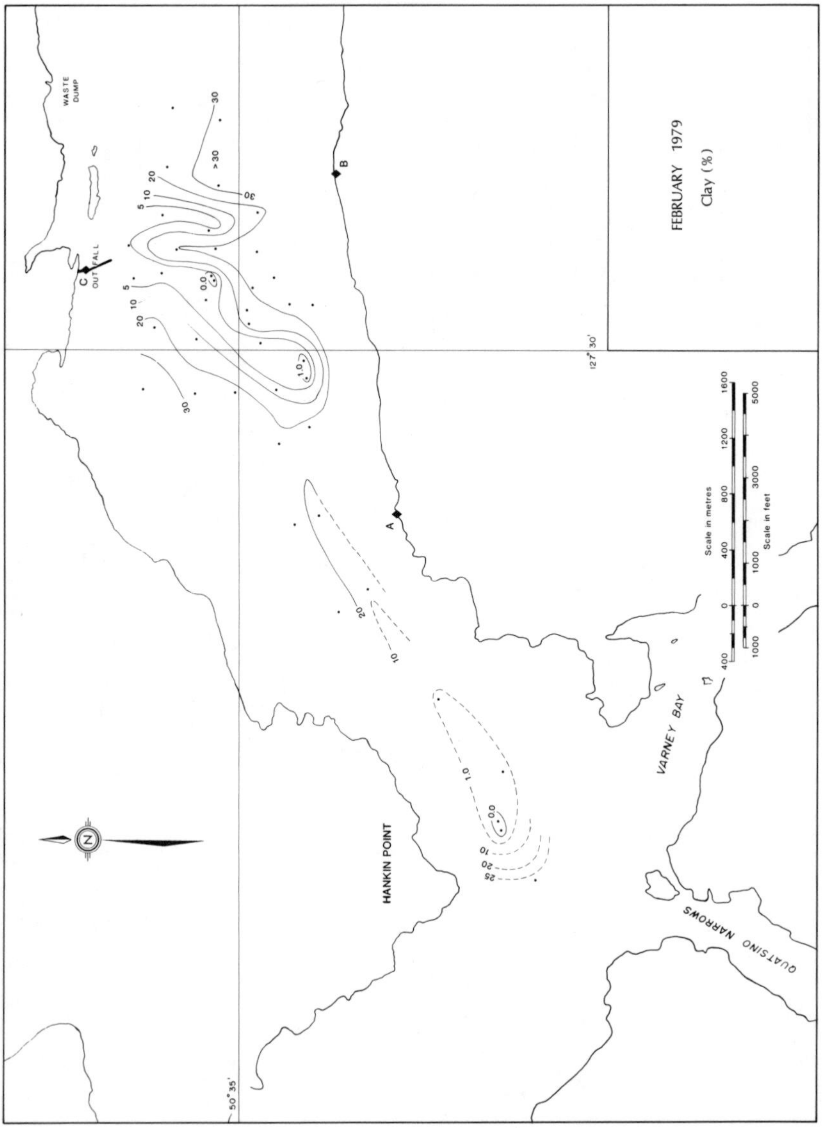

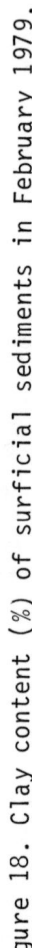

Figure 18. Clay content (%) of surficial sediments in February 1979.

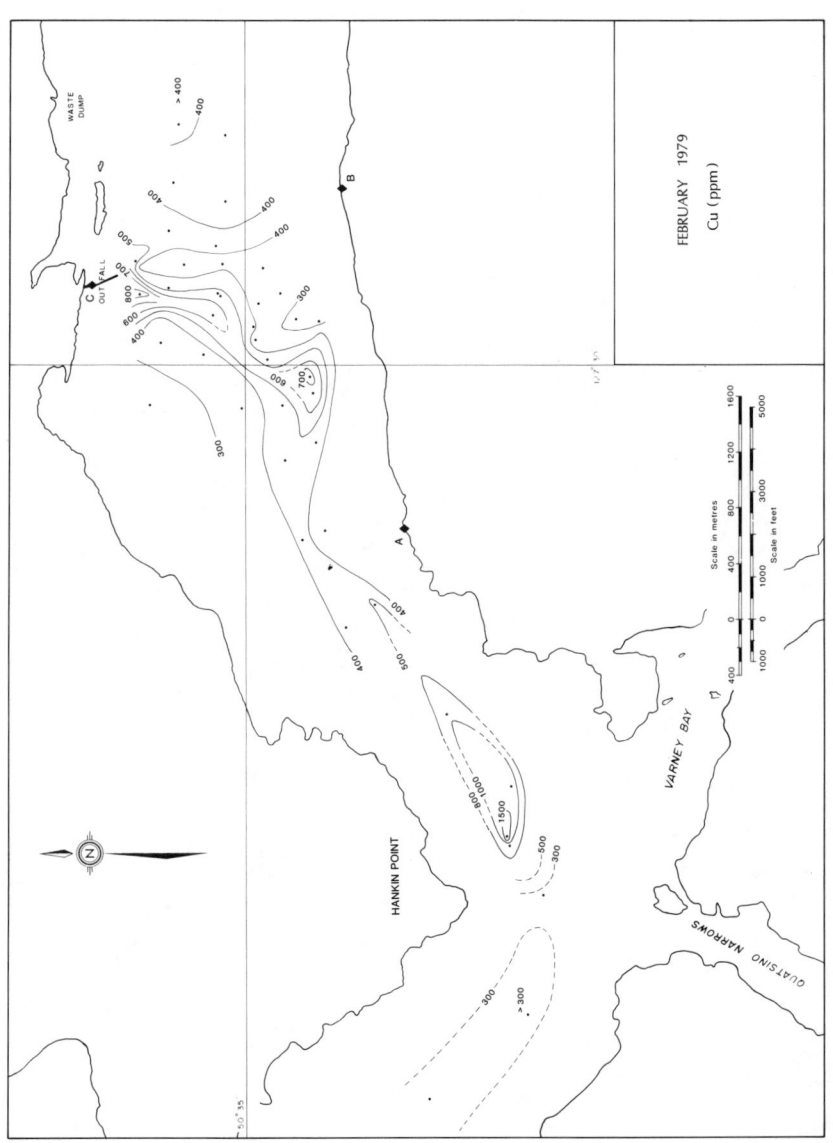

Figure 19. Cu concentration (ppm) in surficial sediments in February 1979.

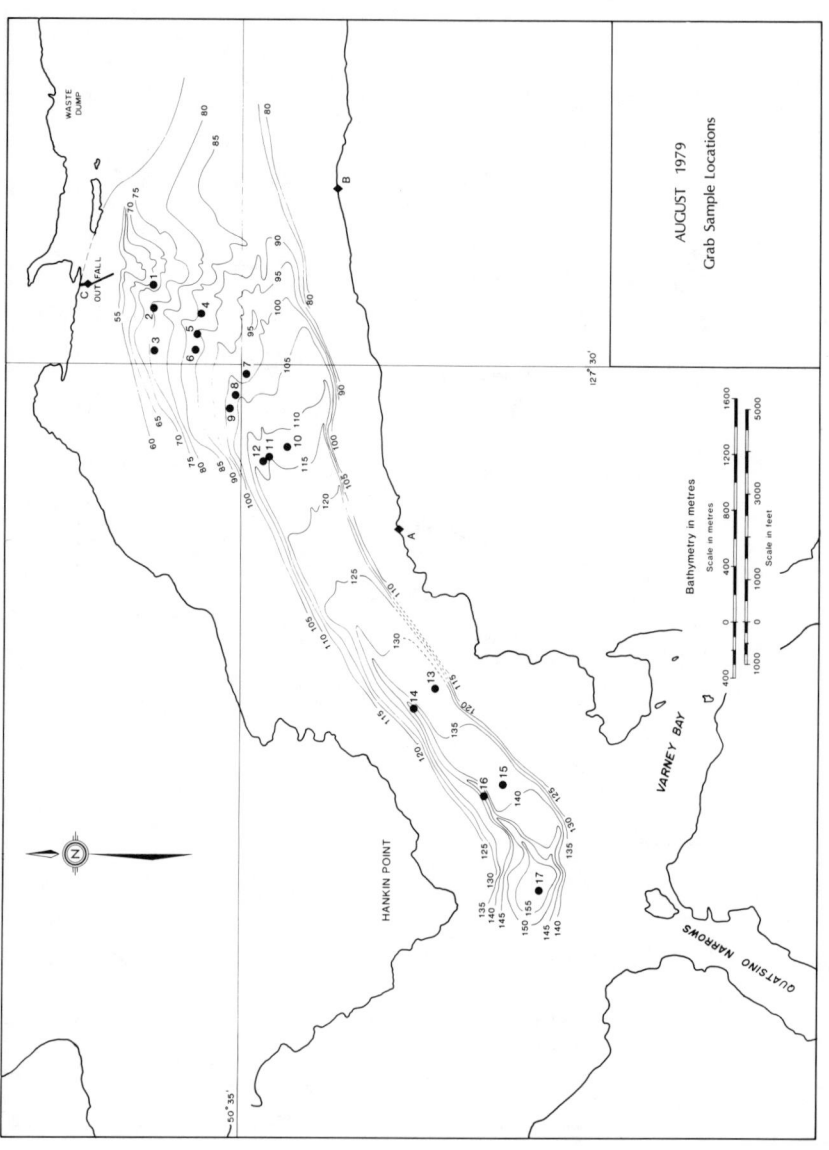

Figure 20. Grab sample locations, August 1979.

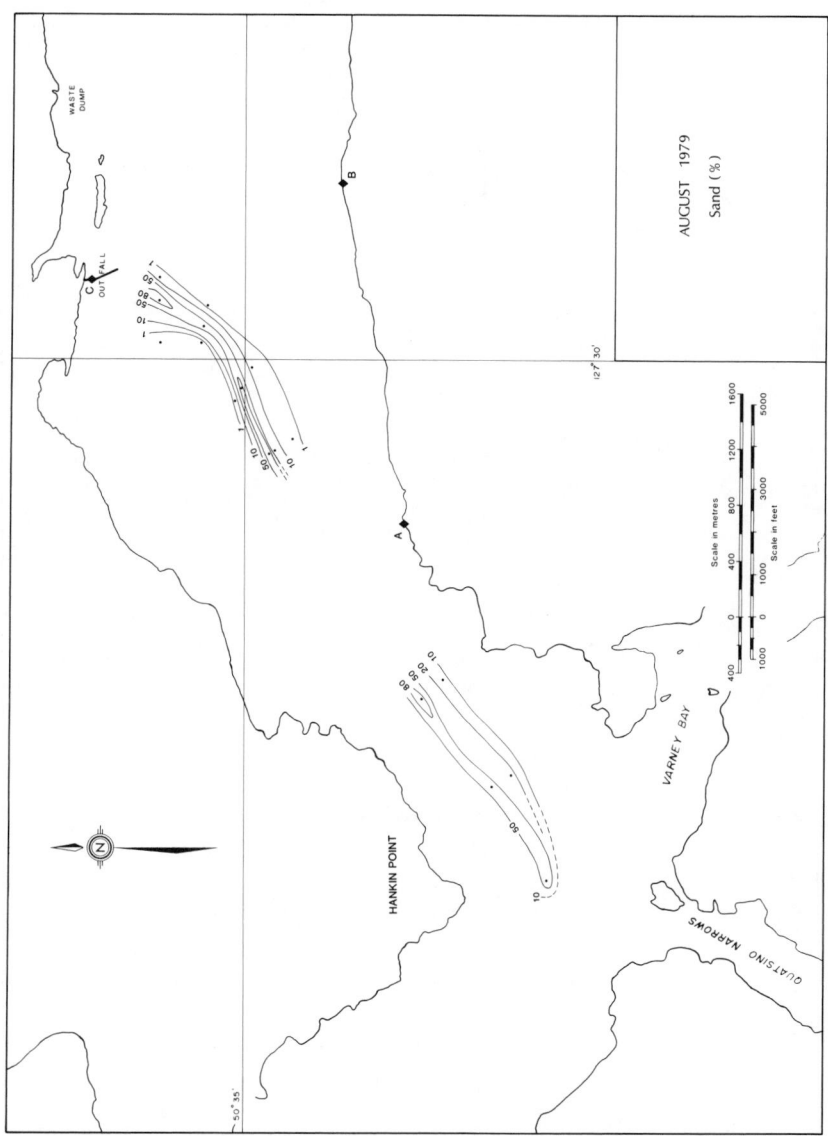

Figure 21. Sand content (%) in surficial sediments, August 1979.

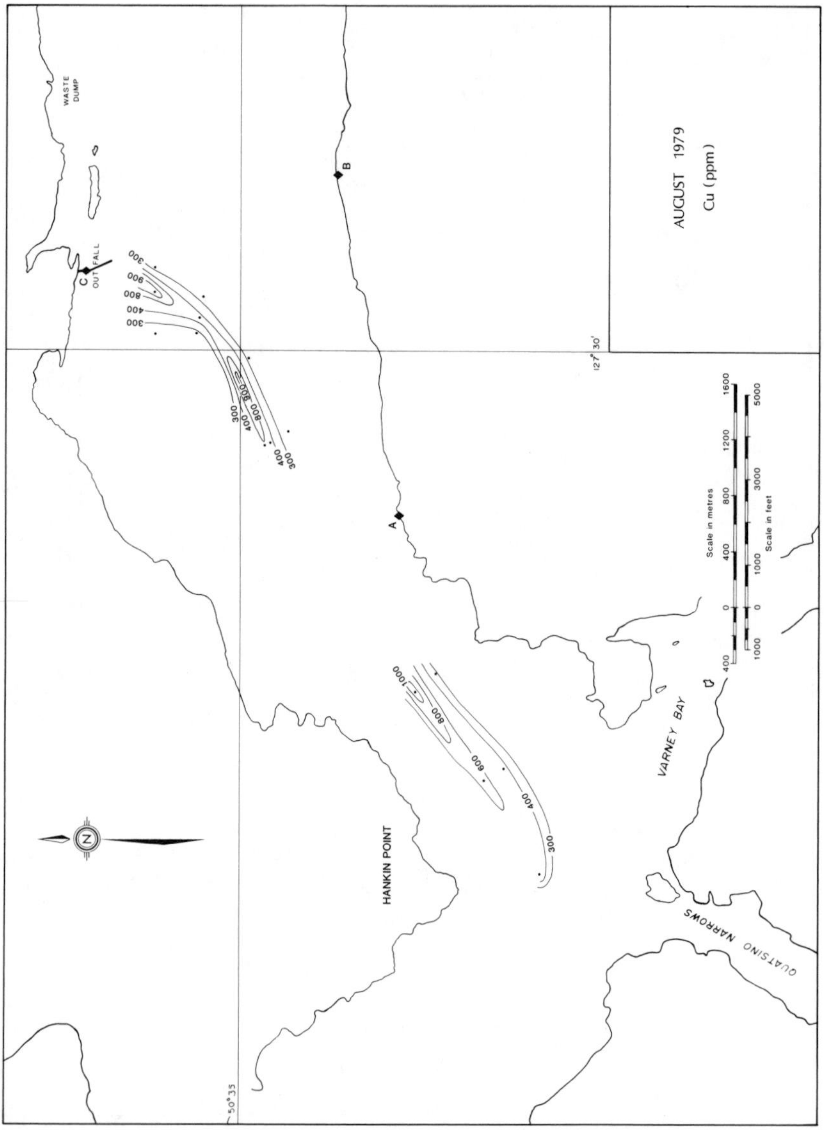

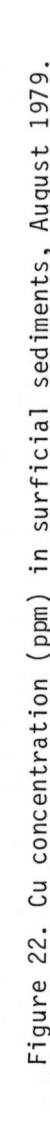

Figure 22. Cu concentration (ppm) in surficial sediments, August 1979.

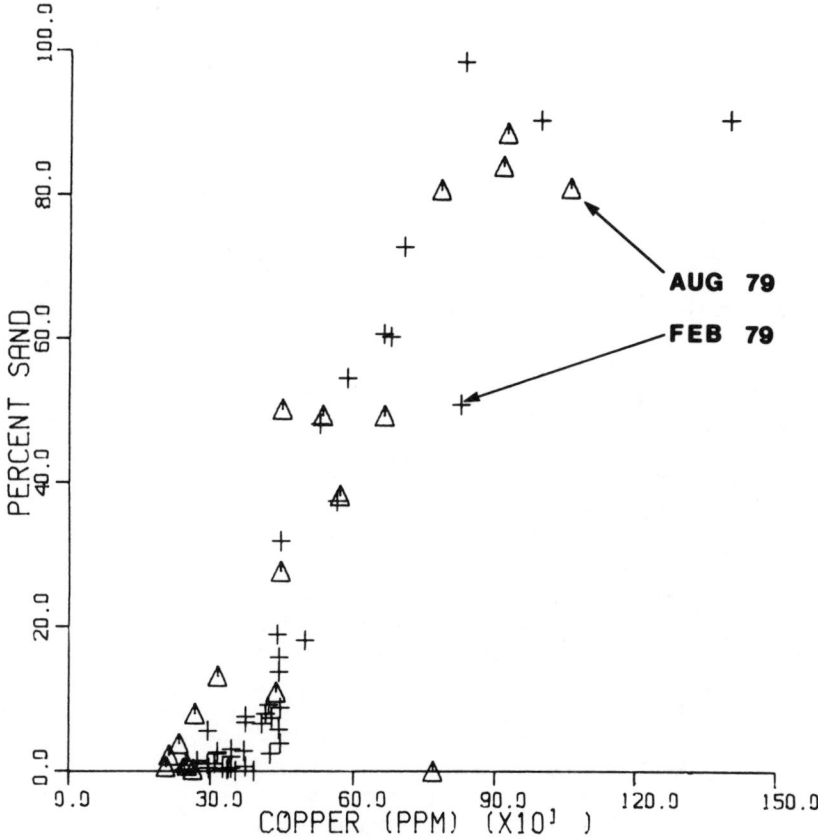

Figure 23. Percent sand versus Cu concentration in the 1979 surficial sediments.

Sediment Column

We return to the meandering channel regime. Since the levee slopes are the steepest ($\sim 25°$) in the system, since the levee crests are sites of rapid deposition (e.g. Fig. 4), and since axial slopes are greatest in the upper reach, the levee walls in the upper reach represent the zones of probable rotational failure (slumping) which are most likely to result in the generation of surge-type turbidity currents. Furthermore, the surficial sediment analyses indicate that these levee walls consist of copper-rich, coarse-grained sediments which, when deposited on the levees downstream as a result of channel overspill, should be both coarser and Cu-rich relative to the background sediments deposited from the discharge plume.

Cores

Summary descriptions of cores taken in November 1976
at locations shown in Fig. 24 are given in Table I. Cores
1 and 2 from the levees near the outfall consisted of
alternating light and dark bands of coarse material. The
banding is thought to reflect the differences in
coloration of the ore-types processed in the mill, since
the dominant type in the feed changes every 2-4 weeks. The
Cu concentration in both cores increased from a low value
of about 300 ppm at the top to 700-900 ppm at the base (75
cm depth). Cores 10 and 11 were taken on the axis of the
Holberg Inlet trough. Each penetrated a layer of tailings
to the underlying non-mine derived material, which had Cu
levels of 60 ppm.
Cores 3 to 9 are of most interest to this discussion.
Fig. 25 shows the results of the analyses of closely
spaced samples from a section of core 7. The layers
sampled are three well-defined turbidites, consisting of
sands fining upwards to silt and a thin layer of mud. The
basal contact in each layer is quite sharp and, in the
case of the lower layer, has been disturbed by removal of
the metal finger core-catcher. The Cu levels decrease from
high levels in the basal sand to low levels in the mud as
expected. Fe concentrations are low in the coarse material
and high in the mud. Iron concentrations, which have not
entered the discussion to this point, increase with the
quantity of silt in the sample, apparently because of
higher levels of fine-grained magnetite in that size
fraction (14). Fig. 26 is a mosaic of photographs of the
freshly split half-cores. The darker, coarse-grained
turbidites increase in thickness and make up an increasing
fraction of the sediment column with distance
down-channel. This suggests that deposition from
continuous flow is not significant in the lower reach, and
this is consistent with theoretical transport estimates
presented elsewhere (14), which indicate that most of the
tailings transported through the lower reach must be
carried by slump-generated surge-type turbidity currents.

Acoustic Profiles of the Discharge Plume

A typical 200 kHz acoustic profile of the discharge
plume in the upper reach is shown in Fig. 27. The bottom
is at the upper edge of the uniform grey band, which is
generated electronically by the high-amplitude bottom

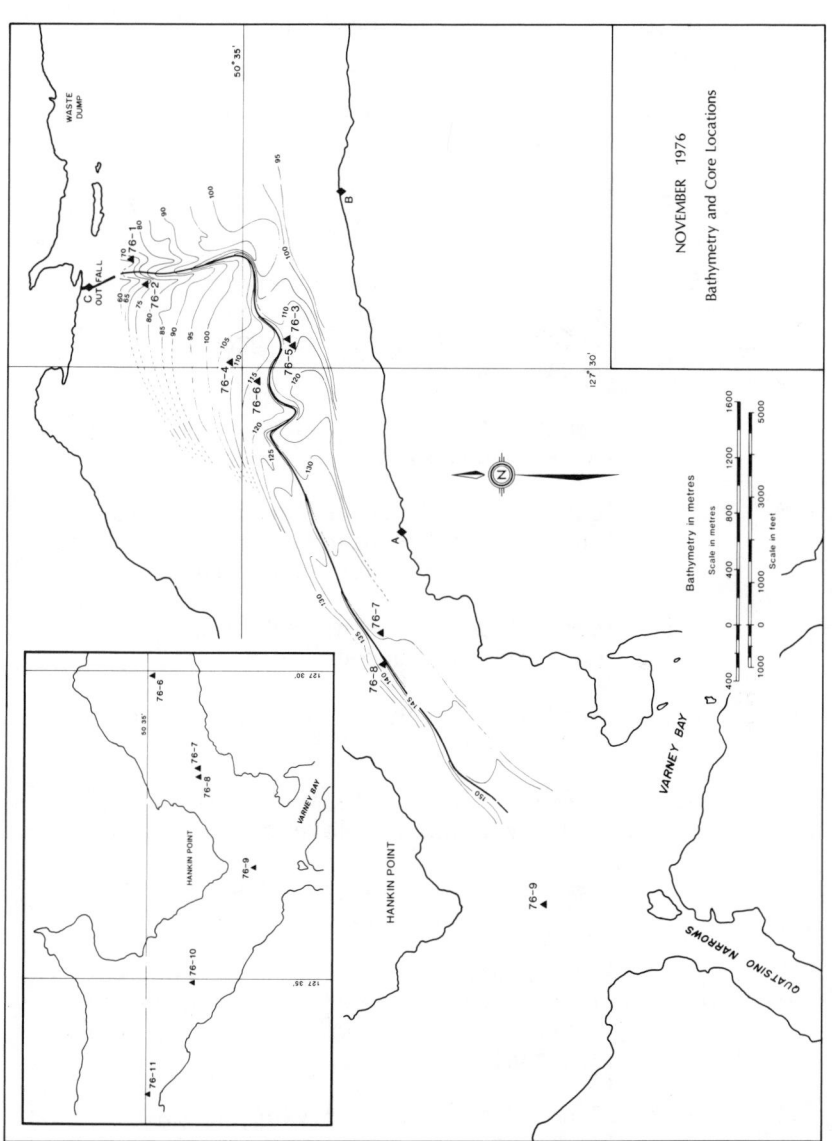

Figure 24. Core locations and bathymetry, November 1976.

Table I. 1976 Cores.

Core	Length (m)	Description
1	0.75	Laminated silts and sand, small amounts of clay.
2	0.83	Laminated silt and sand. More clay than 1.
3	0.2	Fine laminated mud interbedded with dark layers of sand and silt, some of which have erosional or gravity loaded basal contacts.
4	1.55	Laminated mud, thin widely spaced layers of sand and silt.
5	1.35	Laminated mud; thick layers of sand and silt with erosional or gravity-loaded basal contacts; possible sand clast.
6	1.50	Laminated mud; thin sand-silt layers, more closely spaced than 4.
7	0.89	Sand-silt layers separated by thin layers of mud.
8	0.49	Sand-silt layers separated by nearly equally thick mud layers.
9	0.38	Sand bed overlying and grading upward into mud. Sand beds at top and base.
10	0.54	Bed of alternating sand-silt and mud layers overlying mud. Mottling and small burrow holes. 1-3 cm of pre-mine sediments at base.
11	0.42	7 cm of tailing overlying pre-mine sediment.

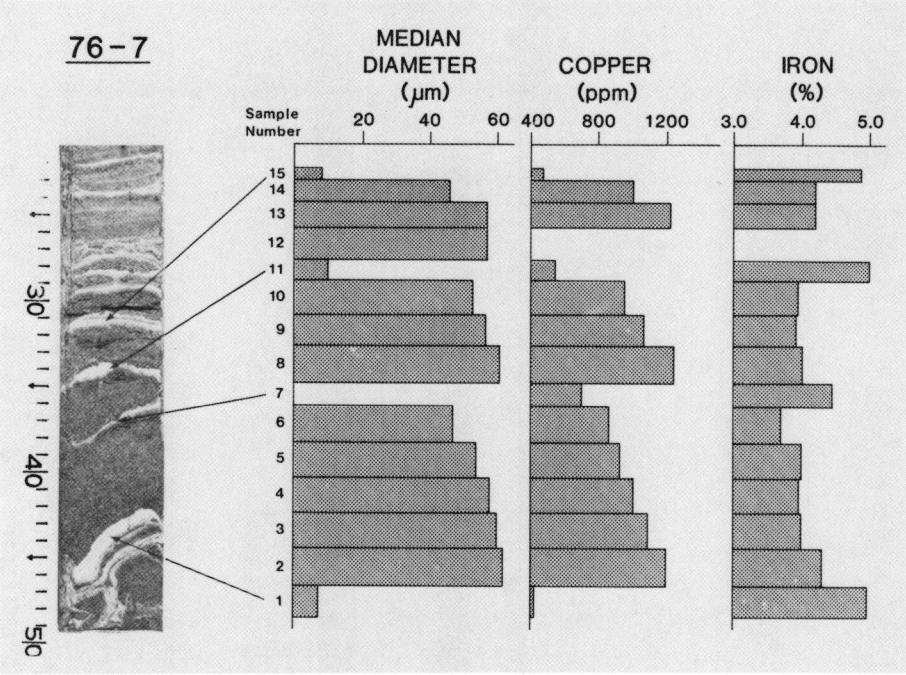

Figure 25. Lower section of core 76-7. Vertical scale of
the bargraph is twice that of the photograph.
The horizontal scales for Cu and Fe are
different for this core. Note disturbed basal
section, and unclassified structure at 27-28 cm.

echo. Backscatter from suspended particulates in the plume
is evident both within the channel and above the higher
west levee. The upper interface of the plume tilts upward
to the west, which is to the right in the
downstream-looking sense, and is due to the combined
effects of the Coriolis and centrifugal accelerations of
the flow in the leftward-curving upper reach (14). The
profile nicely confirms the selectively greater deposition
due to channel overspill on the right levee which occurs
as a result of these inertial accelerations.

Fig. 28 is a profile of the plume during the
rechannelized phase which has been converted to sediment
concentration (14). In this instance the plume hugged the
left levee, because the channel curved to the right (Fig.

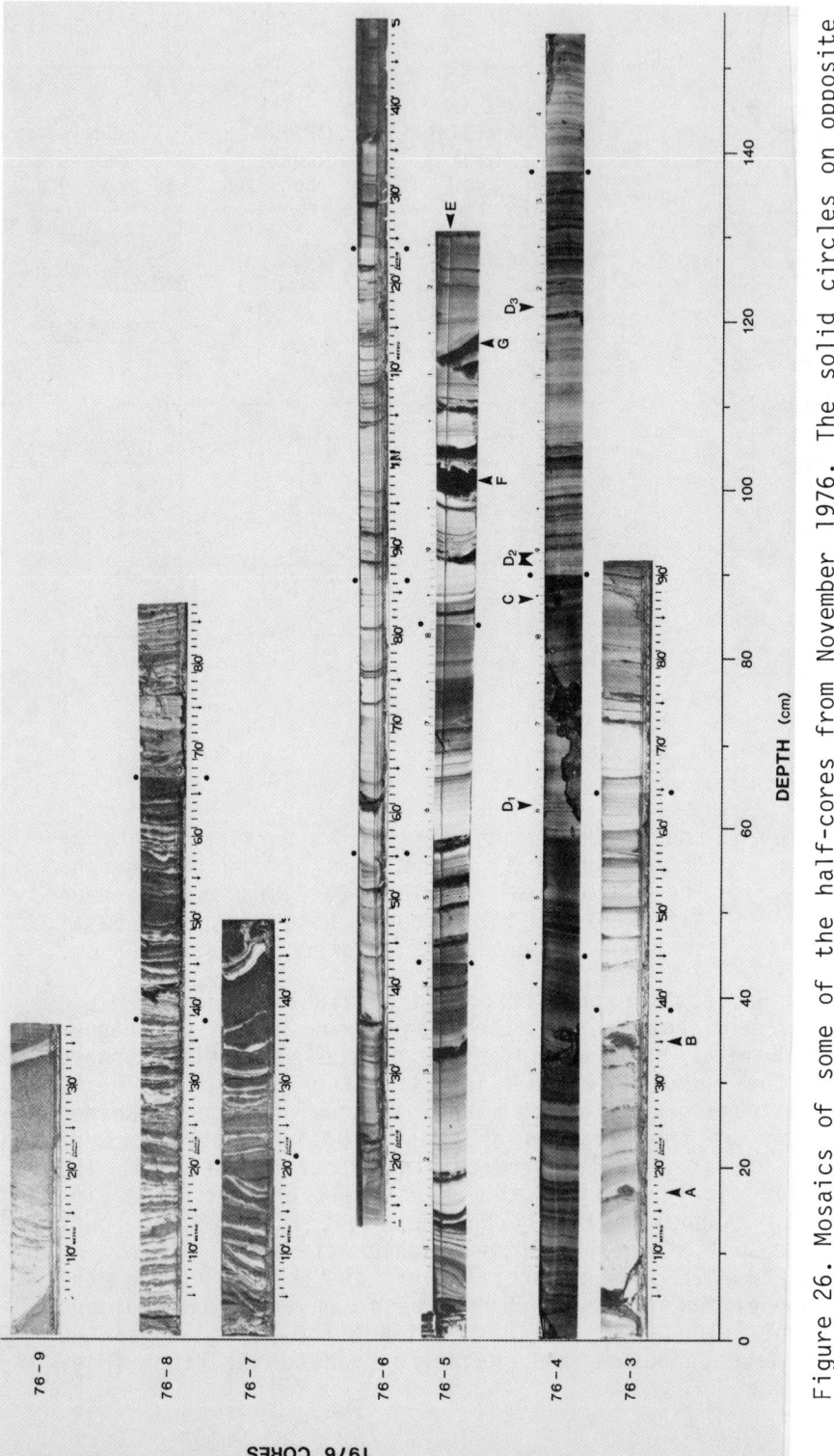

Figure 26. Mosaics of some of the half-cores from November 1976. The solid circles on opposite sides of a core identify the location of a splice between two photographic prints. 76-6 is a quarter-core. Disturbances due to splitting are present at 5 cm in 76-3; 33-35 cm and 60-78 cm in 76-4; and 42 cm in 76-8. Slight breaks in sand layers occur at 22, 36 and 61 cm in 6; 10 and 21 cm in 7; and 2 cm in 8. Core-support and by material adhering to

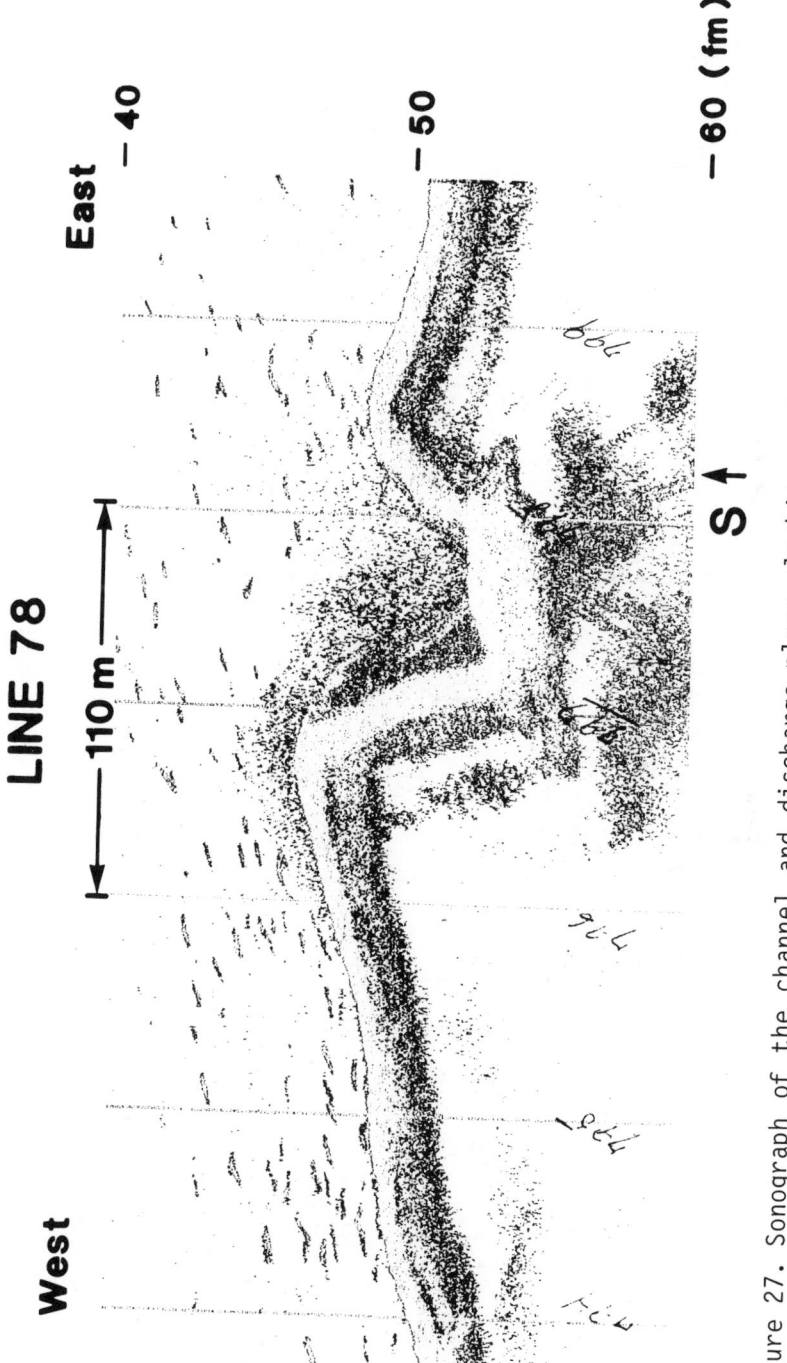

Figure 27. Sonograph of the channel and discharge plume, looking upstream, during flood tide 3.5 h after low water, 22 November 1976, 1050 h PST. S = side-echo, P = plume. Vertical lines indicate times of position fixes taken at 1 min intervals. Line location is 620 m off the outfall (Fig. 6).

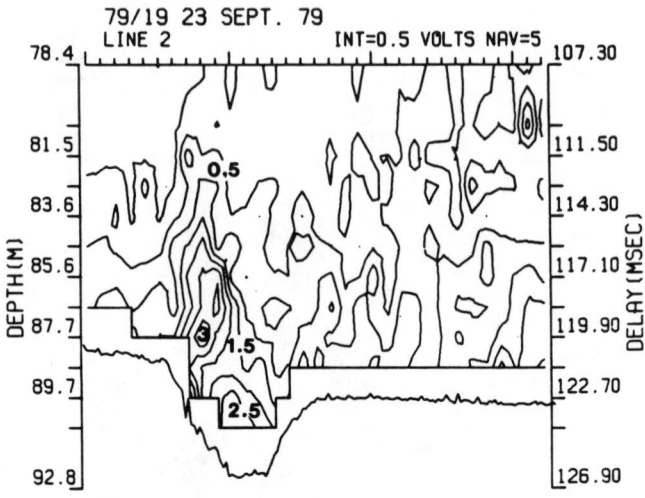

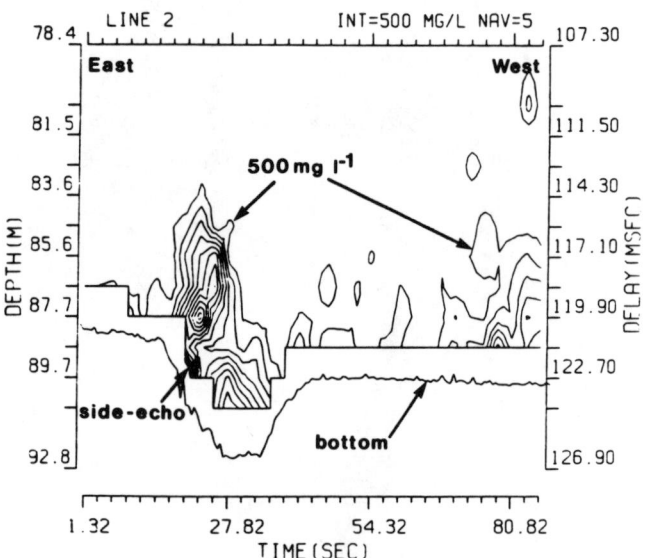

Figure 28. (a) Contours of signal level (in 0.5 volt intervals), uncorrected for spreading or attenuation. Grid points are at the intersections of horizontal and vertical hash marks, and represent averages over 28 points (1 m) in the vertical and 5 points (2.5 s) in the horizontal.

(b) Same as (a), but converted to concentration after applying spreading and attenuation correction. Contour interval is 500 mg/L.

15). Solids concentrations in the plume reach very high values near the bottom and over the left levee (5 g/L). This value may be a factor of two higher than expected, which could be the result of the signal statistics or a weak echo from a fish at the edge of the beam pattern.

SUMMARY AND CONCLUSIONS

The time-series of seismic reflection and bathymetric surveys show that the submarine channel system, once formed, underwent three successive metamorphoses: (a) a meandering channel regime; (b) an apron regime; and (c) a rechannelized regime. This evolution indicates that the leveed channels in Rupert Inlet form in a direction parallel to the steepest accessible slope. Laterally confining features in the bathymetry (such as slump scars) probably play an important role in the early stages because they provide lateral constraints on the rates of mass and momentum loss through lateral divergence. This, coupled with the preferentially higher deposition rates at the edge of the flow and possible erosion of the bed beneath the velocity core, provides a positive feedback mechanism by which the channel could become more pronounced. For outfalls on initial slopes sufficiently steep that the discharge is supercritical, the morphology of the channel in the near-outfall zone will be strongly affected by the occurrence of a hydraulic jump where there is a decrease in axial slope. The levees in the proximal zone are sites of rapid deposition; the levee walls are very steep; the material constituting the levees is primarily sand and silt and therefore non-cohesive; and the axial slopes in the proximal zone are the highest in the system. It follows that the levees in the near-outfall part of the upper reach, below the point of occurrence of the hydraulic jump, represent the most probable source of the slump-generated turbidity surges which deposit coarse-grained, copper-bearing turbidites on the levees in the middle and lower reaches.

Having dealt with the first question posed in the introduction, consider the second: the effect that channels have on tailing dispersal. There are two. The first is that because lateral divergence is inhibited, the velocity and excess density of the continuous flow is maintained to greater distances and depths, and deposition in the near-outfall zone is reduced in areal extent. The second is that because the levees are subject to failure, surge-type turbidity currents will carry material to even greater distances along the comparatively flat floor of an inlet.

This is probably advantageous from the engineering standpoint, since the tailings are carried to depth more quickly, in larger quantity and to greater distances than would be the case in the absence of a channel. This means that build-up in the vicinity of the outfall is not likely to be a problem and that environmental requirements pertaining to the minimum depth of deposition will be more readily met in most cases.

In Rupert Inlet, however, this assessment is complicated by the fact that the presence of a channel means higher rates of transport into the Hankin Point area, where the material can then be resuspended and brought to the surface by the intense turbulence generated by the tidal jet during the annual replacement of deep water. Such annual cycles are not unusual in high latitude, silled fjords in British Columbia and Alaska, because of high seasonal runoff, high-amplitude tides and seasonal variability in the wind stress along the coast (17, 21, 22). If these effects combine to drive an annual deep-water exchange cycle in silled fjords in the vicinity of a mine site, then the possibility that massive resuspension will occur in the vicinity of the sill must be considered and the environmental effects of any resulting transport of tailings into the near-surface zone assessed. If they are not negligible, then there are two alternatives: (a) to prevent transport of tailings into the vicinity of the sill, which would require preventing the formation of submarine channels; or (b) to discharge the tailings into an unsilled basin.

REFERENCES

1. Daly, R.A. "Origin of Submarine Canyons", Am. J. Sci. 31: 401-420 (1936).
2. Menard, H.W. "Deep-sea Channels, Topography and Sedimentation", Bull. Am. Assoc. Petroleum Geologists 40: 2195-2210 (1955).
3. Kuenen, Ph. H. and C.I. Migliorini. "Turbidity Currents as a Cause of Graded Bedding", J. Geol. 58: 1-127 (1950).
4. Johnson, R.D. "Dispersal of Recent Sediments and Mine Tailing in a Shallow-silled Fjord, Rupert Inlet, British Columbia", Ph.D. Thesis, University of British Columbia, Vancouver, B.C. (1974).
5. Normark, W.R. and F.H. Dickson. "Sublacustrine Fan Morphology in Lake Superior", Am. Assoc. Pet. Geol. Bull. 60: 1021-1036 (1976).
6. Normark, W.R. and F.H. Dickson. "Man-made Turbidity Currents in Lake Superior", Sedimentology 23: 815-831

(1976).

7. Carstens, T. and E. Tesaker. "Erosion by Artificial Suspension Current", presented at the Thirteenth Coastal Engineering Conference, Vancouver, B.C. (1972).

8. Tesaker, E. "Modelling of Suspension Currents", Symposium on Modelling Techniques, San Francisco, CA., ASCE: 1385-1401 (1975).

9. Evans, J.B. and G.W. Poling. "Discharging Flotation Mill Tailings into a Coastal Inlet", presented at the 77th Annual Meeting of the Canadian Institute of Mining and Metallurgy (1975).

10. Poling, G.W. "Environmental Considerations in Tailing Disposal", CIM Bulletin 72 (801): 144-153 (1979).

11. Canadian Hydrographic Service. Canadian Tide and Current Tables, Volume 6, Barkley Sound and Discovery Passage to Dixon Entrance, Fisheries and Oceans, Ottawa (1976).

12. Folk, R.L. Petrology of Sedimentary Rocks (Austin, Texas: Hemphill's, 1968).

13. Carver, R.E., Ed. Procedures in Sedimentary Petrology (New York: John Wiley and Sons, Inc., 1971).

14. Hay, A.E. "Submarine Channel Formation and Acoustic Remote Sensing of Suspended Sediments and Turbidity Currents in Rupert Inlet, B.C.", Ph.D. Thesis, University of British Columbia, Vancouver, B.C. (1981).

15. Haner, B.E. "Morphology and Sediments of Redondo Submarine Fan, Southern California", Geol. Soc. Am. Bull. 82: 2413-2432 (1971).

16. Hillis, R. Personal communication.

17. Stucchi, D.J. "The Tidal Jet in Rupert-Holberg Inlet" in Fjord Oceanography, H.J. Freeland, D.M. Farmer and C.D. Levings, Eds. (New York: Plenum Press, 1980), pp. 491-497.

18. Stucchi, D.J. and D.M. Farmer. "Deep-water Exchange in Rupert-Holberg Inlet", Pacific Mar. Sci. Report 76-10, Institute of Ocean Sciences, Patricia Bay (1976).

19. Hay, A.E. "Turbidity and Deep-water Exchange in Rupert-Holberg Inlets, 1973-1976", in Summary Report, Fifth Production Year: October 1975 - September 1976, J.B. Evans, Ed. (1978).

20. Hay, A.E. "A Brief Report on Trace Metal Levels in the Rupert-Holberg Basin" in Summary Report, Fifth Production Year: October 1975 - September 1976, J.B. Evans, Ed. (1978).

21. Mathews, J.B. "The Seasonal Circulation of the Glacier Bay, Alaska Fjord System", Estuarine, Coastal and Shelf Science 12: 679-700 (1981).

22. Heggie, D.T. and Burrell, D.C. "Deep Water Renewals and Oxygen Consumption in an Alaskan Fjord", Estuarine, Coastal and Shelf Science 13: 83-99 (1981).

DISCUSSION

Question: How permanent will the submarine channel be after mine discharge has stopped?

Answer: The levee walls were about 10 m high, and the material underlying them to the depth that our core was able to penetrate tended to consist mostly of sand and silt. There wasn't very much clay which might consolidate over time and form a stable system. Thus I believe that in an area of reasonably high bottom currents or high seismic activity they may not persist for long periods of time. A great deal of material must be moved to destroy them, however, and they could be stabilized by biological growth.

Question: Did the submarine current in Rupert Inlet do something to generate up-welling?

Answer: Any tailings in suspension would normally tend to stabilize the water column against up-welling. One of the things which does occur is transport of shallow water to depth within the turbidity current. This water then becomes buoyant after the sediment has been deposited and will rise to some intermediate depth. This is reflected in the salinity and temperature profiles of the discharge plume. They indicate that the water in the immediate vicinity of the bed has lower salinities than would be at that depth normally.

Question: Why did the submarine channel disappear, and what is its engineering significance?

Answer: We just didn't have enough temporal control in our monitoring of the bathymetry to really make a definitive statement about the cause of disappearance.

Question: Do you have any measure for how long any of the areas of deposit become stable?

Answer: One of the things we did to date the sediment column was to follow the distribution of diatoms within several cores. In one or two cores you can find two or three seasonal layers of large diatoms. The mud was also well laminated. The laminae showed no obvious sign of disturbance. The deposit from such an area probably was stable over a two or three year period. These tended to be on the flanks away from the submarine channel itself; that is, away from areas where turbidity current surges could be expected.

Question: I found it very curious that the channel disappeared and then reappeared again. What happened to the material in between? Why did the channel reappear again?

Answer: The channel reappeared because material is discharged at quite a high rate on a mobile bed that has a reasonably steep slope. Therefore, if anything happens to prevent the lateral spreading of the discharge it is going to build a channel both due to selective local erosion and to deposition on either side. Between the disappearance and reappearance, the material was deposited on the apron.

Question: How far down Holberg Inlet do your tailings go and do they go into Quatsino Narrows?

Answer: Tailings are detected in cores and samples from the bottom taken a substantial distance up Holberg Inlet. While I was doing my study it was 15 km or halfway up the inlet from Quatsino Narrows. There is intense up-welling in the neighborhood of the Narrows and fine material that doesn't settle out as the tide ebbs gets out into the Quatsino Sound area.*

---

*Editor's note: see Ch. 6, p.236 for details on amounts.

ENGINEERING AND SCIENTIFIC PRINCIPLES

D.M. Farmer
    Institute of Ocean Sciences
    P.O. Box 6000
    9860 West Saanich Rd.
    Sidney, B.C.
    Canada V8L 4B2

The first part of these proceedings has included four introductory talks on various aspects of mine tailings and their disposal in the environment. At this point it is useful to summarize the topics that have been discussed and to add a few comments on some aspects, especially those connected with physical oceanography, that seem especially relevant to the purpose of this symposium.

From Mr. Caldwell's discussion of the various schemes available for landfill disposal it is clear that the engineer has considerable control over the way in which tailings can be deposited on land so as to minimize the long-term environmental impact. The cost varies by a factor of about 20, between the simple upstream method and the more elaborate schemes. It is also evident that in deciding what kind of land disposal scheme to adopt, close attention must be paid to the environment in which it is being done, particularly with regard to such features as precipitation, the ruggedness of the terrain, the area of the watershed and so forth. The great variety of environmental conditions make the land disposal engineer's job a complex and sensitive one.

Tailings disposed on land can be reclaimed to a greater or lesser extent. The description of some of these techniques leaves me a little uneasy about their long-term environmental impact. For example, in areas of high precipitation, water must be ducted in such a way that the tailings are not eroded and that toxic elements are not leached from them. These ducts must stand the test of time; in many cases it is quite unrealistic to expect maintenance to continue following closure of the mine. There are also aesthetic considerations and the tailings and ducting must be made to blend into the natural surroundings as far as possible. However, we must accept the fact that in many locations land disposal is the only practicable technique.

How does one go about choosing a particular disposal scheme? Mr. Caldwell showed us a little matrix in which he had placed various numbers representing preferences for and against, depending on various environmental considerations as well as costs and other technical aspects. The real difficulty, as pointed out by Dr. Ellis' questions, is in deciding just what weight to apply to each component of the choice. This is not just a technical or an economic problem; it is also a social and an aesthetic one. We have to understand it thoroughly before a choice is made. Moreover, it is important that not only the experts but also the public understand the basis for the choice.

Dr. Poling discussed the nature of mill tailings. Enough is known about the physics and chemistry of the ore particles so that engineers can decide how to do the grinding so as to optimize the extraction of the metal and also control the way in which the tailings behave after they leave the mill. It is evidently possible to control the electric charge on the tailings so as to alter the way in which they flocculate in the water and settle. Control over this behavior is required so as to minimize the chances of resuspension in the marine environment. A subtle and perhaps counter-intuitive consequence of the grinding processes that needs consideration in marine disposal schemes relates to the size to which the particles are reduced. Very fine milling increases the surface area and thus might be expected to increase the likelihood of chemical interaction or metal release. But a quite different and beneficial effect is adsorption onto the particle surface, which also increases with surface area. The interaction of these two effects is one example of the complex behavior of mine tailings that must be considered in optimizing a disposal scheme. Again we see that mining engineers have substantial ability to control the long-term consequences arising from tailings discharge, and as with land disposal there will be a matrix of choices, each having different economic and environmental costs. How do we weigh those gains and losses and how do we arrive at an optimum solution? And what is meant by an optimum solution, not just from an economic point of view, but from the point of view of those of us concerned about both the economic and environmental consequences of mining? There is no simple answer to this question, but we must at least ensure that we are aware of the full range of choices and implications before we decide on one scheme over another.

Dr. Hay described his acoustic observations in Rupert Inlet and showed us the way in which mine tailings, disposed in the ocean, run down slopes forming levees, thus establishing their own channels which from time to time collapse and change course. Because the mine waste forms a channel with banks, it travels much farther than might otherwise have been expected. An important point emerges from this presentation: once a marine disposal scheme is underway, we should be prepared to monitor it closely and see exactly what is happening.

The example of Rupert Inlet discussed by Dr. Hay is rather unique; there are a few other inlets that have a similar geometry, but they are not common. On the other hand, it is common for strong currents to occur over the sill and those currents are often capable of resuspending sediments that move into the vicinity. This point was raised by Dr. Burling in his discussion of physical oceanography. Since it is so critical to tailings disposal in fjords of the Pacific Northwest, which experience large tidal ranges, I shall amplify Dr. Burling's comments and add to them on the basis of recent research into sill flows.

Tidal exchanges between fjords and the open coast tend to be greatly accentuated near constrictions. The resulting flows can be effective at mixing and involve significant vertical movements of water. Fig. 1 shows a set of acoustic images that I obtained in Knight Inlet. It demonstrates the kind of flow patterns that occur in many of the inlets along our coast. In this example the surface water was drawn down to depths of 100 m or more, bounded by vortices of up to 50 m height. These are the processes then that can mix the freshwater with saltier water beneath it, but can also disturb any suspended material near the sill. A related but different process described by Dr. Burling is the exchange of deep water with the ocean. The fjord "breathes" the ocean water and at different depths it breathes at different speeds. Perhaps in the deep water an exchange may only occur once a year, or even once every ten years. These exchanges have an important bearing on the movement, resuspension and deposition of mill wastes. An analysis of both tidal and deep basin exchange processes can lead to an understanding of mechanisms likely to influence the deposition of this material.

Another point that needs to be made is that data collection and data reduction do not in themselves constitute a meaningful environmental study. If the

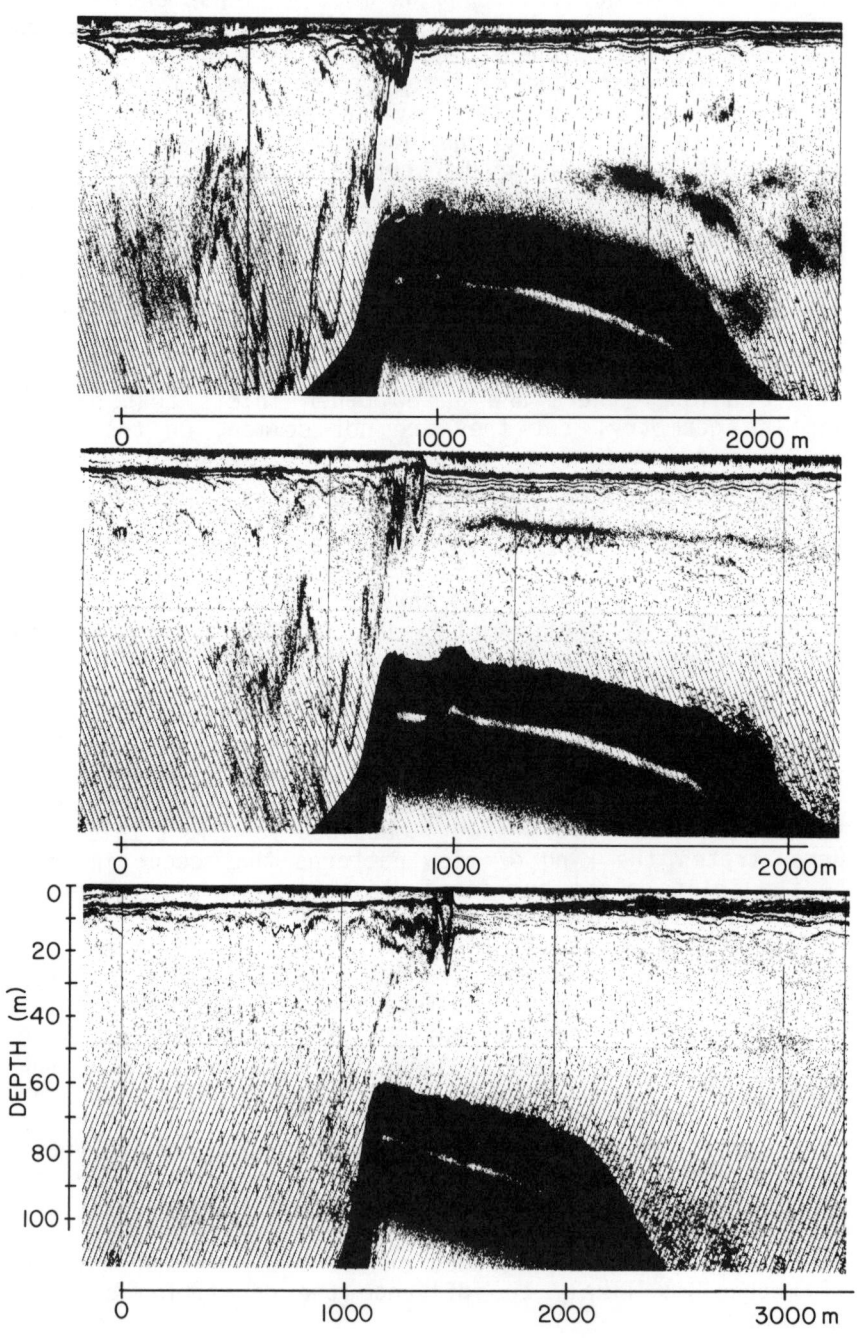

Figure 1. Three acoustic images showing the formation and relaxation of an internal hydraulic jump over the sill of Knight Inlet, B.C. (1). In the first image, the tide is ebbing and the flow is from right to left. Deformation of the flow is made visible by acoustic scatterers, principally biological particulates, which move with the water. As the tide slackens, the disturbance of the flow downstream of the sill relaxes and travels back up the channel as a sequence of internal waves or an intrusion. Active flows over sills such as this are responsible for much of the mixing and circulation in the inlet. They will also be important in determining the movement or possible resuspension of mine tailings that may approach the sill.

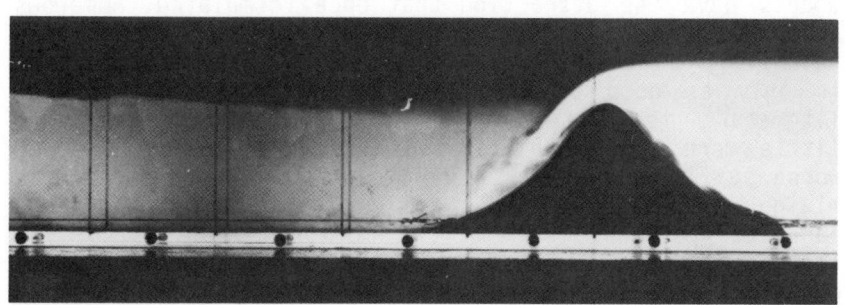

Figure 2. A model of tidal flow over a sill. The sill is moving to the right so as to represent a tidal current from right to left. The surface layer (dark) is freshwater, the deeper layer (clear) is salty. Notice the jet of water that runs down the downstream slope of the sill. Simple laboratory models of this type can provide insight into the physical processes that occur near constrictions and the way in which they might influence mine tailings in the vicinity.

measurements are worth the trouble of obtaining in the first place, then they are also worth the trouble of analyzing properly. This includes a careful evaluation of the problem prior to observation, development of a sound plan for data gathering and then, following the field program, a careful analysis and interpretation of the results. For a physical oceanographic study one should allow about 1/3 of the total cost for analysis. The final result must not be limited to a compilation of data reports (these can be helpful, though many users will prefer data on magnetic tape), nor a superficial summary or purely descriptive account of the relevant processes. Rather, the goal should be to focus on specific physical processes that govern the circulation of the water and might influence the disposal scheme, and to interpret the data quantitatively in terms of simple models of these processes.

Models can assist in the interpretation in various ways. They may be physical models or mathematical models of various complexity, from simple analytical representations to complex computer schemes. Mathematical models have the attraction that once formulated, numerous experiments can be run using different parameters with relatively little trouble. However, a mathematical model is only as good as the physics that goes into it. The literature is littered with numerical models that are little more than sophisticated interpolation formulae, or worse still, purport to describe complex phenomena such as mixing on the basis of wholly inappropriate parameterizations. Mathematical models can be useful for testing concepts and developing insight, but it is important to see them clearly for what they are: a mathematical representation of certain physical concepts which may or may not be correct.

Laboratory models can also provide insight and offer a means by which complex flows can be directly observed. A consideration with such models is the accurate scaling of the flow, which must be carried out such that the important physical processes are adequately reproduced. While complex topographies are often constructed for modelling specific flows in particular locations, experience suggests that a deeper insight is usually gained with a much simpler model that allows investigation of the underlying physics, uncluttered by the complicating effects of real topography. There would appear to be several aspects of marine tailings disposal that are particularly suitable to such modelling, including interaction of material in the vicinity of strong tidally induced flows such as have already been discussed (see Fig. 2).

Ideally, we should strive for interaction between experimental work, data analysis and theoretical or laboratory modelling. The modelling can be used to test ideas that can be checked against the data. Often questions will arise that suggest further experimental work and the process can be repeated. Thus in planning an experimental program it is desirable to avoid the temptation of having one large and expensive field project followed by data analysis and interpretation. The same amount of money is often better spent on a more limited pilot study, followed by careful evaluation of results, followed again by a second more selective field study.

It is evident from the discussion thus far that a real difficulty in tackling problems of marine disposal of tailings, is that there are very many technical subjects involved, in engineering, physics, chemistry and biology, but few people who are competent in more than any one of these. Thus, it is essential that the experts make their knowledge as easy to understand as they can and pay close attention to the interactions of the different disciplines. The technical problems we have addressed at this meeting are the details of mining and extraction, the civil engineering problems of tailings disposal, the physics and chemistry of particulate matter behavior, the physics of circulation in inlets and of course, the biology and chemistry of the water into which the tailings are disposed. For each of the possible disposal options, we must also be in a position to balance the technological difficulties, the economic considerations and the aesthetic and environmental costs. This process must involve not only experts, but also lay people who have particular concerns about the effects of a mine on their community and their environment.

REFERENCES

1. Farmer, D.M. and J. Dungan Smith. "Tidal Interaction of Stratified Flow with a Sill in Knight Inlet", Deep Sea Research 27A: 239-254 (1980).

DISCUSSION

Question: Please comment on water velocities and stability associated with the hydraulic jumps and very rapid flows generated at fjord sills.

Answer: The flow speeds are typically of the order of 2-4

m/s, perhaps more than that close to the sill crest. This is the range of speeds that we have observed in one particular case. Other cases, however, will be very different. One can make a reasonable estimate of the kind of velocities to expect with quite limited information, on the basis of tidal velocities and knowledge of the stratification. One can then model at least the initial phase of the hydraulic jump and estimate the overall structure of the resulting flow. Many sill flows are going to be a lot more energetic than the one in Knight Inlet. In Observatory Inlet just near the mouth of the Nass, we have seen evidence of penetration of water from a hydraulic jump to 150 m or more and high velocities must be expected, quite a bit higher than in Knight Inlet. As for the stability, that of course depends on the shear. If the water at one depth is flowing very fast and at another depth is not flowing very fast, instabilities will develop and quite a lot of mixing can occur. Such shears are always generated over sills and can be a source of turbulence.

The Knight Inlet studies were conducted throughout the year. One of the more striking observations is that processes occurring at one time of the year are very different from those occurring at another time of the year. The reason is clear enough: when there is a high river discharge, the stratification is changed and the behavior of the flow over the sill is altered. In hydraulics terminology, the sill flow is controlled by the densimetric Froude number and this depends on both the tidal current and also on the stratification and hence the river discharge.

Question: Do you have any feel for how far up the inlet the internal waves generated by this process will travel, and also how far up the hydraulic jump will travel?

Answer: What happens when you have very energetic processes on a sill is that after each half-tidal cycle, both the mixed water and the potential energy stored in the lee waves or hydraulic jump spread away from the sill. They can spread in various ways. One of the more subtle of these ways has no surface manifestation; it spreads at depth. But it can be followed with sonar devices.

Fig. 3 illustrates such a process of subsurface intrusion. The surface freshwater of the inlet lies above the intrusion; the acoustic image shows the distortion of the flow as a great mass of mixed water rolls along underneath the surface layer. So if you just measured what was in the surface layer, you would never know this was

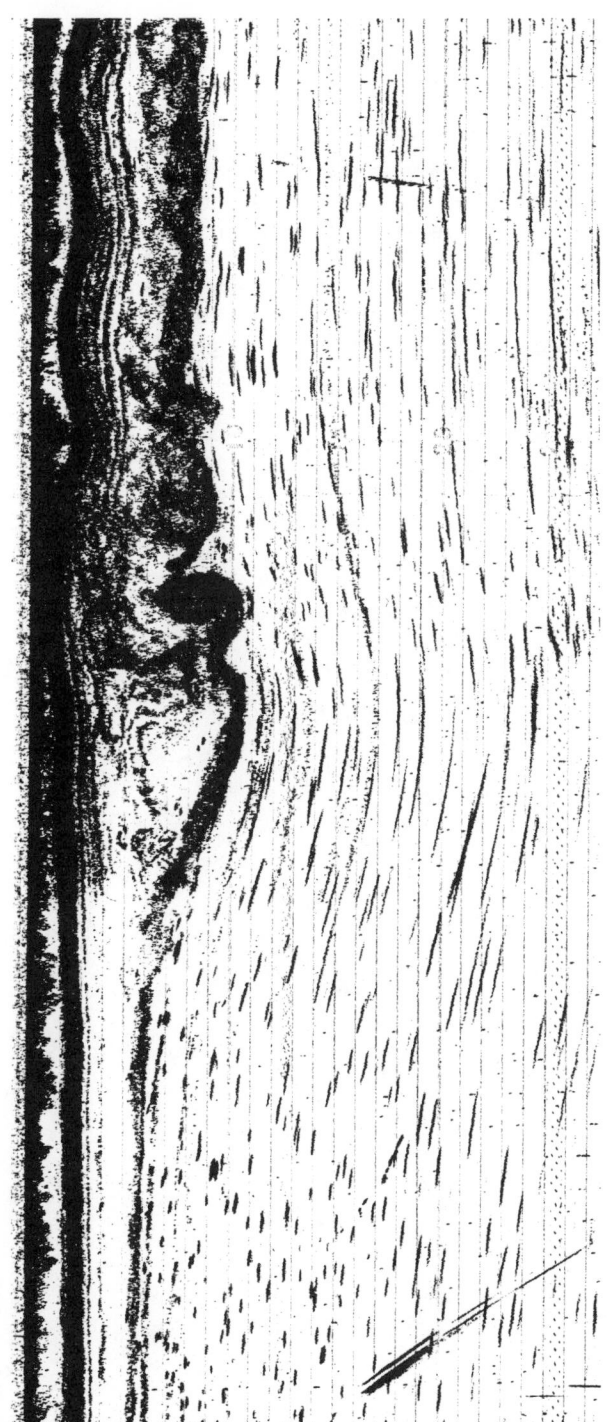

Figure 3. An acoustic image obtained from a stationary vessel of an intrusion or "internal bore" in Knight Inlet, B.C. The intrusion consists of water intermediate in density between the surface layer (lying above the dark line) and the deeper water; it was formed by tidal mixing near the sill and spreads out beneath the surface layer for 10-20 km. Such processes can occur at different depths and are the far-field manifestation of tidal mixing. Note the rolling vortex at the base of the intrusion. Fluid instabilities of this type are sites of active mixing and can arise from the shear at the interface between the two layers of water having different density. In this image the vertical scale is 50 m, so the thickness of the intrusion is 15-20 m. The acoustic record was obtained over a period of approximately 5 min.

happening, except that you might see a few ruffles on the surface. Active mixing is occurring here with vortices being formed as water travels up the inlet. We have observed flows like these and also a slightly different effect resulting in interfacial waves that travel some 20 km or so up the inlet; eventually they interact with the sides of the inlet and are dissipated. When they give up their energy, they also do some more mixing and they are an effective means for redistributing water mixed near a sill. So they are, if you like, a mechanism by which nature takes energy from the tide and redistributes some of that energy over a considerable distance along a fjord.

SECTION II

CASE STUDIES

Section II Frontispiece. A submarine tailings outfall. The seawater mix tank and its control valves can be reached by a walkway. A float in the distance marks the discharge point at 50 m depth.

There is a reason for including case studies in a symposium ranging from general considerations to local specifics. We can draw upon experience at other areas, bring the information together, try to develop some generalizations about patterns which have occurred elsewhere, and then predict for new developments. This is the essence of scientific method - to gather data, hypothesize or predict, and test. The organizing committee for the symposium has considered case studies which illustrate typical problems. They selected three for presentation at the symposium, all of which come from Canada. There is a reason for this. Dr. Poling has touched upon it in Chapter 2. All three have considerable interest, in part because they are unique. They all provide data which simply is not available from other localities. Canada has a long history of marine tailings disposal.

The first case study was the Polaris lead-zinc mine in the arctic. It has a most unusual topographic situation for tailings disposal. The speaker, Mr. W. Kuit, was unfortunately unable to prepare his material for a book chapter, but there are some comments in this section's summary.

The second case study is the Kitsault molybdenum mine just 35 miles east of Quartz Hill. It is important partly because of the regional similarity with southeast Alaska. But more important than that, there is a conceptual reason for including it. Kitsault is engaged at the moment in what can be called outfall performance checks. How is the outfall behaving in dispersing the tailings which are being discharged? It is the only mine to date which has gone about outfall performance checks systematically. In this sense it is a very important case study for this meeting and for other mines which will develop in the future. Unfortunately the speaker, Dr. Littlepage, also could not prepare his material as a book chapter, although a summary report of his baseline studies is available (1).

The third case study is Island Copper Mine, which produces copper and molybdenum. It is some 400 miles southeast of Quartz Hill on Vancouver Island. It is important for yet another reason. Island Copper has more than ten years of monitoring data available about its submarine tailings disposal system. This is an older data bank than is held by any other mine which is discharging waste to the sea. Island Copper is now in the phase of what can be called trend monitoring: to determine if any unexpected long-term consequences are arising and whether the long-term impacts which were predicted are still within levels which are acceptable to the community.

The Frontispiece to Section I locates both Kitsault and Island Copper on the Canadian west coast.

The book Frontispiece illustrates the most important engineering feature of a tailings disposal system: the outfall, with its seawater mix chamber. Both the original and a replacement outfall are shown. The later outfall at Kitsault is similar (see the Frontispiece to this section).

REFERENCES

1. Littlepage, J.L. "Oceanographic and Marine Biological Surveys, Alice Arm and Hastings Arm, B.C. 1974-1977", Report to Climax Molybdenum Corp. of B.C. Ltd. (1978).

CHAPTER 6

ENVIRONMENTAL DATA HANDLING AND
LONG-TERM TREND MONITORING AT ISLAND COPPER MINE

C.A. Pelletier*
    Rescan Group Ltd.
    Suite 1105, Board of Trade Tower
    1177 West Hastings St.
    Vancouver, B.C.
    Canada V6G 2K3

INTRODUCTION

Island Copper mine is located at the north end of Vancouver Island, British Columbia (see Section I Frontispiece). It is an open pit, low grade copper and molybdenum mining operation with ore averaging 0.52% copper and 0.018% molybdenum (1). In 1980 the concentrator handled approximately 38,000 tonnes per day, while the mine moved an average of 145,000 tonnes of ore and waste rock. Since the beginning of production in October 1971 to the end of September 1980, 106,000,000 tonnes of tailings have been discharged through a submarine disposal system into the adjacent fjord, Rupert Inlet (Fig. 1). The mill concentrator system has been described by Poling in Chapter 2 and the outfalls are illustrated in the Frontispiece.

A novel submarine outfall system was designed for the mill (2) and consisted of a tailings line discharging to an open seawater mix chamber. The seawater intake line originated at 20 m depth and provided a self-maintaining flow mixing with the tailings slurry with subsequent gravity flow to the outfall discharge point at 50 m depth. In 1974 the original outfall was converted to standby use,

---

*The author was formerly the Environmental Manager for Utah Mines Ltd., Vancouver, B.C., Canada V6E 3S7.

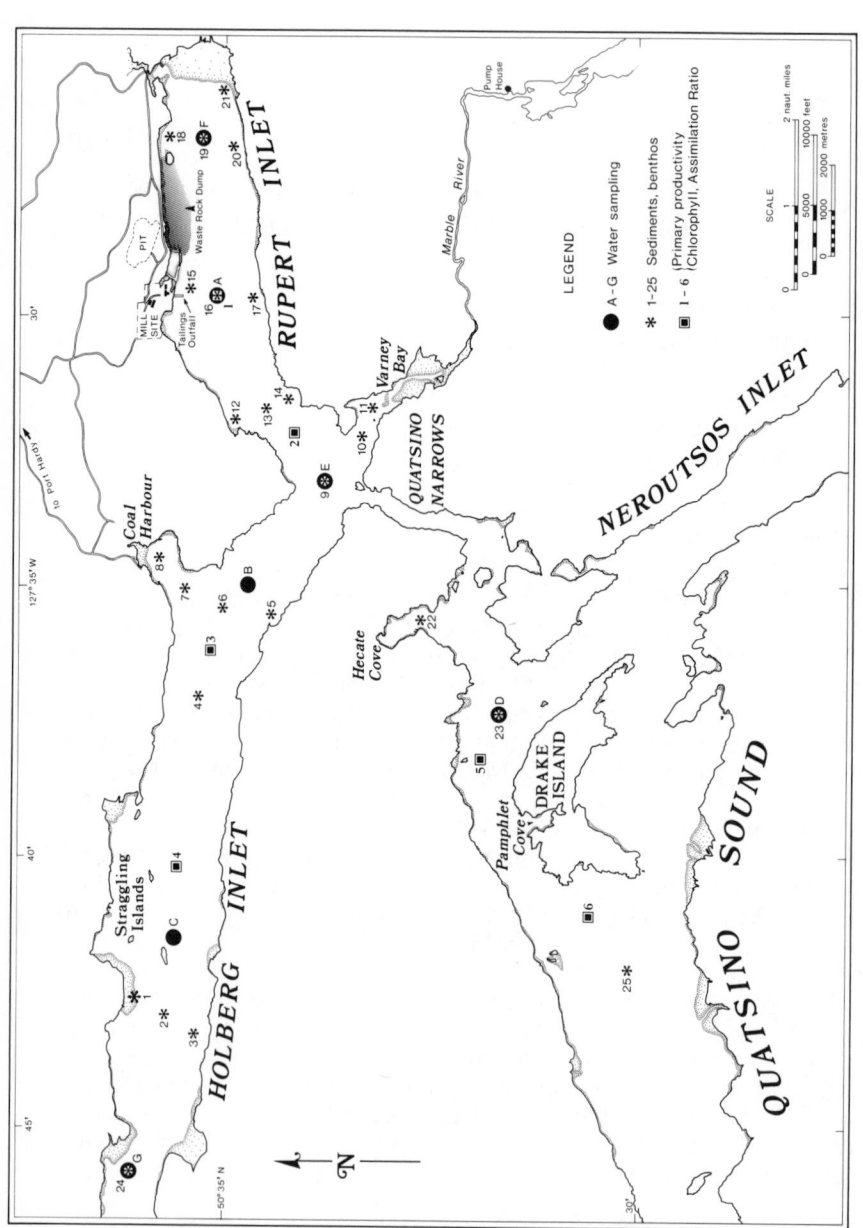

Figure 1. Sampling locations.

as a redesigned system became operational. The tailings line discharged the slurry below the surface of a new seawater mix chamber and thereby minimized air entrainment.

The raw effluent contains 35-45% solids by weight prior to mixing, with a final seawater-slurry ratio of 2:1. Subsequent behavior of the tailings as a density plume is described by Hay in Chapter 5. Composition of the slurry prior to seawater mixing is shown in Table I for 1978-79, along with constraints set by the Discharge Permit issued by the British Columbia Waste Management Branch.

The provincial regulatory agency also requires that Island Copper maintain a comprehensive marine monitoring program, the details of which are specified for 1980 in Table II. Methods are described in a series of annual reports, e.g. (1), which include copies of data and interpretive reviews. These are submitted to both provincial and federal regulatory agencies. The monitoring program is extensive, has been modified several times since 1970, and presents an increasingly demanding need for effective data storage, retrieval and analysis. The environmental staff at Island Coper Mine have had to develop a computerized system of data handling. This was initiated in 1973 as a variety of hard and software packages became available.

By September 1978 the data bank consisted of some 22,000 observations with anywhere from 2 to 28 parameters per observation. Such a large data base is beyond the scope of manual analysis. A computer-based data system was available to provide univariate statistics and some multiple mean comparison procedures: one-way ANOVA and Newman-Keuls multiple range test (3). However, the design of this system did not adequately provide for the time-series application of these tests.   Such site-specific, time related comparisons are particularly important in the analysis of sampling regimes such as the one particular to this monitoring program where control comparisons are not applicable. To mitigate this deficiency somewhat, interstation comparisons by ANOVA and multiple range testing were repeated for a number of truncated time periods. (The logistics for this type of approach were straightforward, given this system design.) These repeated analyses were then compared. If some trend was suggested the pertinent data were then analyzed again by desk top mini-computer (HP 9830A) using an orthodox time-series application of the multiple mean comparison procedures. This was an inefficient and time consuming process. However, it probably revealed most of the salient trends in the data.

Since 1979 Island Copper has been using a relatively

Table I. Effluent physical and chemical analyses summary, 1978-1979.

Parameter[x] (Units)	N	Mean	Standard Deviation	Minimum	Maximum	Permit or Provincial Objective+
Volume (10^6 gallons)	365	15.3	2.8	0.0	21.4	16 daily mean* 22 maximum*
% Solids (by wt)	52	39.8	2.4	13.9	44.9	50*
Temperature (°C)	52	21.3	4.9	15.0	31.4	-
pH (log)	52	10.2	0.6	8.4	11.1	7.5 - 11.5*
Copper (μg/L)	52	8.2	3.4	2.9	17.0	50*
Molybdenum (μg/L)	52	154.0	30.0	82.0	213.0	500*
Cadmium (μg/L)	52	0.1	-	-	-	5*
Chromium (μg/L)	52	1.1	0.5	0.3	3.3	50 - 300
Cobalt (μg/L)	52	1.0	-	1.0	4.6	500 - 1000

Iron (µg/L)	52	30.8	15.9	8.0	110.0	300 - 1000
Lead (µg/L)	52	2.6	1.2	2.0	5.0	50*
Manganese (µg/L)	52	1.9	1.3	0.7	4.3	100 - 1000
Zinc (µg/L)	52	3.0	1.9	0.9	10.2	200 - 1000
Arsenic (µg/L)	52	46.8	17.0	14.0	92.0	100*
Cyanide (µg/L)	52	25.9	25.0	4.0	150.0	250*
Mercury# (µg/L)	52	0.1	0.05	0.1	0.2	1*

x Heavy metals are dissolved fractions only.
+ Permit level indicated by asterisk.
# Analysis on decanted, unfiltered sample from August 1979.

Table II. Outline of environmental control monitoring program at Island Copper Mine.

	Description	Frequency	Objective
EFFLUENT DISCHARGE PROGRAM	From weekly composites of daily samples taken from the thickener U/F determine pH, % solids, temperature, total cyanide, total mercury, dissolved Cu, Mo, Cd, Cr, Co, Fe, Pb, Mn, Ni, Zn, As.	Daily samples analyzed	Monitor physical and chemical characteristics of effluent.
	Samples of final-effluent sent out for 96 hour TLM bioassay tests.	Monthly	Record tailings acute toxicity.
	Determine effluent volume.	Continuous	Record volume of effluent discharged to sea.
	Study composition of tailings, settling rates and leaching potential.	Ongoing research	Research activity.
	Outfall performance program, i.e. seawater dilution, wear characteristics, etc.	Ongoing research	Physical performance of system.

MARINE PROGRAM

PHYSICAL

Bottom Sediment
Distribution

Bottom Coring	Cores at 26 stations – log and measure tailings thickness on bottom.	Quarterly	Visually determine tailings distribution less than 2 ft in thickness.
Bottom Grab Sampling	Sediment sample at 26 stations for metal analysis; Cu, Mo, Pb, Zn, Cd, Ni, Co, Cr, Mn, Fe, As and Hg.	Annually	Chemically delineate the spread of tailings.
Echo Sounding Survey	Bottom profiling.	Monthly in Rupert Inlet, semi-annually throughout system	Monitor changes in sedimentation regime.
Seismic Survey	Bottom profile and sediment distribution.	Annually	Record tailings greater than 2 ft in thickness.

Table II, continued

	Description	Frequency	Objective
MARINE PROGRAM			
PHYSICAL			
Suspended Sediment Distribution			
Discrete Water Sampling	Water samples collected in profile at 7 stations for measurements.	Monthly	Delineate suspended sediment distribution by scattered light measurement.
Discrete Water Sampling	Water samples collected at 7 stations for suspended sediment analysis.	Quarterly	Quantify suspended sediment load.
Transmissometer Survey	Continuous in situ profiling of water column.	Bi-weekly in Rupert Inlet	Delineate turbidity field in Rupert Inlet.
		Annually throughout system	Delineate extent of turbidity, throughout the water column.

Fixed Suspended Sampler Survey	At 4 stations and 2 depths (30' & 60') monitor sedimentation rate and source of sediment, i.e. chemical analysis, particle size distribution, organic to inorganic ratio, SEM and microprobe identification.	Quarterly	To identify and quantify sedimentation at 30' & 60' throughout the system.
CHEMICAL			
Chemical Water Analysis			
Discrete Water Sampling	At 7 stations profile temperature, salinity, turbidity, color and surface transparency.	Monthly	Record water column physical properties.
Discrete Water Sampling	At 7 stations profile dissolved oxygen, pH, alkalinity, salinity, spent sulphite, cyanide, dissolved and particulate Cu, Mo, Fe, Mn, Ni, Pb, Zn, Cr, Co, Cd and total As, Hg.	Quarterly	Record water column chemical properties.

Table II, continued

	Description	Frequency	Objective
MARINE PROGRAM			
BIOLOGICAL (PLANTS)			
Primary Production Study			
Euphotic Depth Survey	Measure amount of light attenuation and scatter at depth.	Monthly	Record depth of photozone in the water column.
Discrete Water Sample for Nutrients	Collect water samples at 6 stations for nutrients (silicates, nitrates, phosphates).	Monthly	Record nutrient levels in water column.
Carbon[14]	At 6 stations monitor the assimilation rate of primary producers.	Monthly during Spring, Summer and Fall	Record assimilation rate of primary producers.
Chlorophyll "a"	At 7 stations collect sample for chlorophyll "a" standing crop.	Monthly	Record standing crop of primary producers in water column.

Study	Description	Frequency	Purpose
Macrophyte Study	At 3 sites detailed samples of flora and fauna.	Annually	Monitor plants and animals in the littoral communities and sub-littoral zone.
Intertidal Plate Study	At 16 sites artificial substrate samplers set for continuous monitoring of flora and sediment deposition.	Monthly	Monitor settling flora and fauna in intertidal zone.
Metal Analysis of Fixed Algae	At 16 sites Fucus and Zostera are collected and analyzed for metal analysis (Cu, Mo, Cd, Pb, Zn, As, Hg).	Quarterly	Monitor metal concentration in sessile macrophytes.
BIOLOGICAL (ANIMALS)			
Zooplankton	At 4 sites collect zooplankton for density, diversity and metal analysis. Identify various larvae of crabs, clams, mussels, etc. Metals include Cu, Mo, Cd, Pb, Zn, As & Hg.	Quarterly	Monitor population changes and metal concentration.
Benthic Organisms	Collect benthic organisms at 26 stations; sort to polychaetes, molluscs and others aboard ship for live counts and biomass.	Quarterly	Monitor benthic population and biomass.

Table II, continued

	Description	Frequency	Objective
MARINE PROGRAM			
BIOLOGICAL (ANIMALS)			
Benthic Organisms	At 26 stations collect benthic samples for first order identification and diversity study.	Quarterly	Monitor benthic population and diversity.
	At each of 26 stations collect 3 samples for detailed identification and biomass.	Annually	Monitor in detail, change in benthic communities.
Crabs	At 6 stations collect to determine body condition and metal concentration Cu, Mo, Cd, Pb, Zn, As and Hg.	Quarterly	Monitor crab population and metal content.
Shrimps	At 3 locations collect shrimps with standard commercial shrimp traps. Metal analysis includes Cu, Mo, Cd, Pb, Zn, As, and Hg.	Annually	Monitor metal content and assess population variations.

Intertidal Invertebrates

Clams	Collect various species of clams at 9 sites; identify, weigh, measure and analyze for Cu, Mo, Cd, Pb, Zn, As, and Hg.	Quarterly	Monitor body condition and metal content.
Mussels	Collect blue mussels at 3 sites. Measure, weigh, and analyze for Cu, Mo, Cd, Pb, Zn, As and Hg.	Quarterly	Monitor body condition and metal content.

Fish

Intertidal Fish	At 7 intertidal sites sculpins are collected by beach seining. Identify, weigh, measure and analyze for Cu, Mo, Cd, Pb, Zn, As and Hg.	Semi-annually	Monitor metal content.
Bottom Fish	Longlines set at 4 stations to collect bottom fish. Identify, weigh, measure, sex and analyze for Cu, Mo, Cd, Pb, Zn, As, and Hg.	Semi-annually	Estimate population and monitor metal content.
	At 6 sites jigging is done to collect reef-dwelling and bottom species. Identify, weigh, measure, sex and analyze for metals.	Semi-annually	Monitor metal content.

Table II, continued

	Description	Frequency	Objective
MARINE PROGRAM			
BIOLOGICAL (ANIMALS)			
Fish			
Pelagic Fish	At 5 sites gill nets are set to collect pelagic species. Identify, weigh, measure, sex and analyze for metals.	Semi-annually	Monitor metal content.
SPECIAL STUDIES			
Chemical Study Interstitial Water	Collect samples of pore water within various zones in the tailings and in natural sediments.	Short-term studies	Assess the pore water dissolved metals concentration and the potential metal fluxing.
Suspended Solids	Dye studies. Coulter counter particle size studies. Remote sensing from aircraft.	Short-term studies	Monitor the amount of up-welling at Hankin Point.
	Time-series studies in Quatsino Narrows.		Monitor amount of mine derived suspended solids deposited outside Quatsino

Salmon Fry Assessment	Spring beach seining for salmon fry. Identify to species, examine stomach contents, and assess school's residence time in various habitats.	Annually in Spring	Narrows. Monitor annual outward migration of fry and examine feeding habit.
FRESHWATER PROGRAM			
Water Sampling	At 9 mid-stream locations collect samples for temperature, pH, alkalinity, dissolved solids, suspended solids, turbidity, color, hardness, dissolved oxygen, sulphates, nitrates, total extractable Hg and As, dissolved and particulate Fe, Cd, Cu, Co, Cr, Mo, Pb, Zn, Ni, Mn.	Quarterly	Record chemical characteristics of water flowing into Rupert Inlet.
METEOROLOGICAL PROGRAM			
	At the mine site record temperature, wind, precipitation, cloud cover, barometric pressure, incident sunlight levels.	Hourly and continuous	Record of meteorological conditions.

sophisticated statistical analysis system package, SAS (4). The statistical procedures used are described in the relevant sections of the annual reports (1).

The majority of the statistical procedures are distribution dependent, requiring that the data be normally distributed for a valid analysis (3). To assess the normality of the data and therefore its compliance with this requirement, a Kolmogorov-Smirnov goodness-of-fit test (3) was done on representative data blocks. This analysis indicated that deviations from normality were minimized if the data were converted to natural logarithm equivalents. The resultant minimal deviations from normality, coupled with the robust nature of the distribution statistics used, reduced the likelihood of erroneous interpretations to acceptable levels.

The mine initiated its environmental baseline program in 1969 two years prior to start-up (5). This was expanded by the regulatory agency one year later. The mine has also added other components such as primary productivity measurements and juvenile salmon migration investigations within the fjord. Data gathering and analysis has been by a combination of staff environmental scientists and technicians, and consultants retained for specialist components, e.g. zooplankton and benthos species identifications and counts. The program is designed and is under the general review of an environmental advisory group. This consisted initially in 1970 of engineers and oceanographers from the University of British Columbia, supplemented in 1972 by a benthic biologist from the University of Victoria. After a review of the program in 1978, the formal management agreement with the University of British Columbia was no longer required by the regulatory agencies. However, Island Copper continues to retain the services of an environmental advisory group consisting of university and consulting scientists and engineers. This group maintains the previous practice of reviewing the monitoring program annually, making on-site inspections of procedures, and suggesting changes when appropriate.

In 1974 at a meeting with government scientists the first need for a trend analysis appeared. Some levels of trace metals in organisms gave the appearance of being high. It was possible at short notice to respond to a regulatory agency inquiry by an Analysis of Variance on clams and mussels. The analysis did not indicate whether the elevation was related to the tailings discharge or was a natural seasonal variation (6). Subsequently it has been desirable to review most of the monitored parameters, and the data handling procedure developed progressively as needs appeared.

At present all data, as assembled from field sampling or subsequent laboratory chemical tests or organism identifications and counts, are reviewed by hand tabulation or plots by Island Copper's environmental scientists. In some cases these have been compared against levels which function as "alerts" or "signals", e.g. the Canadian Food and Drug allowable metal limits for edible species (Table III). Where the visual inspection indicates a trend beyond or approaching the limits of prior variations in the data, a statistical analysis can be implemented, or the data gathering and laboratory analyses repeated for confirmation. If necessary reagent control action can be taken, as in one case of bioassays indicating that a new reagent was synergizing with others to produce mortality to tested juvenile salmon. On one occasion repeat testing showed faulty monitoring procedures which were corrected (7).

Table III. Canada Food and Drug Act limits for metals in marketable seafoods (ppm wet weight).

Metal	Limit
Arsenic	5
Copper	100
Lead	10
Mercury	0.5
Zinc	100

Significant features of Island Copper's monitoring program include the comprehensive coverage of parameters able to monitor potential impacts, quality control of the program through annual reviews by an independent group of university scientists and engineers, and development of procedures for handling the enormous bank of data now stored. The latter is the principle subject of this chapter.

TREND MONITORING BY GRAPHS AND TABLES

Data are assembled in figures and/or tables, depending on the nature of the test, and the extent and complexity of the data (2,3 or 4 dimensions, i.e. regional divisions, regional divisions with different depths, and regional divisions with depths and through time).

As an example, for the 1980 annual report (1) water

column dissolved metal levels were averaged by year and by regions. Plots for copper and manganese are shown in Figs. 2 and 3 (sampling stations are shown in Fig. 1). Copper levels from 1971 to 1976 show an increase trend which was the source of some concern at the time even though involving very low levels of the metals. However, continued monitoring showed a decrease to and below the original levels, followed by a return in 1980 to the 1970 levels. There is no sign in 1980 from these data that dissolved copper levels have risen since 1970 as a result of the mine's activities.

In contrast to copper, dissolved manganese levels (Fig. 3) reflect an increased liberation of the element as predicted by the pre-operational leaching experiments (Chapter 2) and confirmed by later tests. The elevations are greater near the bottom than near surface, and have been accepted as reflecting a real but harmless increase. The significance of the higher levels in the more remote Holberg Inlet than in Rupert Inlet still needs exploring through a further data analysis or through redesigned sampling.

A combined graphical-tabular review procedure has been used where appropriate, and is shown for rate of sedimentation in collection trays. The sampling stations are shown in Fig. 4. The graphical presentation (Fig. 5) shows an annual oscillation with possibly an increase trend from 1975 through 1980 at Hankin Point where it is known that resuspended tailings deposit (8). The additional tabulated data (Table IV) show that Hankin Point sedimentation was substantially higher than elsewhere. The yearly averages suggest that shallow sedimentation there (30 ft depth) may have increased from 1977 to 1980, but deeper water sedimentation (60 ft depth) has not. There appears to have been a much higher proportion of volatiles (biological material) in the shallow deposits in 1979-80 than previously. Finally Table IV, by providing standard deviations, shows that the data is probably non-normal and that statistical testing will be a complex process requiring data transformations. The 1980 decision concerning the results was to review further years' data by statistical analysis before deciding whether the results merited a changed or extended program to determine causes for the variations in the complex of parameters involved. One difficulty to be overcome is missing data due to sampler overturn, and bioperturbation of the deposits.

A simple tabular presentation can on occasion be sufficient for reviewing some parameters, such as the copper content of deep sediments. Table V gives results for seabed stations shown in Fig. 1. Copper levels greater

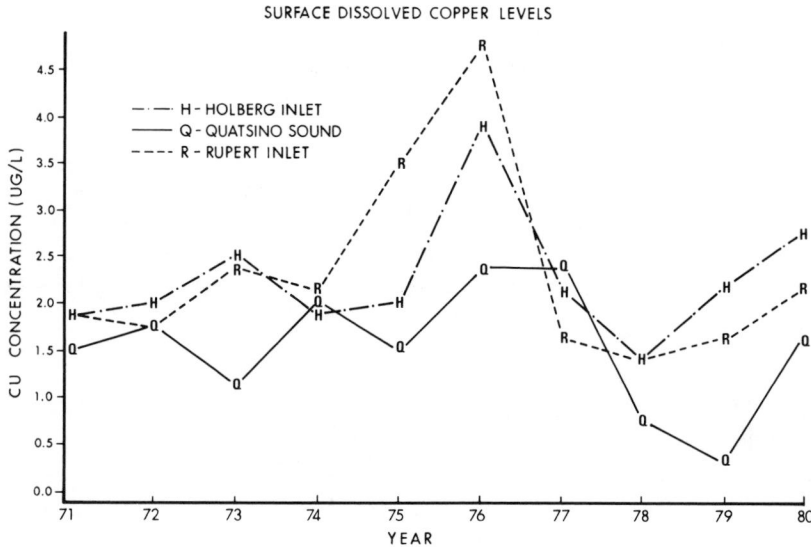

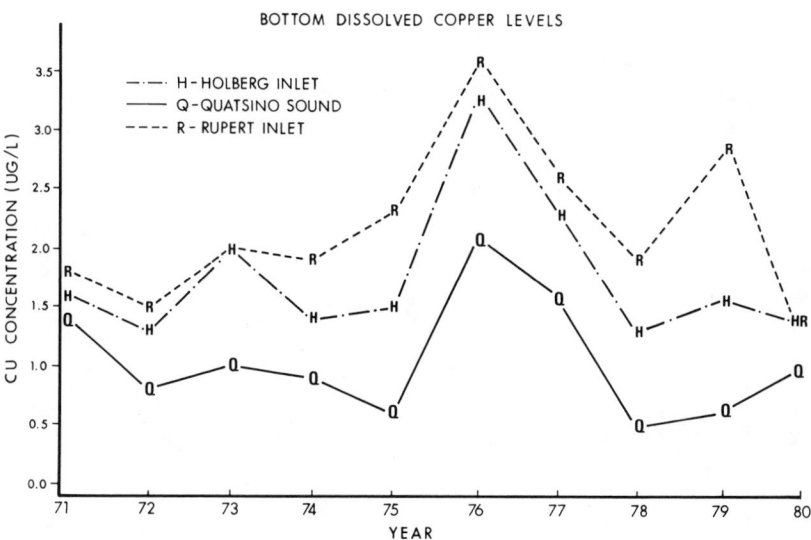

Figure 2. Mean inlet dissolved copper levels.

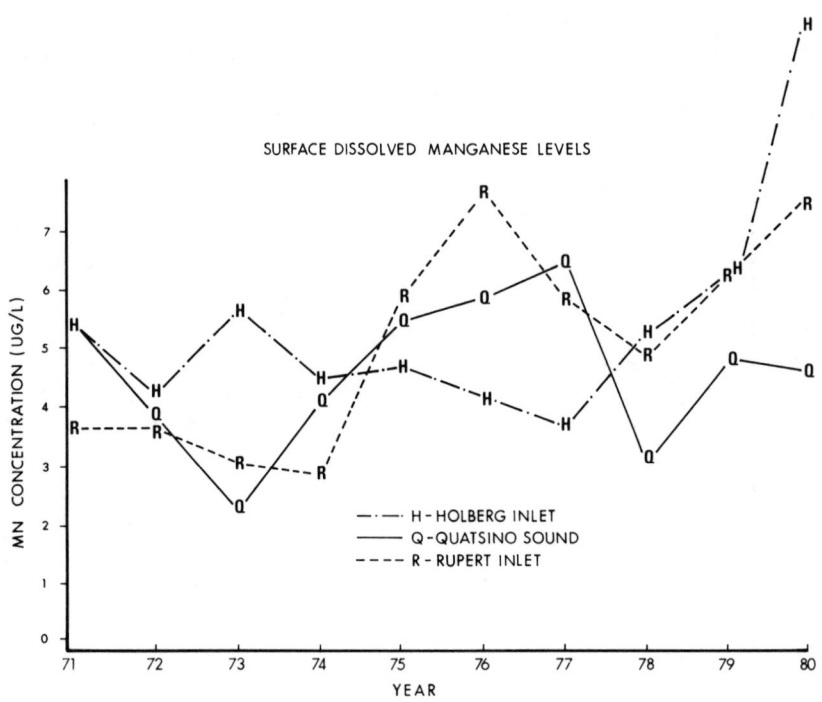

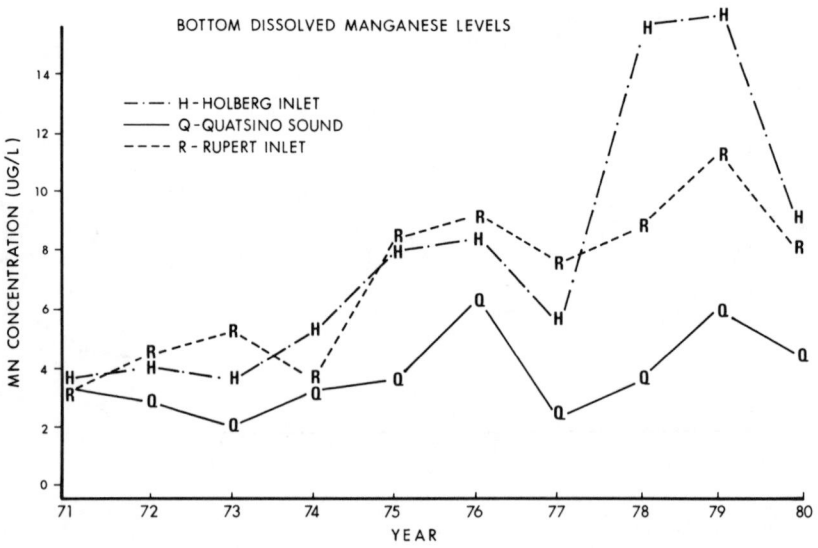

Figure 3. Mean inlet dissolved manganese levels.

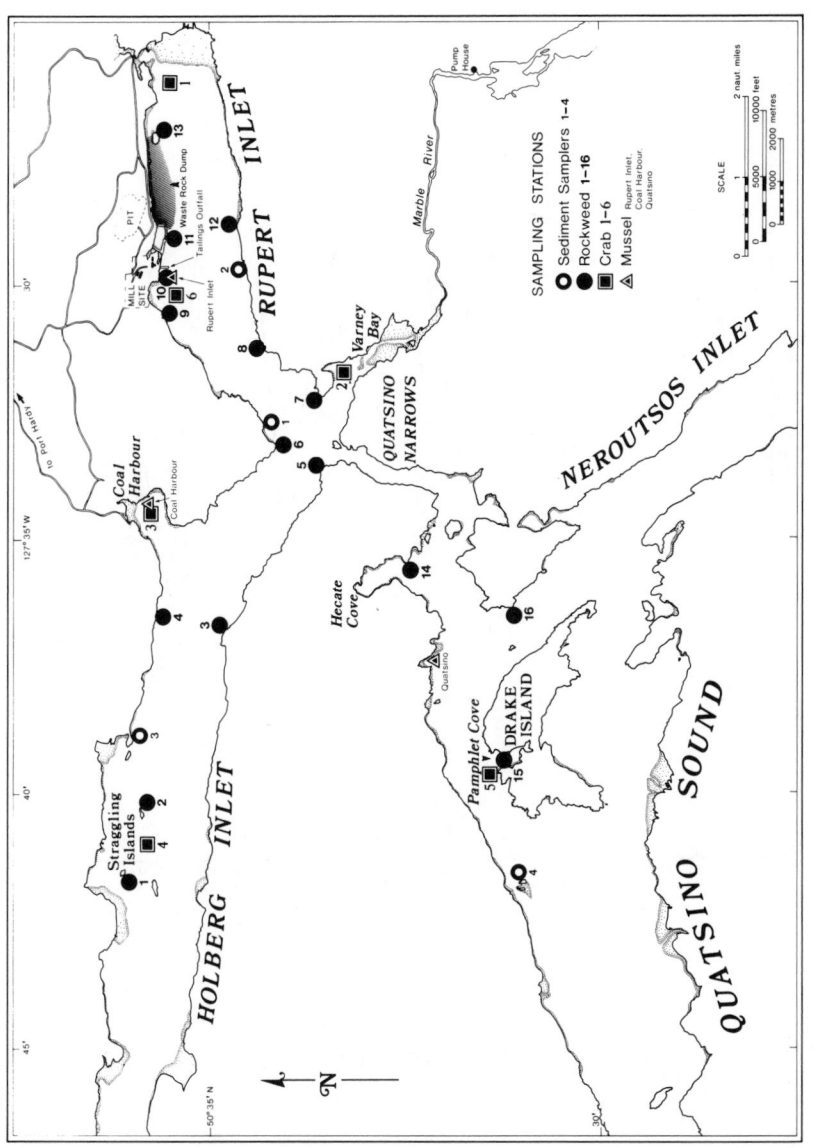

Figure 4. Additional sampling locations.

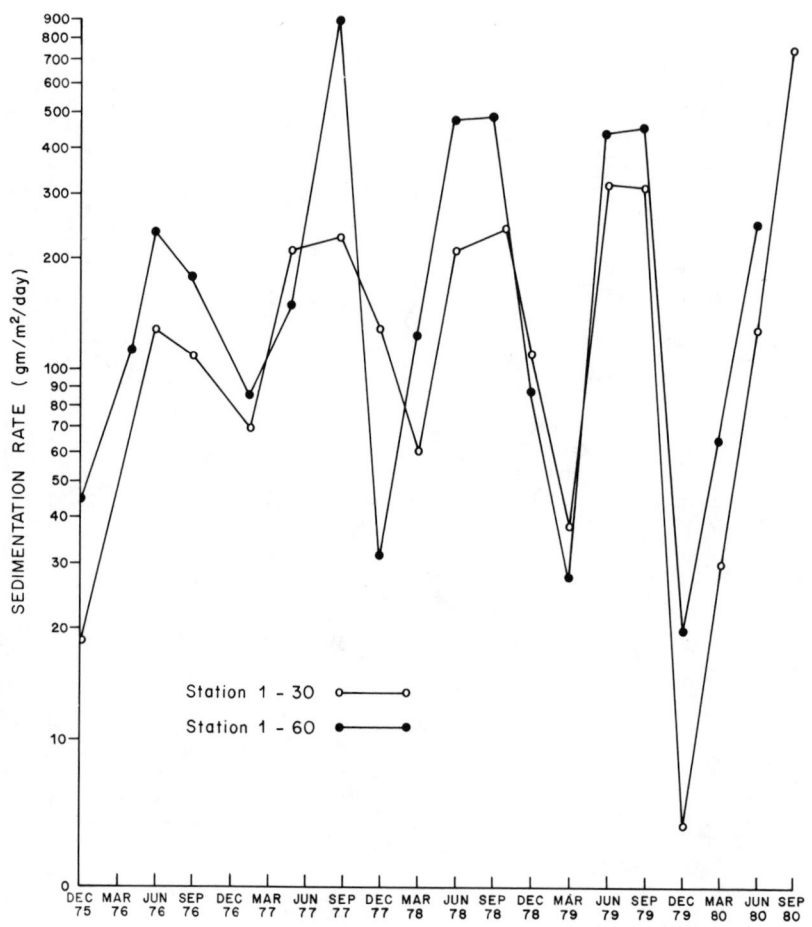

Figure 5. In situ sedimentation rates at Hankin Point, 1975-1980.

than approximately 100 ppm can be taken as an indicator of tailings deposits. By 1980 tailings were settling at most stations, although not to the extent of being visible in core samples most remote from the discharge point. This represents an irregular expansion of deposition since

Table IV. Mean in situ sedimentation rate (g/m^2/day) and percentage volatile component, 1977-1980.

Station #	Total Sedimentation Rate			Mean % Volatile		
	1977/1978 $\bar{x} \pm$ S.D. (n)	1978/1979 $\bar{x} \pm$ S.D. (n)	1979/1980 $\bar{x} \pm$ S.D. (n)	1977/1978	1978/1979	1979/1980
1 - 30'	174.6 ± 103.1 (4)	201.0 ± 148.7 (4)	240.85 ± 377.71 (4)	2.1	1.8	11.1
1 - 60'	278.9 ± 235.3 (4)	243.1 ± 217.5 (4)	115.27 ± 122.82 (3)	1.9	1.9	2.1
2 - 30'	15.9 ± 9.4 (4)	15.3 ± 8.2 (4)	2.60 ± 1.19 (3)	4.8	3.7	4.9
2 - 60'	12.6 ± 7.3 (4)	18.8 ± 19.2 (3)	18.85 ± 16.79 (3)	6.6	11.6	5.0
3 - 30'	2.3 ± 1.6 (4)	3.6 ± 0.3 (2)	3.37 ± 1.03 (2)	14.3	14.2	8.2

Table IV, continued

Station #	Total Sedimentation Rate			Mean % Volatile		
	1977/1978	1978/1979	1979/1980	1977/1978	1978/1979	1979/1980
	$\bar{x} \pm$ S.D. (n)	$\bar{x} \pm$ S.D. (n)	$\bar{x} \pm$ S.D. (n)			
3 - 60'	4.8 $\pm$ 2.6 (4)	7.3 $\pm$ 1.8 (3)	9.47 $\pm$ 2.92 (3)	9.7	7.7	5.0
4 - 30'	5.5 $\pm$ 5.7 (4)	12.9 $\pm$ 8.6 (3)	1.37 (1)	9.1	5.7	7.3
4 - 60'	15.8 $\pm$ 14.2 (4)	5.8 (1)	N/A	6.1	5.8	N/A

N/A (Not Available) = stations not sampled.

Table V. Copper content (ppm) of deep marine sediment, from 1971-1980.

Station	1971	1972	1973	1974	1975	1976	1977	1978	1979	1980
1	23	19	33	43	45	61	56	50	33	40
2	42	87	95	181	167	308	276	78	143	254
3	42	54	59	117	139	76	140	52	117	157
4	45	68	130	227	286	359	350	251	268	235
5	34	62	255	284	505	515	449	408	342	279
6	39	73	148	302	579	581	471	489	373	306
7	22	37	103	232	318	460	356	284	256	230
8	18	29	35	51	116	211	187	144	141	199
9	32	407	330	391	600	693	381	452	628	406
10	39	27	78	157	331	318	267	237	264	131
11	25	39	36	64	114	113	147	123	128	156
12	23	41	160	135	208	421	329	247	270	298
13	45	98	940	361	632	1124	442	352	335	434

Table V, continued

Station	1971	1972	1973	1974	1975	1976	1977	1978	1979	1980
14	35	122	320	387	747	729	549	639	607	545
15	34	785	17300	27800	3480	1602	1505	369	418	433
16	40	624	940	382	789	688	279	340	382	304
17	35	97	500	358	331	637	260	284	359	311
18	32	47	144	115	82	159	196	218	167	263
19	35	111	134	203	589	293	315	269	251	327
20	27	73	43	162	249	298	231	175	204	266
21	30	33	65	68	57	82	74	81	75	72
22	21	26	45	45	78	139	224	227	169	159
23	30	71	70	115	235	282	255	271	281	253
24	N.A.	N.A.	N.A.	56	100	67	101	80	123	108
25	N.A.	N.A.	N.A.	N.A.	N.A.	113	115	136	137	146

N.A. (Not Available) = stations not sampled.

1972. Note also how very high 1973-74 levels of copper at station 15 near the outfall have declined, possibly through reactivation of coarse deposits by the channelization processes documented by Hay in Chapter 5.

TREND ANALYSES

Data for some parameters are too complex for easy graphical presentation, and trend analyses have proceeded directly from tabular collations. Table VI compares tissue metal levels in the edible shoreline softshell clam <u>Mya arenaria</u> (sampling stations, Fig. 6). This is a species which is gathered commercially, recreationally and for native use in British Columbia. The format of arranging levels in order, and underlining to denote no significant difference at the 5% level allows visual checking for consistent trends between regions, between individual stations and between years. No mine-related trends appear to be derivable from the data.

Data analyses of this type in the mid-1970s (9) indicated a need to extend the tissue metal monitoring program. As a result regional coverage was extended in

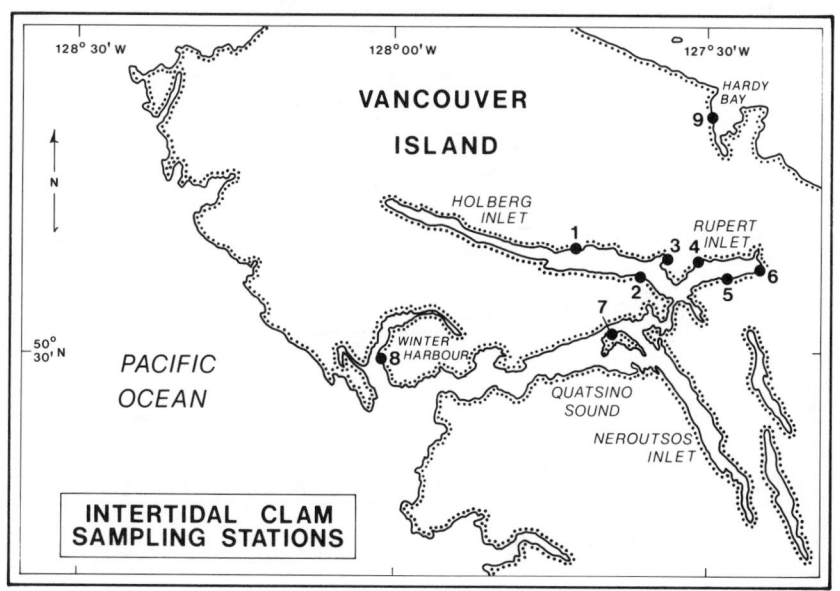

Figure 6. Intertidal clam sampling stations.

Table VI. Relative metal content of Mya arenaria, 1971-1980.

Metal	Grouping	Higher								Lower
Copper	-Inlet	Holberg		Quatsino		Rupert		Port	Hardy	
	-Station	4	3	2	7	5	6	1	8	9
	-Year	77	74	73	75	78	79	72	71	80
Molybdenum	-Inlet	Port	Hardy	Quatsino		Rupert		Holberg		
	-Station	9	7	6	8	4	1	2	5	3
	-Year	72	73	74	71	80	76	75	79	77	78
Lead	-Inlet	Quatsino		Port	Hardy	Rupert		Holberg		
	-Station	5	7	4	1	9	2	6	8	3
	-Year	73	72	74	71	79	80	77	75	78	76
Zinc	-Inlet	Holberg		Rupert		Quatsino		Port	Hardy	
	-Station	4	5	3	2	1	6	7	9	8
	-Year	80	71	78	77	79	76	72	75	74	73
Cadmium	-Inlet	Port	Hardy	Quatsino		Rupert		Holberg		
	-Station	9	7	8	6	3				
	-Year	73	72	74	80	71	79	75	77	76	78

Table VI, continued

Arsenic	-Inlet	Port		Hardy		Holberg		Quatsino		Rupert
	-Station	9	3	2	7	1	8	4	5	6
	-Year	71	77	74	76	75	79	72	78	80	73

Mercury	-Inlet	Rupert		Quatsino		Port		Hardy		Holberg
	-Station	5	6	7	1	4	8	9	3	2
	-Year	73	74	71	72	75	76	79	77	80	78

Note: Underlining denotes no significant difference at the
      5% level.

---

some cases, and new species were brought into the program
testing for bioaccumulation and biomagnification.
     The edible mussel Mytilus edulis is known to be an
efficient bioaccumulator (10,11). Tests were initiated in
1973 and results are presented in Tables VII and VIII
(sampling stations are shown in Fig. 4).   Copper in
mussels at the mine loading dock in Rupert Inlet is not
significantly higher than at the remote control station in
Port Hardy, nor is there a time trend. The significantly
higher values for molybdenum at the mine loading dock are
not supported by a time trend, and are difficult to
ascribe to the mine in view of the practice of shipping in
closed containers. Some actual levels (for 1979-80) are
shown in Table VIII. They are substantially below levels
set by the Canada Food and Drug Act (Table III) for
marketing seafoods, and in most cases lower than in most
of the earlier years (Table VII).
     Crab tissue metal levels have presented a number of
peculiarities during the monitoring program, in particular
relatively high levels of arsenic (8) close to or above
those set by the Canada Food and Drug Act. Accordingly, a
reduction in monitoring frequency during the mid-1970s
from quarterly to annually was reversed in 1977.   Remote
control stations from the west and east coasts of
Vancouver Island were also introduced.   Table IX shows
statistical results from data accumulated from stations
occupied over the ten years from 1971 to 1980. There is no
support for there being mine-related differences between

Table VII. Relative metal content of <u>Mytilus edulis</u>, 1974-1980.

Metal	Grouping	Higher						Lower
Copper	-Inlet	Rupert	Port Hardy		Holberg		Quatsino	
	-Year	74	79	77	78	80	75	76
Molybdenum	-Inlet	Rupert	Holberg		Quatsino		Port Hardy	
	-Year	74	75	77	78	76	80	79
Lead	-Inlet	Rupert	Quatsino		Holberg		Port Hardy	
	-Year	74	79	80	78	77	76	75
Zinc	-Inlet	Port Hardy		Rupert		Holberg		Quatsino
	-Year	79	80	77	75	78	76	74
Cadmium	-Inlet	Quatsino	Port Hardy		Rupert		Holberg	
	-Year	74	78	77	79	80	75	76
Arsenic	-Inlet	Quatsino	Holberg		Rupert		Port Hardy	
	-Year	77	79	74	78	76	80	75
Mercury	-Inlet	Holberg	Port Hardy		Quatsino		Rupert	
	-Year	74	77	76	75	79	80	78

Note: Underlining denotes no significant difference at the 5% level.

Table VIII. Mean metal content of Mytilus edulis, 1979-1980 (ppm wet weight).

Location	Date	Copper	Molybdenum	Lead	Metal Zinc	Cadmium	Arsenic	Mercury
Coal Harbour	79-12	1.7	0.1	0.6	31	0.42	1.7	0.04
Quatsino	79-12	1.3	0.2	0.5	21	0.42	2.0	0.02
Utah Dock	79-12	2.1	0.2	1.2	47	0.52	0.9	0.01
Coal Harbour	80-03	1.1	0.1	0.1	24	0.28	1.0	0.02
Quatsino	80-03	1.1	0.1	0.9	7.8	0.75	1.6	0.04
Utah Dock	80-03	1.7	0.1	0.7	41	0.55	1.3	0.02
Coal Harbour	80-06	1.1	0.1	0.6	16	0.19	1.2	0.03
Quatsino	80-06	1.2	<0.1	1.4	26	0.52	0.9	0.02
Utah Dock	80-06	2.5	0.2	1.5	38	0.58	1.4	0.01

Table VIII, continued

Location	Date	Copper	Molybdenum	Lead	Metal Zinc	Cadmium	Arsenic	Mercury
Coal Harbour	80-09	1.1	< 0.2	0.6	10	0.17	0.6	0.07
Quatsino	80-09	1.3	0.2	0.4	11	0.23	0.5	0.01
Utah Dock	80-09	1.7	0.2	0.9	21	0.40	0.6	0.01
Coal Harbour	1979-1980	1.2	0.1	0.5	20	0.27	1.1	0.04
Quatsino	1979-1980	1.2	0.1	0.8	16	0.48	1.2	0.02
Utah Dock	1979-1980	2.0	0.2	1.0	37	0.51	1.0	0.01

Note: "Less than" values averaged at reported detection limit.

Table IX. Relative metal content of Cancer magister by inlet, station and year, 1971-1980.

Metal	Grouping	Higher								Lower
Copper	-Station	1	6	3	2	4	5			
	-Inlet			R	H	Q				
	-Year	77	78	79	76	80	74	75	71	72	73
Molybdenum	-Station	2	6	3	5	4	1			
	-Inlet			R	H	Q				
	-Year	73	71	72	74	77	80	75	78	76	79
Lead	-Station	5	1	6	2	3	4			
	-Inlet			Q	R	H				
	-Year	73	71	72	80	79	78	77	76	75	74
Zinc	-Station	3	6	2	1	5	4			
	-Inlet			R	H	Q				
	-Year	75	74	71	77	80	79	78	76	72	73
Cadmium	-Station	5	6	2	3	4	1			
	-Inlet			Q	R	H				
	-Year	73	72	71	77	74	78	76	79	80	75

Table IX, continued

Arsenic	-Station	5		1	2	6	3		4
	-Inlet				Q	R	H		
	-Year	77	75	76	73	71	80	78	74	72	79

Mercury	-Station	3	6	4	2	1	5
	-Inlet			H	R	Q	
	-Year	73	77	71	76	75	72	78	74	80	79

Note: R = Rupert, H = Holberg, and Q = Quatsino.
Underlining denotes no significant difference at the
5% level.

the fjords or between years.

Finally, in this context of statistical reviews and revising programs to better monitor potential impacts, the rockweed alga Fucus distichus was brought into the testing program in 1976. A combined graphical and statistical tabulation format has been developed. Fig. 7 groups stations (see Fig. 4) 6, 8, 10 and 11 closest to the mine or the tailings resuspension area, and compares results with remote stations 1, 2, 3, 4, 15 and 16, and an intermediate group. There appeared to be an increase in copper levels in the close-in stations showing through seasonal oscillations, but the low 1980 results did not support this. The time trend analysis of Table X also did not support a consistently increasing trend, although the zinc data might merit further analysis.

CONCLUSIONS

The sequential procedure adopted at Island Copper Mine of reviewing data by tables and graphs, and then by various statistical tests if considered necessary, allows rapid review of quantitative data provided by chemical analysis, or biological species identifications. As the

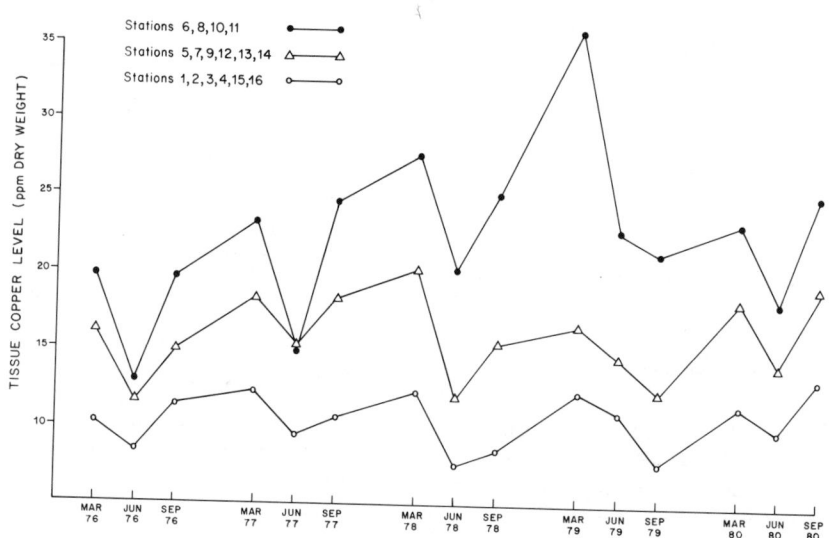

Figure 7. Copper content of <u>Fucus</u>, grouped means, 1976-1980.

data bank becomes even more enormous than at present, the data handling becomes more complex. Staff and outside reviewers must be careful to retrieve all the relevant data and not pre-select small data blocks unless it is on a randomized basis. In practice reviewers need to specify their review objectives precisely, format their data collations in a relevant manner, and undertake appropriate statistical tests.

Island Copper's environmental program reflects constraints in data collection and handling imposed by the original 1970 baseline survey design, and the then operational procedures considered acceptable. Nevertheless

Table X. Rupert and Holberg Inlet <u>Fucus</u> relative metal content by year.

Metal	Higher				Lower
Copper	<u>79</u>	<u>77</u>	<u>80</u>	<u>78</u>	<u>76</u>
Molybdenum	<u>76</u>	<u>77</u>	<u>78</u>	79	80
Zinc	<u>80</u>	<u>79</u>	<u>77</u>	78	76
Arsenic	<u>77</u>	<u>80</u>	<u>79</u>	<u>78</u>	<u>76</u>
Mercury	<u>78</u>	<u>76</u>	<u>77</u>	<u>79</u>	<u>80</u>

Note: Underlining denotes no significant difference at the 5% level. No relationships could be statistically demonstrated for lead or cadmium.

there is periodic review of the program with modifications designed to maintain comparability of data. Similar programs at other mines under development should be able to shortcut some of the early development difficulties encountered by Island Copper, e.g. limited computer software available. In addition, sampling designs can be specific to detecting such relevant potential impacts as bioaccumulation of locally predictable trace metals, and eliminate tests for improbable impacts, or others which early testing shows have not been realized.

REFERENCES

1. Island Copper Mine. "1980 Annual Environmental Assessment Report", Utah Mines Ltd. (1982).
2. Evans, J.B., D.V. Ellis, J. Leja, G.W. Poling and C.A. Pelletier. "Environmental Monitoring of Porphyry Copper Tailing Discharged into a Marine Environment", Proc. XIII International Mineral Processing Congress (Warsaw, Poland, 1979), pp. 650-690.
3. Sokal, R.R. and F.J. Rohlf. Biometry (San Francisco, CA: W.H. Freeman and Co., 1969).
4. SAS. Statistical Analysis System (SAS Institute Inc.: North Carolina University, 1979).

5. Evans, J.B., D.V. Ellis and C.A. Pelletier. "The Establishment and Implementation of a Monitoring Program for Underwater Tailing Disposal in Rupert Inlet, British Columbia", Tailing Disposal Today: Proc. First International Tailing Symposium, C.L. Aplin and G.O. Argnall, Eds. (San Francisco, CA: Miller Freeman Publications, 1973), pp. 512-552.

6. Ellis, D.V. and L. Bissel. "A Preliminary Report on Increasing Copper Levels in Intertidal Clams near Island Copper Mine", Report to Island Copper Mine in Appendix 3 to Island Copper Mine "Summary Report Second Production Year, Oct 1972 - Sept. 1973 Environmental Control Program,. The University of British Columbia (1974).

7. Island Copper Mine. "1978 Annual Environmental Assessment Report", Utah Mines Ltd. (1979).

8. Waldichuk, M. and R.J. Buchanan. "Significance of Environmental Changes due to Mine Waste Disposal into Rupert Inlet", Fisheries and Oceans Canada, B.C. Ministry of Environment (1980).

9. Ellis, D.V. "Pollution Controls on Mine Discharges to the Sea" in Proc. International Conference on Heavy Metals in the Environment (Toronto, 1975), pp. 677-686.

10. Goldberg, E.D., Ed. The International Mussel Watch (Washington, D.C.: National Academy of Sciences, 1980).

11. Phillips, D.J.H. Quantitative Aquatic Biological Indicators" (London: Applied Science Publishers, 1980).

DISCUSSION

Question: How effective were your suspended sediment samplers?

Answer: We found that the samplers actually trapped sediment (i.e. higher sediment rate indicated by sampler than on the seafloor). This could indicate that sediment on the seafloor was moved by currents while the trapped material in samplers remained intact. The sampling cup, being 3 m off the bottom, is a good indicator of the suspended sediment "fallout".

Question: How much of your analysis was by computer?

Answer: The computer system now in place at Island Copper was developed over a number of years. Initially, the data was handled by hand (i.e. preparing tables and graphs).
   In 1974, we set up a "RAMIS" data storage and

retrieval system. A Hewlett Packard 9830 programmable calculator was used to do some basic statistical analyses. By 1978, the data bank had expanded to the point where a complete computer and statistical system was warranted. A terminal was installed in the environmental laboratory which gave us access to a large IBM computer system. A software package called "Statistical Analysis System" was purchased, which now permits elaborate statistical analysis right at the laboratory.

Most of the trend analyses are now being done by computer. Two and three-way ANOVAS and various other statistical tests can be done routinely.

Question: Have you noted cyclic changes in the data on a seasonal or annual basis?

Answer: At Island Copper, we are in the fortunate position of having over ten years of operating data to review. We have observed a number of seasonal, annual and even 5 to 10 year cycles. An example of long-term cycles is evident in the benthic data. In the pre-operational data, a large polychaete worm Heteromastus filobranchus was prominent throughout Rupert and Holberg Inlets. Coincident with the start of tailings discharge in 1971, this large polychaete disappeared and was replaced by a smaller species, Ammotrypane aulogaster, as the dominant species. In 1977, the Heteromastus filobranchus started coming back and by 1978 they were again present in large numbers. In 1979, at a station with over 30 ft of tailings, a new large polychaete Lumbrinereis luti appeared. In 1980 we found that the Ammotrypane aulogaster was being replaced as the dominant polychaete by the large polychaetes Heteromastus filobranchus and Lumbrinereis luti. This is an example of population trends in the benthic community. There are a number of other trends that have been noted, such as in zooplankton diversity, and metal content of various shellfish.

Question: Why did you not use liver tissue instead of muscle tissue as an indicator of metal population? And secondly, why do you report your results in wet weight instead of dry weight?

Answer: There is no question that the liver tissue concentrates metals more readily than muscle tissue.

The reason we chose to use muscle tissue is that it represents what people consume in fish or shellfish. Also, the livers of most species monitored are so small that an adequate sample for analysis is hard to obtain. There are enough difficulties with tissue metal analysis without having to work at microgram levels.

An example of problems with small sample size is evident at Kitsault where by permit, they are required to monitor the metal uptake of a small clam Yoldia. This species can be very small and thus very difficult to analyze with reasonable accuracy, and reproducability. Yoldia is not a commercial species and may have limited value in assessing the impact of mining on the commercial fishery. It is appreciated that a brown king crab may feed on these animals, but difficulty depurating these small forms combined with the analytical problems, make the use of Yoldia as an indicator species questionable. The problem here is the same as the liver problem, not enough sample to analyze for the metals.

To answer your second question, we chose to analyze the tissues on a wet basis because it saved one step in which contamination could occur. Second and most important, the Canadian Food and Drug limitations for edible tissue are expressed on a wet weight basis. There is also the possibility of volatilizing, particularly mercury and arsenic, during the dry ashing process.

Question: In addition to monitoring for soluble metals in effluent, did you conduct any basic work on leaching from the tailings?

Answer: Yes, a series of tailings/seawater leaching studies were done at Island Copper. Different types of studies were done to simulate different conditions on the bottom of the fjord, and the behavior of the tailings from the thickeners to the discharge point. A series of static tank tests were done for 60 days, during which the release of 12 metals was monitored. The chemical characteristics of the seawater used in the study were monitored very closely. Manganese was the only metal that was significantly released from the tailings over the 60 day study. Agitation tests under highly oxidized conditions were done to simulate the behavior of tailings in suspension in seawater. Again the results indicated the manganese came into solution in seawater. After 10 years, of operations and discharging over 100 million tons of tailings to Rupert Inlet, the dissolved manganese in the water columns of Rupert and Holberg Inlet show an average increase of approximately 3-4 parts per billion. There is no apparent increasing trend with time. This could be due to the flushing effect caused by the tidal changes, which would not permit the buildup within the fjords.

Question: Does the company intend to discontinue
monitoring now that they have demonstrated that the
tailings are having minimal impact on the environment?

Answer: The company has been operating one of the most
comprehensive environmental programs associated with a
mining operation for 12 years. The only way to continue to
know and appreciate the health of the inlets is to
continue to monitor. Rupert and Holberg Inlets are used
extensively by the local people for recreational fishing.
It behooves the mine to continue to monitor in order to
protect the multiple use concept of the inlet. The
long-term monitoring program is part of the permit;
therefore some basic monitoring is required by law.

Question: The Canadian Federal Government released a
report in 1978 that suggested that conditions in Rupert
Inlet and surrounding area are significantly different
than what you have been telling us this afternoon. In
particular, it indicated that the bottom habitat had been
obliterated and that a significant amount of tailings were
being carried through the Narrows and into Quatsino Sound,
impacting an area much larger than had been predicted.

Answer: First a comment on the tailings going through
Quatsino Narrows on an ebb tide. Dr. Hay also mentioned in
his talk this morning that there is photographic evidence
to show that there is turbidity extending out of Quatsino
Sound. We did a number of time-series studies in Quatsino
Narrows during large tidal cycles to monitor the net loss
of tailings during a period of up-welling off Hankin
Point. Our studies indicated that 30 - 50 tons of material
was lost out of the system during a period of high
up-welling. Fifty tons (the maximum recorded) is 0.1% of
the daily discharge of tailings. Although 50 tons seems
high, the occurence is not continuous and the amount lost
from the system is not highly significant. The evidence of
tailings in Quatsino Sound cannot be seen by coring or
bottom dredging. Evidence is indicated by sediment
chemistry only. "Landsat" satellite photography shows a
lot more turbidity exists in the mainland fjords such as
Knight Inlet than in the Rupert-Holberg, Quatsino Sound
system. There has been a tendency to only show the system
at its worst. Maybe I could be accused of only showing the
positive, but I do believe that I have given both sides of
the story in my presentation.
     As a result of the controversy the federal and
provincial governments jointly appointed a two-man
commission to review the situation in Rupert Inlet. I can

assure you that they did a thorough review of our program
and the situation in the inlets. They reviewed all
available information and spent one week on Rupert and
Holberg Inlets examining the various forms of life
firsthand. We went through all our sampling procedures and
they retained samples for cross checks on our results. In
my opinion they left very few stones unturned. One
statement that I recall from their report was that the
environmental impact or consequences of the present means
of tailings disposal into Rupert Inlet does not warrant a
change to another procedure.*

---

*Editor's note: See Section III, Chapter 7, by R. Buchanan
for results of this government review.

D.V. Ellis
    Department of Biology
    University of Victoria
    Victoria, B.C.
    Canada V8W 2Y2

We had three speakers. Walter Kuit gave us a world premiere since the environmental program of the Polaris lead-zinc mine had not been presented in any detail to a technical audience previously. Considerable care has been given to the potential toxicity problems known to be involved with lead extraction, as opposed to less toxic metals, and to baseline surveying a unique and apparently highly favorable natural tailings impoundment from a relic arm of the sea.

Dr. Littlepage gave us not quite a world premiere of the baseline environmental surveys at Kitsault, and the detailed program still in operation to determine outfall performance relative to government regulations. This was his second opportunity to present his results. His first was at a public inquiry a year ago, but this time around he could supplement with data from the continuing monitoring.

Our third case study by Mr. Pelletier from Island Copper mine is the latest of a series of data and review presentations from that mine. They include public inquiry submissions, professionally published papers, and many limited distribution reports to government agencies. There is a great deal of information available on Island Copper; although it does take a bit of effort to recover that information. Much of it is in the form of limited distribution reports, in government files, but available also from the mine. Restricted reporting is a common problem in dissemination of environmental information, and one that can only be solved by persistence in correspondence with the data owners, regulatory agencies or attendance at conferences such as these where information is exchanged verbally.

I want to make a few comments and I will break them down into three different components.

The first component is the general concept that I raised earlier, that of collating data, generalizing and then making site-specific deductions. Mr. Caldwell in Chapter 1 showed us that the process of generalizing from case studies of land tailings disposal is reasonably far advanced. There are patterns of design that engineers can follow in deciding where to put the tailings (on land or in the sea) and other waste products from a mine. These well-known patterns indicate that marine disposal is a legitimate option, which has to be considered seriously in coastal regions where there is high rainfall and an earthquake risk. High rainfall and the earthquake risk are the reasons why in Canada the decision has been made over the past ten years to continue, knowingly, the longstanding practice of using this disposal option. The Philippines is another major area where marine disposal is a legitimate option and there are one or two other locations around the world where this option also appears to be needed.

We also learn from the monitored marine disposal systems that we now have a lot of information about tailing behavior in seawater. There are also continuing basic oceanographic studies in progress which provide us with a good deal of information about fjord dynamics. The two kinds of investigation when put together, allow us to deduce what will happen in new areas, and have led to the specific studies at Kitsault, for instance.

But in spite of increasing generalized knowledge, Polaris in the Canadian arctic is a good example that site-specific assessments must be done. There is always the possibility of developing atypical ways of solving waste disposal problems, which may be improvements on the standardized procedures of discharging into deep embayments or land impoundments. Polaris has a very unusual combination of environmental conditions which hopefully will facilitate the disposal of the waste without disturbance to the fragile arctic ecosystems. It is a case study which should be watched with very considerable interest. It should provide information about a unique situation from which we may get presently unpredictable ideas which can be applied in other parts of the world. The basic principle from Polaris, as I see it, involves site-specific assessments. Always check to see if there is a unique set of environmental conditions which may provide better disposal options than the standard procedures.

I must comment that Cominco were fortunate in that when Garry Lake was explored by the baseline surveys no arctic snail darter was found there.

The next major point that I want to raise is that we must have in our minds a schematic flow chart for data gathering relevant to a mine waste disposal program. We must schematize somehow for easy understanding the complex data gathering sequence that is possible. I suggest a five stage sequence of data gathering for environmental impact assessment (Fig. 1). This schematic allows us, for example, to compare how Polaris, Kitsault and Island Copper have progressed and how Quartz Hill must proceed if a marine discharge system is developed there.

First of all, in any data gathering, there has to be a baseline survey. Some people, I know, object to this concept as a largely irrelevant number gathering exercise. But what a baseline survey does is document the state of the environment over the years in which the baseline is run. It documents the resources which are at risk, and hence for the specific area are potentially impactable. When we put that local information together with the basic concepts of ecosystem dynamics (which are available from fundamental research) then we can develop predictive modelling, or predictive hypothesizing, of impacts for that particular area. The baseline survey is the basis for predicting impacts. From such predictions then assessors and engineers together can design mitigation before concrete gets poured. It is important to predict potential impact and introduce mitigative measures into hardware design and blueprinting before construction goes ahead.

After the baseline surveys, there has to be a stage of pre-operational monitoring. The baseline surveys allow decisions on the ecosystem parameters which are most efficient for monitoring so that the post-discharge situation can be compared with the pre-discharge situation. The pre-operational monitoring of these parameters exposes the natural variations which are important in the local area. Pre-operational monitoring allows the amount of noise in these parameters to be measured so that the eventual statistical comparisons can make necessary allowances. Pre-operational monitoring of parameters which are important for before and after comparison, is vital for efficient results. Polaris has just completed this stage.

After start-up, Kitsault-type outfall performance checks are needed. What these do is allow fine-tuning the outfall system - the waste disposal system. They feed back information to allow process controls so that the system will work optimally. Dr. Littlepage has shown us that Kitsault is well ahead in this particular kind of program.

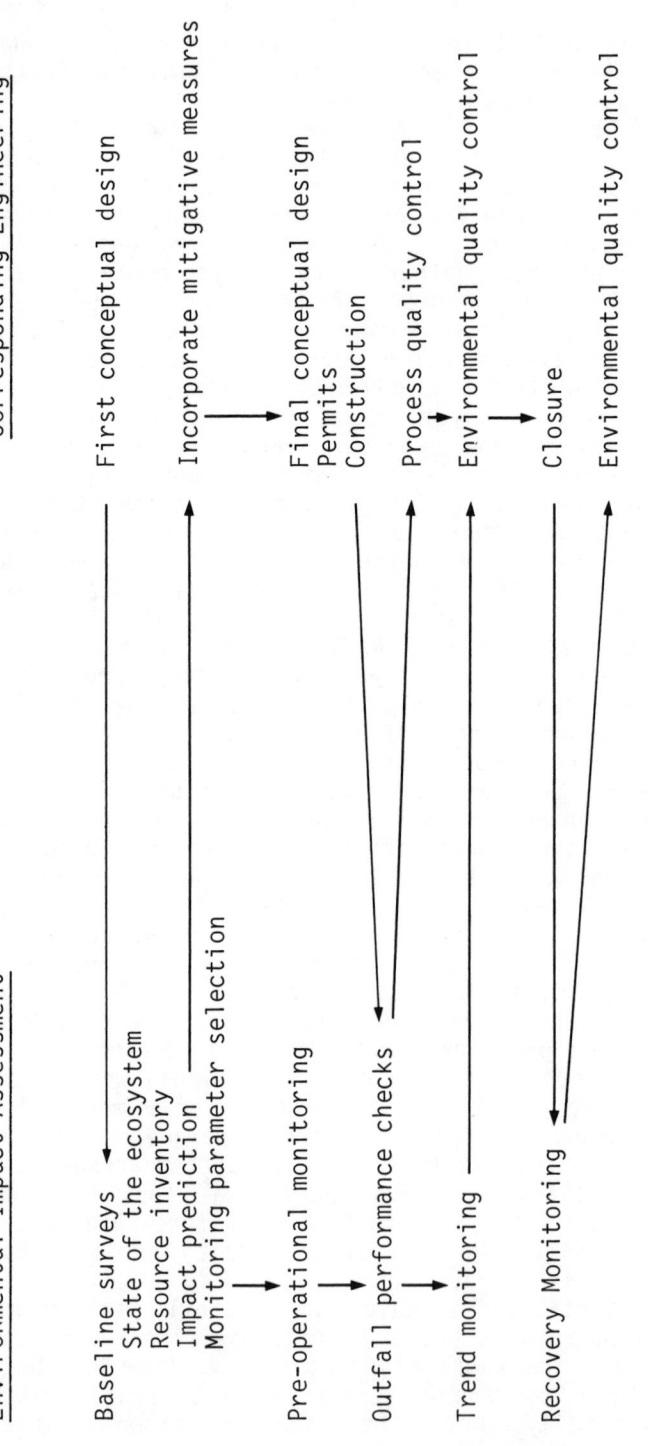

Figure 1. The process of environmental impact assessment for a marine waste discharge and relationships with project engineering.

Once these outfall performance checks are complete, and plume behavior is known, such monitoring can be reduced to occasional checks to show that the outfall is still performing the way it was designed, and performed on commissioning. These outfall checks should be made through a variety of discharge rates and seasonal environmental conditions. The pattern of outfall performance can be shown over a year, or over other environmental and process variables. Once the patterns are known the tests do not need repeating to the same intensity. Outfall performance checking is a short-term monitoring program to show that the design criteria have been met.

The fourth part of the EIA process also needs initiating soon after start-up. This is long-term trend monitoring, or monitoring for early warning of slowly developing and at times unpredictable ecosystem responses. The parameters and their sampling design will probably be very different from those in the outfall performance testing. Trend monitoring tests, for example, whether arsenic or copper or molybdenum is insidiously elevating in the water, primary producers or higher up the food chain. That involves a very different testing program from determining the behavior of the outfall plume or density current. We need to know plume behavior for designing long-term trend monitoring, but the two programs are quite different. We can expect the long-term monitoring to go on for the lifetime of the mine - be it ten years, twenty years, or even seventy years. It is an early warning system, like the Alaskan-Canadian DEW-line radar stations. Island Copper is now well into this stage.

Finally, it must not be ignored that ecosystem recovery monitoring is also needed. It is important to find out for the mines now discharging to marine receiving areas, how the ecosystem recovers from those impacts which have been allowed. Island Copper has initiated such monitoring. It is seeking to determine how quickly an underwater tailings deposit recovers. Does it reclaim naturally? Will it reclaim at all? How do patterns compare with land-mine reclamation practice? This information is going to be important in management. If it can be input into decision making about the allocation of a fjord for waste disposal - if it can be input how quickly a fjord will recover or establish some other beneficial balance of the marine ecosystem - that is very important information for decision making.

I have a few extra incidental points to note. One of these is the matter of computerization. Mr. Pelletier mentioned computerization of Island Copper's environmental data bank. He mentioned that Island Copper started computerizing after the monitoring program got started.

Island Copper was unfortunate in that sense. They started their comprehensive data assessment before computer facilities became widely available, before many biologists and environmental scientists were trained in the necessary kind of programming, the concepts of data retrieving, averaging, time trend analysis and so on. Kitsault and other new developments have come on stream at a time when computerization is much more readily obtainable. No comprehensive EIA should go ahead now without considering the statistical analyses which are going to be needed and how the data can be banked, retrieved, analyzed and formatted. Ideally it should be possible to obtain real-time presentations of environmental results. In practice, of course, that may be rather difficult for biological monitoring due to the need for time-consuming and non-computerizable species identifications. However, the Kitsault suspended solids monitoring is very close to real-time analysis and information feedback. We are getting close to our goal of real-time environmental quality control through data gathering, information feedback and process controls.

Finally, a concept that I must present to you is that progress in EIA has been almost unbelievably swift in the past decade. We talk in general terms about the pace of technological advance, and the hot pace applies to waste disposal and its quality control, particularly in the disposal of mine wastes. Today's results and today's understanding largely did not exist even five years ago when the Kitsault development was already conceived. This means that a decision in Alaska on the Quartz Hill development will be based on information far better than any decision made previously in B.C. and elsewhere. There is simply much more information available in 1982 than even 1981 and we understand it better. The information is also more retrievable than it used to be. It also means that decision makers and citizens who are concerned with the kinds of impacts that might occur, need to check dates of the data that are available to them. They need to check dates when prior decisions were made and prior impacts noted. The consequences of decisions taken ten years ago may well not be the consequences of similar decisions now. Circumstances have changed. More information is available. Particularly, there are different mitigative technologies available for handling the problems that we now can predict might but not necessarily will occur.

In that context, I want to close with some illustrations. Fig. 2 is a tailings outfall of the kind that was conventional about ten years ago. You see disposal of tailings to a beach. Fig. 3 shows a shallow subtidal tailings disposal system creating a turbidity

Figure 2. Surface disposal.

Figure 3. Shallow water disposal.

plume at the surface. These two obviously impacting discharges are the kinds that have occurred in the past but there is absolutely no reason why they should be built today.

The present state of the art in disposal to the marine environment is a submarine outfall (Frontispiece, and Section II Frontispiece). In both cases the bulk of the tailings remain well below the surface. My concluding point is, think time, and the rapid changes that are ongoing in technologies and understanding.

DISCUSSION

Question: I would like to make a point. Three different outfall systems have been presented as a measure of technological progress, but in fact as far as I can make out, this research was available ten years ago.

Answer (D. Ellis): Submarine outfalls have long been used for sanitary sewage and disposal of pulp mill effluent, but they had a very different set of design criteria. There they wanted a plume with initial dilution in order to facilitate assimilation. The new kind of mine outfall is designed to generate a density current and get the tailings down deep.

Comment (J. Littlepage): In the past an outfall was built and that's it. The ministry would say, "Well we tried, we can't do anything about it". Now we are at the point where we build an outfall according to some plan as to what we want it to do, then look and see if it is doing it. If it is not doing it, then a program is initiated to find out why it is not performing the way it was designed and correct it. I think we now have enough background so that we can make these corrections and given a reasonable but short time to make adjustments we can make outfalls function in a way which, five years ago, was impossible.

Question: not recorded.

Answer (D. Ellis): Beneficial use of tailings was proposed for the Jordan River copper mine on Vancouver Island. The intent was to build up a waterfront area of coarse sands which could be used for recreational purposes. It might have worked but the mine went bankrupt and closed down.

Texada used tailings for beach sands and for building

a causeway. At one particular mine in the Philippines, the coarse fraction was used for a causeway. The causeway was planted with mangroves and there is a potential for aquaculture nearby. So there are some possibilities.

Question: Dr. Littlepage was commenting on the permit requirements for milled solids above a certain level, Really, when we are talking about 1 ppm solids, there is some doubt that such a low level has an impact. I think the issue here is that mining companies really are honestly prepared to come to grips with genuine environmental concern. But is there any environmental impact from suspended solids at 1 ppm? Is that a real issue or is it not?

Answer (D. Ellis): I'm not going to answer that one. Would anyone from the audience care to answer it?

Comment (R. Burling): I wouldn't mind commenting. During discussions of the panel on the distribution of tailings in the Amax mine, we became very clearly aware that during the previous year or two instruments had become available that would allow you to measure the kind of concentrations just mentioned: 1 mg per kg of seawater. No doubt in the future we will do a lot better than that. Our immediate reaction was, "What is meaningful?" I don't think anyone has answered the question yet.

One mg/kg of suspended sediments is an extremely small amount. Community water supplies may contain more than 1,000 or even 10,000 times this amount following a mud slide into a catchment reservoir. A reasonable criterion for setting an upper limit on concentrations of tailings desirable in shallow water might be based on it being a reasonable fraction of an average concentration of the natural suspended sediments from rivers. In determining the ratio the relative amounts of potentially harmful constituents in the natural and tailings sediments could be considered.

Comment (J. Littlepage): I think another thing we have to do on that is go back to the point previously raised: that these limits have to be site-specific. If you were talking about a coral atoll where perhaps you had a 50,100 - 150,200 m photic zone, then one would have to be much more careful about the limits which one allowed in the water. On the other hand, when you have an inlet of the type of which there are so many in southern Alaska and northern B.C., where the surface turbidity naturally is running in the tens to hundreds of ml/L and the photic zone, during the summer months is in the order of probably 2-3 m, then this certainly would require a totally different set of limits. I think this is one of the areas where we should be putting some of our research interest. Determination of a level of no effect has to be tied in with other environmental factors that you measure simultaneously. "No effect" limits can't be taken completely out of context.

Question: not recorded.

Answer (C. Pelletier): In my experience I have noted that the different types of pipes have different uses. The steel pipe, rubber lined, is the best wearing pipe that we have used. It wears slower (the rubber lining) than the plastic pipe. So it becomes a matter of costs. It costs more money to have a steel pipe 1" rubber lined than to use the plastic pipe, but we have found that a plastic pipe was better than a conventional steel pipe unlined. It is strictly a cost matter why people use different types of pipe. There are different ways of disposing of tailings and some of the operations are discharging tailings from a high elevation down to the sea through a pressurized line. If you use a pressurized line, you can't use plastic. You have to use metal. High density polyethylene pipe is also reasonably new in application. I think Island Copper was the first to use it in 1974. Amax is now using it underwater, something which Island Copper did not do. Island Copper used the first 100 ft of line made from machine-lined polyurethane, which is another way of protecting pipe from wear.

Comment (G. Poling): Certainly there are differences in behavior that can be quite dramatic between various plastic pipes. There are quite significant advances being made in pipe characteristics. Previously some pipelines would wear out in a matter of weeks; but now some of these high density polyethylene pipes can last for years and pass perhaps 1,000,000 tons. But of course, they won't withstand as high a pressure as steel pipe.

Question:* In areas of strong, prevailing wind has wind

---

*Editor's note: The following questions were directed at W. Kuit following his paper on the Polaris mine. They are included here since they raise the issue of wind effects on tailings.

dispersal of tailings ever been considered? Is there a means to take sufficient moisture out so that the tailings could become wind-borne?

Answer (W. Kuit): In our land disposal concept, even though it is cold in the arctic, tailings deposits will dessicate in strong wind and disperse. That was of great concern to us. It was one of the environmental reasons why land disposal was rejected. This would have spread tailings over considerable distance and impacted what little wildlife there is.

Question: Has wind dispersal actually been tested to the point that it has been rejected?

Answer (W. Kuit): On environmental grounds, wind dispersal of tailings is out of the question.

Question: With the wind blowing towards the sea instead of over the land, would it still be out of the question?

Answer (W. Kuit): If tailings were to be disposed of in the sea, they would have been discharged underwater in the nearby Crozier Strait, not on the surface by wind dispersion.

Question:* Will you comment on the orientation of the outfall line at Kitsault with respect to the longitudinal access of the inlet and any leveed channel formation?

Answer (J. Littlepage): The outfall pipe terminates at 50 m on a reasonable slope. It extends essentially at right angles into the inlet. The tailings plume moves out from the outfall pipe, turns to the left and heads down the inlet. We have attempted to look at this with submersible cameras, but there is a layer of turbidity over the bottom, so we haven't been able to identify any channelization. We will do our first soundings this summer to see if we can detect it. The outfall hasn't been in

---

*Editor's note: The following questions were directed to J. Littlepage following his paper on the Kitsault mine. They are included here since they summarize some basic information about outfall design, operation and tailings related environmental monitoring.

operation long enough yet to allow us to do this. I asked a question earlier about how long you would expect these channels to exist. There was no evidence from our surveys in 1980 or 1979 (when we did a high resolution echo-sounding of the area) of any prior channelization. So if there was channelization from the older outfall, it had been obliterated by that time.

Question: Bearing in mind the low suspended solids levels and low metal levels in the solid, how do you confirm that your remote sensed records are of tailings?

Answer (J. Littlepage): There are two ways that you can do this. When you see the development of the upper field plumes in record after record, and you can trace this very low level material back to the outfall region by the records, then there is little doubt what they are. Also the material settles out of suspension very rapidly and when the outfall is down for one reason or the other, the record disappears. Also, one can do metal analysis on the material. The problem that we are encountering here is that because they are in such dilute suspension, we have to filter in the order of 200-300 L of water to get enough to get a reliable estimate. This is very difficult. It is basically the continuity of measurements that allows you to confirm that the records are of tailings.

Question: Do you have any idea what the natural thickness of the tailings is on the bottom, and also do you have any idea of the impact of tailings on the seabed?

Answer (J. Littlepage): Unfortunately, I have to answer in the negative to both those questions. We have not been in operation long enough to do repeat surveys in the inlet. These will be coming up this summer and we hope to have some answers then. From the amount we are discharging, I have doubts whether we will be able to measure the amount of suspended material on the bottom away from the outfall, because it is in 350 m of water. It is an extremely small amount of accumulation. We do know from some evidence from local fishermen that turbidity currents do exist, but whether they are natural or from tailings we don't know. There are turbidity currents in the area and they have caused some disruption of equipment that was put down on the bottom.

SECTION III

REGULATORY ACTION

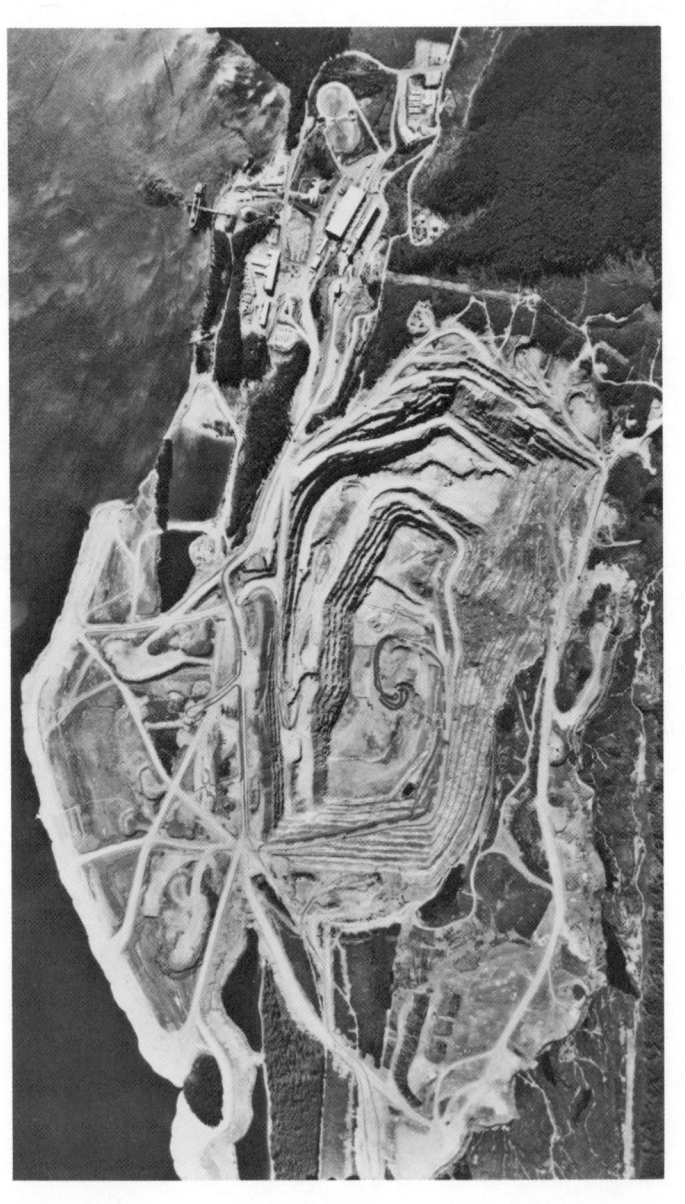

Section III Frontispiece. Aerial view of Island Copper Mine showing the pit, and to the right the mill complex. At the water's edge there is a bulk freighter at the loading wharf. Left along the shoreline the outfalls form a small projection and further is first the emergency tailings pond, and then the shoreline waste rock dump (B.C. Government air photo).

INTRODUCTION

REGULATORY ACTION

The single paper in this section presents the results of a government review of mine waste disposal by Island Copper Mine. It was not the first time that a regulatory authority considered the mine's discharge system and environmental impact. Island Copper was the subject of a public inquiry held in 1970 at the time of issuing a Discharge Permit. Two public inquiries were held by the province in 1972 and 1978 into the setting of discharge standards for the mining, mine-milling and smelting industries. Submissions from and about the mine were presented.

Dr. Buchanan's chapter is thus of particular interest in that it summarizes the results of the fourth of a ten year series of government inquiries which have considered the environmental impact of a particular mine. The environmental performance of this mine has been scrutinized like no other mine before it.

It should also be noted that the Waldichuk-Buchanan review was a joint federal-provincial action. In Canada, environmental protection is an overlapping responsibility, dealt with largely through the authority of the federal government over fisheries, and the provincial governments over lands and resources.

Provincial policy in British Columbia concerning environmental impact assessment is that the discharger will normally be responsible for monitoring and its interpretation. Island Copper Mine was therefore required to initiate such a program in 1971, as part of the terms of its Discharge Permit. It had earlier initiated less extensive investigations in 1970. The mine has undertaken these investigations using a rather unusual procedure. Sampling and analysis has been conducted primarily by in-house environmental scientists and technicians (working from the mine's custom designed small sampling vessel), but under the advice of a university engineering and scientific consulting board. These scientists also have initiated, on occasion, graduate student research training projects in the receiving area. Some of the biological sampling and analysis was contracted out to specialist consultants, e.g. the benthic transect investigations of 1974, zooplankton identification and quantification.

There are several chapters in this book which arose from the Island Copper Mine environmental program. There is Chapter 5 (university research) and Chapter 6 (in-house monitoring). In addition, some of the material presented in Chapter 2 by G. Poling, Chapter 3 by R. Burling and the Summary in Section II, are derived from investigations at and on behalf of Island Copper Mine.

For an understanding of the complexity of the marine disposal problem at Island Copper Mine, and the enormous amount of monitoring and assessment data that has been logged, archived and can now be retrieved, it is important that Chapters 2, 3, 5 and 6 be read in association with the following chapter. To some extent, there is disagreement between the authors concerning the conclusions drawn and opinions expressed, and these should be noted. On the other hand, there are substantial areas of agreement on the extent and intensity of the actual impacts.

It should be noted that the tailings disposal system of at least one other marine discharging mine has been reviewed by government inquiry. This is the open pit molybdenum mine at Kitsault. Although Kitsault was the subject of a paper presented at the Ketchikan symposium, it has not been prepared for publication at this time. However, descriptions of the government review process and results are available (1, 2).

REFERENCES

1. Burling, R.W., J.E. McInerney and W.K. Oldham. "A Technical Assessment of the Amax/Kitsault Molybdenum Mine Tailings Discharge to Alice Arm, British Columbia", Report to Minister of Fisheries and Oceans Canada (1981).
2. Ellis, D.V. "Kitsault, B.C. - Technical Communication in a Non-technical World", Mar. Poll. Bull. 13(3): 78-80 (1982).

CHAPTER 7

A GOVERNMENT INQUIRY INTO MINE WASTE DISPOSAL
AT ISLAND COPPER MINE

R.J. Buchanan
B.C. Ministry of Environment
Parliament Buildings
Victoria, B.C.
Canada V8V 1X5

The author and M. Waldichuk were appointed by federal and provincial inter-agency agreement to review and report on the dispersal pattern of tailings from Utah Mines in Rupert Inlet, the derived environmental changes, and their significance. Selected photographs of conditions in Rupert Inlet in July 1978 are presented and discussed.

Our review recommended improvements in waste disposal systems and monitoring. These included:

·investigation of the relationship between turbidity and slumps of accumulated tailings near the outfall pipe, and of the feasibility of changes in the outfall;

·sampling water at several depths in the water column and in the surficial sediments and analysis by the most sensitive available technique for selected dissolved metals to determine the availability of these elements to marine organisms;

·investigation of the effect of mine tailings on benthic organisms along transects from deep water to the intertidal;

·regrading inactive waste rock dump faces to a shallow slope in the intertidal and subtidal area to promote colonization by algae and invertebrate animals;

·monitoring the bioaccumulation of metals by standard organisms such as blue mussel in several locations;

·continual upgrading of analytical techniques used in

the monitoring program; and
‘improvements in monitoring the biota and metal content.

The panel also recommended short and long-term research related to mine tailings, in five subject areas.

INTRODUCTORY REMARKS

The Minister of Environment for Canada and the Minister of Environment for British Columbia requested in March 1978 a joint review of the effects of sea disposal of mine tailings from the Island Copper Mine at Rupert Inlet, British Columbia, owned by Utah Mines.

The author and M. Waldichuk of Fisheries and Oceans Canada were appointed by the Deputy Ministers of Environment for Canada and British Columbia to "review and report on the facts with respect to (a) the dispersal pattern of tailings from Utah Mines in Rupert Inlet; and (b) the environmental change which is taking place, and its significance". The review was carried out in collaboration with those scientists who had been working on the subject and in consultation with other scientists and sources of knowledge about relevant subjects. The review included a lengthy series of personal and telephone interviews during 1978 and 1979, a site visit in July 1978, extensive literature review, examination of unpublished data, and checks on chemical analyses (1).

OBSERVATIONS

Island Copper Mine is located immediately beside the north shore of Rupert Inlet, which is the eastern extremity of the Quatsino Sound/Neroutsos Inlet/Quatsino Narrows/Holberg Inlet/Rupert Inlet fjord complex located near the north end of Vancouver Island, British Columbia (Section I Frontispiece and Chapter 6, Fig. 1). It is served by and most employees live in Port Hardy, the nearest major community.

The mine milling complex and concentrate storage are adjacent to the shore, and the concentrate is loaded onto seagoing ships nearby (Section III Frontispiece). The waste rock is deposited mainly in a marine dump along the shore of Rupert Inlet to the east of the mill complex and adjacent to the pit, in an area leased for that purpose from the province. The mine tailings are thickened in circular tanks, the clarified water is returned to the milling process and the thickened tailings are piped to a submarine discharge system located beside the milling complex (Frontispiece). The tailings are mixed with an equal volume of seawater in a tank which provides for removal of gas bubbles prior to discharge, and the mixture is discharged through a pipe at a depth of 50 m. There is an emergency tailings pond located along the shore nearby.

The shoreline of Rupert Inlet, including the pilings of the copper concentrate shiploading dock and the works of the tailings disposal system (Fig. 1), has a very dense and diverse community of algae and invertebrate animals. It is postulated that the exceptional density of growth in this inlet system is made possible by the relatively low freshwater discharge which normally occurs. In years of unusually high freshwater runoff, local residents report substantial mortality of intertidal and shallow-water organisms. The intense mixing caused by tidal flows through Quatsino Narrows assures an atypically small density gradient from the surface to the bottom in Rupert and Holberg Inlets. This small density gradient, coupled with the mixing forces of the tidal flows, probably imposes the most important limitation on planktonic primary production: light limitation owing to deep mixing relative to light penetration. In these circumstances, the relative importance of primary production by attached plants (Fig. 2) and of organic matter carried into Rupert

Figure 1. Shoreline growth by the mill extends onto the tailings line where it penetrates underwater.

Figure 2. Attached plants grow at the Rupert Inlet end of
the Quatsino Narrows sill. (D. Ellis photo).

and Holberg Inlets by tidal exchanges will be greater than
is typical for fjord systems having large discharges of
turbid freshwater at their head ends and deeper sills at
or near their mouths.

The waste rock dump contains a large amount of fine
material which is subject to erosion and sorting by water
along the face of the dump. The steepness of the dump face
also contributes to the instability of the shore materials
even when the face is not being extended. The result of
this chronic instability of the surface materials is an
inability of marine organisms to colonize the rock dump
face as effectively as in other places. The mine
experimentally regraded a short segment of the dump face
to approximately 10 degrees slope, and the resultant
improvement in the colonization was dramatic in spite of
the high proportion of fine grain materials present (Fig.
3).

The shore directly across Rupert Inlet from the waste
rock dump and the tailings discharge works is a broad
cobble and gravel beach. This beach was obviously very
productive, being covered with dense colonies of various
small seaweeds (Fig. 4). There was an apparent abundance
of invertebrate animals as well.

The steeper rocky shore segments showed very dense
colonization by algae and attached invertebrates in the

Figure 3. Experimental regrading of the waste rock dump face added growing area for shoreline plants and animals forming the ecosystem base for juvenile salmon.

Figure 4. Dense growths of algae cover the broad cobble and gravel beaches of Rupert Inlet opposite the mine site, and form habitat for invertebrate animals.

intertidal zone. There were numerous starfish, crabs and other macroinvertebrates among these attached organisms (Fig. 5).

On one short stretch of beach adjacent to the east side of Hankin Point, which is directly opposite Quatsino Narrows and exposed to the strong tidal currents and associated up-welling of suspended tailings, a deposit of tailings was accumulating and being colonized by eelgrass, a species not previously found at this site, and by the small green seaweed Enteromorpha (Fig. 6). Other organisms indigenous to this site did not appear to have been displaced where tailings were not greater than a few centimeters. The greatest depth of tailings observed was approximately 15 cm over the original cobbles and coarse gravel. I am informed that the accumulated tailings are intermittently carried away and subsequently redeposited along this section of beach.

During the site tour, there was visual evidence of the resuspension of tailings in the Quatsino Narrows area. Nevertheless, the intertidal and subtidal community of seaweeds was very impressively dense (Fig. 7). There was no difficulty catching several rockfish for use as crab bait, either.

Two standard crab pots were set out a short distance from the north shore of Rupert Inlet at two separate sites

Figure 5. Rocky shore in Rupert Inlet also shows abundant growths of algae zoned into bands of different species typical for the area.

Figure 6. Eelgrass and seaweeds colonize deposits of tailings at Hankin Point in Rupert Inlet opposite Quatsino Narrows.

Figure 7. Shoreline growth in Quatsino Narrows through which resuspended tailings pass. (D. Ellis photo).

west of the mine site, the nearest being about one kilometer (0.6 mile) from the mine dock. These were recovered approximately 24 hours later, and they averaged 22 Dungeness crabs in each (Fig. 8). Specimens from among these were taken for chemical analysis at an independent laboratory on behalf of the review panel. The analyses did not reveal any unusual accumulation of metals in the crabs.

Examination of the bottom sediments of Holberg Inlet in a site not subject to tailings deposition and in a site subject to moderate deposition rate revealed that the invertebrate animal community in the latter was more dense than in the former. The sediments in the unaffected site were much coarser, consisting of sand, fine gravel, shells of bivalve molluscs and some wood debris. The sediments in the site exposed to a moderate tailings accumulation rate were apparently pure tailings, and were light in color (Fig. 9). This comparatively dense population of benthic organisms in the tailings was surprising in light of the low organic matter content that could be expected in tailings. It is postulated that the settling tailings particles adsorb dissolved organic matter from the water column, thereby transporting a food supply to the bottom from the overlying water at a rate exceeding that which would occur without the "rain" of tailings particles. The descending particles may also aggregate with suspended

Figure 8. A crab pot with its catch after fishing for 24 hours about 1/2 mile west of the mine's loading dock.

Figure 9. Corer liners containing gray tailings (left) and
darker natural sediments (right) from south of
Coal Harbour and Station 24, respectively (see
Chapter 6, Fig. 1).

organic particles and enhance the settling rate of
particulate organic matter within Rupert and Holberg
Inlets. This would enhance the ability of these two inlets
to act as traps for dissolved and particulate organic
matter carried in by tidal exchange fron Quatsino Sound.
      During our site visit, the resuspension of tailings
in the vicinity of Hankin Point, opposite Quatsino
Narrows, was evident, as it was in Quatsino Narrows itself
(Fig. 10). It was apparent that the resuspension of
tailings is not always as pronounced as shown by Goyette
and Nelson (2) in photos taken in August 1973 and May
1974. It was also apparent that resuspended tailings,
particularly the finest sized fraction, can be transported

Figure 10. The junction of Rupert Inlet with Quatsino Narrows where resuspension occurs. The mine is visible in the distance. (D. Ellis photo).

substantial distances by the tidal currents that are so strong in this particular inlet system.

The Rupert Inlet/Holberg Inlet/Quatsino Sound system is an extremely productive system for seaweeds and other littoral zone marine organisms. The Island Copper Mine apparently has not altered this obvious fact.

CONCLUSIONS

We arrived at several conclusions, as follows:

1. It was not possible to reach a firm conclusion as to whether the resuspension of tailings in Rupert Inlet was due primarily to erosion of deposited material by tidal currents or to sloughing of banks of accumulated material near the outfall pipe and associated turbidity currents.
2. Mine tailings from the Island Copper Mine are altering the Rupert Inlet/Holberg Inlet ecosystem, primarily by smothering benthic communities on the bottom.
3. The recolonization of the tailings-covered bottom should be quite rapid when tailings discharge ceases.

4. Resuspension of tailings is most pronounced on the flood stage of tide during periods of large tidal range. Some of the resuspended tailings is deposited in the intertidal and shallow subtidal areas of Rupert and Holberg Inlets.

5. Tailings deposited in the littoral zone do not appear to be toxic to the marine plants and animals which colonize these deposits.

6. The waste rock dump extending along the north shore, and out into the deep waters of Rupert Inlet, is occupying a significant fraction of the intertidal and shallow subtidal habitat formerly available within Rupert Inlet as a rearing area for juvenile salmonid fish. The unstable actively growing face of the rock dump is not well colonized by organisms, and contributes suspended particulate matter to the inlet.

7. The effects of the mine's operations, if any, on the commercially important fish and shellfish populations or catches could not be documented by available records.

8. There was no evidence of appreciable leaching of copper, molybdenum, cadmium, lead or zinc into the water column, and there was a lack of data on the concentration of these metals in the interstitial water of the tailing deposits on the bottom. There was some indication of leaching of manganese from the tailings, from both laboratory and in situ measurements.

9. Metals were not being bioaccumulated appreciably except in some localized areas of high exposure.

10. The environmental monitoring program, though remarkably comprehensive, did not enable determination of the effects of tailings and waste rock on the commercial and recreational fisheries.

11. The long-term effect of the tailings in the Rupert Inlet/Holberg Inlet system cannot be predicted, particularly with regard to the diagenic changes that will result from additions of organic matter after deposition ceases, and the consequent effects on leaching of metals.

12. There was no basis for changing to any of the other alternatives for tailings disposal.

13. It might be possible to design an outfall system which would reduce the incidence of tailings resuspension.

14. The impact of the turbidity from suspended tailings and fines from the waste rock dump on light penetration and primary production is

uncertain.
15. It is not known whether turbidity affects
    migratory behavior of salmonids and other species.

RECOMMENDATIONS

We recommended improvements in waste disposal systems
and in monitoring. These included:

- investigation of the relationship between visible
  surface turbidity and slumps of accumulated tailings
  near the outfall pipe, and investigation of the
  feasibility of reducing the incidence of suspended
  tailings through changes in the outfall;
- careful sampling of water at several depths in the
  water column and in the surficial sediments and
  analysis by the most sensitive available technique
  for at least dissolved Cu, Pb, Zn, Cd, Mo, As, and
  Mn, to determine the availability of these elements
  to marine organisms;
- investigation of the effect of mine tailings on
  benthic organisms along transects from the deepest
  part of the inlet to the intertidal;
- regrading inactive waste rock dump faces to a slope
  no steeper than about 10 degrees in the intertidal
  and subtidal area to promote colonization by algae
  and invertebrate animals (Fig. 11);
- monitoring the bioaccumulation of metals by standard
  organisms such as blue mussel in several locations
  exposed to high, intermediate and low amounts of
  tailings;
- continual upgrading of analytical chemical
  techniques used in the monitoring program; and
- improvements in the examination of sediment cores
  and settling plates used for monitoring the biota
  and metal content.

We also recommended short- and long-term research
related to mine tailings, including:

- studies of the direct and indirect effects of mine
  tailings on important organisms;
- studies on the long-term effects of tailings,
  particularly after discharge ceases, to determine
  changes in metal content of particles and
  interstitial water and effects of subsequent
  sedimentation of organic matter;
- studies on the release of metals from tailings into
  seawater under various combinations of conditions

Figure 11. Regraded inactive dump face with reclaimed flat
surface, and marine growth below the high tide
level. (D. Ellis photo).

such as salinity, temperature, pH, and chemical
pretreatment of tailings;
·measurements of ambient light intensity and primary
production at various depths in areas affected by
suspended tailings and in unaffected areas with
otherwise comparable conditions; and
·studies on the effect of tailings turbidity on adult
salmon migration in comparison with the effects of
natural turbidity.

REFERENCES

1. Waldichuk, M. and R.J. Buchanan. "Significance of
   Environmental Changes due to Mine Waste Disposal into
   Rupert Inlet", Federal/Provincial (British Columbia)
   Review of the Mine Waste Disposal Problem in Rupert
   Inlet, B.C. Fisheries and Oceans Canada, and British
   Columbia Ministry of Environment (1980).
2. Goyette, D. and H. Nelson. "Marine Environmental
   Assessment of Mine Waste Disposal into Rupert Inlet,
   British Columbia", Surveillance Report EPS PR-77-11,
   Environmental Protection Service, Pacific Region,
   Environment Canada (1977).

DISCUSSION

Question: You made a lot of suggestions and recommendations. Would you suggest that the monitoring program you investigated in 1978 was actually quite a sophisticated one that had been going on for eight years prior to your suggestions? You recommended that mussels be tested for metals. Wasn't this already being done for five or six years prior to your study?

Answer: It escaped my attention. That is something I didn't know. If I knew, I had forgotten.

Question: Did you do any data collection during your review?

Answer: We relied on information provided to us by other people who had done a fair amount of work on the subject. We did not actually carry out any of our own measurements. The only exception was analyses performed on crab samples collected during our site visit, at the Fish Inspection Laboratory of Fisheries and Oceans Canada and by the Geological Survey of Greenland, for heavy metals.

Question: Are you saying that the discharge of tailings into Rupert Inlet does no harm to Rupert Inlet productivity?

Answer: I can't say it hasn't affected it, but it certainly hasn't been calamitous.

Comment (D. Ellis): I would just like to comment on that last question. There have been three reviews now on the phytoplankton in Rupert Inlet. Two of these were done by government scientists and one by a group of university environmental scientists. So far it has not been possible to tell whether there has been a reduction in primary production or not, nor to relate it to the natural variation. If there is an impact it is less than the natural variation from year to year. That is the kind of review that is going on to check out the problems that might develop.

Question: My question is that when there is a public inquiry with some recommendations, does the provincial government intend to take any action as a result?

Answer: I understand that the provincial Waste Management Branch has been in communication with the company, as has the federal Environmental Protection Service. From the point of view of the Waste Management Branch, I understand that the mine intends to carry out a number of the recommendations. Perhaps they have already implemented some of them. They are not able to grade the whole waste dump face now, but they have graded something like 12% (Fig. 11). They will get about 85% of it regraded by the end of 1985. I don't think the company has come to any agreement with the Waste Management Branch yet on what is actually going to be done. It was not too long ago that they began to discuss the matter. I don't know what the federal agency is saying at all.

Question: One of the conclusions that you can't help but draw when you look at the deposition of tailings is that they stabilize and recolonize very rapidly. There seems to be some sort of a natural underwater reclamation. Would you like to comment on that?

Answer: Yes, I was impressed by that too. It looked like barren sand, but there was a remarkable population of creatures in it. Some of them, of course, may be living on material in suspension in the water overlying the sediments, but some of them would not be. I speculate that the tailings, especially the finer fraction, may be adsorbing organic matter from the water column and carrying it to the bottom, thus providing a greater supply of carbon in the sediments than would be the case if there were no rain of fine tailings.

REGULATORY ACTION

G. W. Poling
Department of Mining and Mineral Process Engineering
The University of British Columbia
Vancouver, B.C.
Canada V6T 1W5

It is a pleasure for me to be asked to review Dr. Buchanan's presentation.

Several of us have been involved in the Rupert Inlet Island Copper tailings disposal system. I believe this to be one of the most highly studied marine disposal schemes any place in the world, and it certainly deserves a great deal of serious scientific study.

The review that Drs. Buchanan and Waldichuk undertook in March of 1978, which resulted in the report that some of you have seen, was really sparked, I think, by headline charges throughout newspapers in British Columbia in January of 1978, to the effect that the mill tailings being discharged into Rupert Inlet were killing the inlet. A very positive aspect of this investigation was that it was, to the best of my knowledge, one of the first chances for cooperation between Canadian federal and provincial regulatory authorities, at least in British Columbia. I think this was a very positive step and I hope that this cooperation will improve in the future.

Dr. Buchanan and Dr. Waldichuk extensively grilled the mining company. They also grilled all of the scientists and engineers who were involved in the monitoring program. Some of us who were working on a research contract within the University of British Columbia were not happy to have our research results questioned by the regulatory agencies, let alone in the public domain. This is something that a scientist and an engineer should think about. To have to explain your written reports, published papers and theses to someone else in an environment that is similar to a court of law, is quite an experience and I highly recommend it. After you get over the initial shock of having to go through the routine, it really makes you stop and think about having to explain your results in terms that other people can understand.

The review substantiated the fact that the results being accumulated by this mining company were highly reliable. In fact, the laboratory in use at Island Copper is probably one of the best in marine monitoring. Examination of the data on this marine inlet indicates that once in a while there is turbidity in the upper reaches of the inlet. Initially, this was not expected. Some question remains about how significant that turbidity is and how often it does occur. I can assure you that there have been significant efforts to establish these kinds of data.

One of the most recent projects that I was personally engaged in was an attempt to get, during ten years of operation, satellite data which would indicate statistically how often turbidity does occur in the upper reaches of Rupert Inlet. After a good attempt, we concluded that the cloud cover over the north end of Vancouver Island mitigated against statistically significant data from the satellite. Several reports of turbidity in the local newspapers were based on infrequent occurrences. These gave the public a real misconception of the situation in Rupert Inlet.

We still do not have answers to all these questions, but what you are hearing about the Rupert Inlet case from the mine operators, the consultants for the mine operators, those of us in academia who have been involved in research, and the regulatory people, is that a diverse group of people are working together trying to find out the true facts about this inlet. To date the study of Rupert Inlet by Drs. Buchanan and Waldichuk shows that although there is an environmental impact on the inlet system, this impact is not severe enough to consider a forced change to an on-land impoundment system.

Drs. Buchanan and Waldichuk are to be congratulated for the review that Dr. Buchanan has presented here today.

SECTION IV

QUARTZ HILL

Section IV Frontispiece. Quartz Hill is surrounded by a triangle of snow-capped ridges in the mid-ground of this photograph. In the distance the two potentially tailings receiving fjords can just be seen. Boca de Quadra is to the left of Quartz Hill and Wilson Arm/Smeaton Bay to the right.

This final section provides some of the relevant environmental baseline information that has been obtained to date by the developers of the Quartz Hill ore-body. A selection has been made, based on the present understanding of the greatest or most probable of the potential environmental risks. These are not the same. The severity of the risk depends very much on the sensitivity of the area and its resources. The probability of a risk depends in part on the nature of the ecosystem and ore-body, but also on sound operational controls, well implemented.

Our present understanding of environmental risks at mines can be grouped in four categories: water column turbidity; seabed smothering; toxicity; and trace metal contamination.

Water column turbidity, apart from being displeasing to those who remember previous clarity, can affect primary biological production with derived impacts on fisheries. Fortunately, water column turbidity can be minimized to virtually nil in calm water, i.e. invisible, and detectable only by the most sensitive of new instruments. The tailings resuspension and up-welling at Island Copper Mine appears to be a localized and site-specific problem due to an irregular descending tidal jet over the shallow sill impinging on a rockface in its path.

Seabed smothering impacts fish and shellfishery resources either directly or through an influence on their feedstocks or habitat. If too many tailings are deposited, the physical oceanography of the fjord may be changed, causing wider ecosystem effects. Fortunately, we know now that smothering need not be massive, that natural reclamation occurs underwater and possibly can be assisted (Chapters 6 and 7). We can also calculate volumes of receiving fjord basins and tailings to be discharged so as not to overload the natural fjord pit.

Toxicity is a recurring fear, but is probably the

275

least probable risk of all. Bioassay tests of locally important species such as salmon can be made, daily if necessary, to ensure that the discharge process chemicals and tailings are not acutely poisonous. There are an infinite number of long-term bioassays that can be developed to test whether persistent effluent components may have an impact on important species. Finally, biological monitoring close to the point of discharge checks to see if species are dying or not growing healthily.

Trace metal contamination is an underlying concern, since its effects can be insidious. Yet, as explained in Chapter 4, there have been no recent cases of public health or fishery contamination associated with marine discharging mines. The known health catastrophes resulted from wastes of chemical plants, or smelters discharging to river water. Potential contamination needs monitoring properly based on an understanding of seawater chemistry and interactions with tailings, and the potential food chain routes for biomagnification to levels impacting on humans, fishery stocks or wildlife stocks.

The selection of the topics for the Quartz Hill session was based on the high priority assigned to providing information related to these potential impacts. First a chapter describes the mineralogy of the Quartz Hill host rock, since this will determine the chemical composition of the mill tailings. The next chapter deals with the physical oceanography of the two fjords, either of which might be designated to receive the discharge if marine tailings disposal is permitted and adopted. The physical oceanography, in conjunction with an understanding of the behavior of tailings in seawater, will lead to predictions of tailings dispersion, good outfall location and design, and minimization of water column and seabed impact. The third chapter in this section discusses chemical and geochemical oceanography, which are important in assessing the potential for trace metal activation in the fjord, and subsequent dispersion. Finally, a chapter on the seabed organisms (the benthos) leads towards an understanding of food chain interactions, and hence potential bioaccumulation and biomagnification of trace metals, but also towards an assessment of fishery losses.

It must be noted that there is considerable ongoing investigation of water column life: the phytoplankton, zooplankton, and fish (especially salmon and shellfish). These have not been included, in part due to lack of room in this book for the large amount of biological data already collected (even when summarized), and in part because the basics are either well-known (salmon stocks) or the studies are not near completion. Also, considerable investigation of the terrestrial and river ecosystems is taking place in an attempt to assess the potential impact of the mine on wildlife, and on salmon during their river spawning and residential stages. These subjects also are not reported here, since they are only likely to be impacted by tailings if land disposal is adopted.

CHAPTER 8

PETROLOGY AND PETROGRAPHY OF THE QUARTZ HILL
MOLYBDENITE DEPOSIT

J. R. Snook
    Department of Geology
    Eastern Washington University
    Cheney, Washington 99004
    U.S.A.

The purpose of this paper is to set the stage for a
discussion of potential tailings disposal problems related
to the Quartz Hill molybdenum deposit. Tailings disposal
is concerned primarily with the grain size, composition,
and quantity of the discarded mine waste material. It is
necessary, therefore, to understand the composition of the
rocks and their origin in order to predict what the
tailings are most likely to be during the mining
operation. In an ore deposit averaging less than 0.3% ore,
large volumes of waste rock are generated and thus even
small quantities of uneconomic elements can accumulate in
the waste rock and be potentially hazardous. Quartz Hill
can be compared to many other molybdenite deposits and
even the crust of the earth itself and be shown to be
exceptional in both tonnage of molybdenite and relative
lack of toxic elements.

Quartz Hill is located approximately 70 km east of
Ketchikan, Alaska, along the western margin of the Coast
Range (Fig. 1). Until about twenty years ago, most mining
properties were found by little old bearded gentlemen with
burros, i.e. prospectors. It has only been in the last
fifteen to twenty years that science has really played a
significant role in finding ore deposits. Quartz Hill is
one of these (1). U.S. Borax began a stream sediment
sampling program for southeast Alaska in 1974. They found
a stream sediment sample containing 168 ppm molybdenum in
the Quartz Hill area. This was very high compared to other
areas. Later in the season they moved two miles upstream
from where the sample was taken and made the discovery of

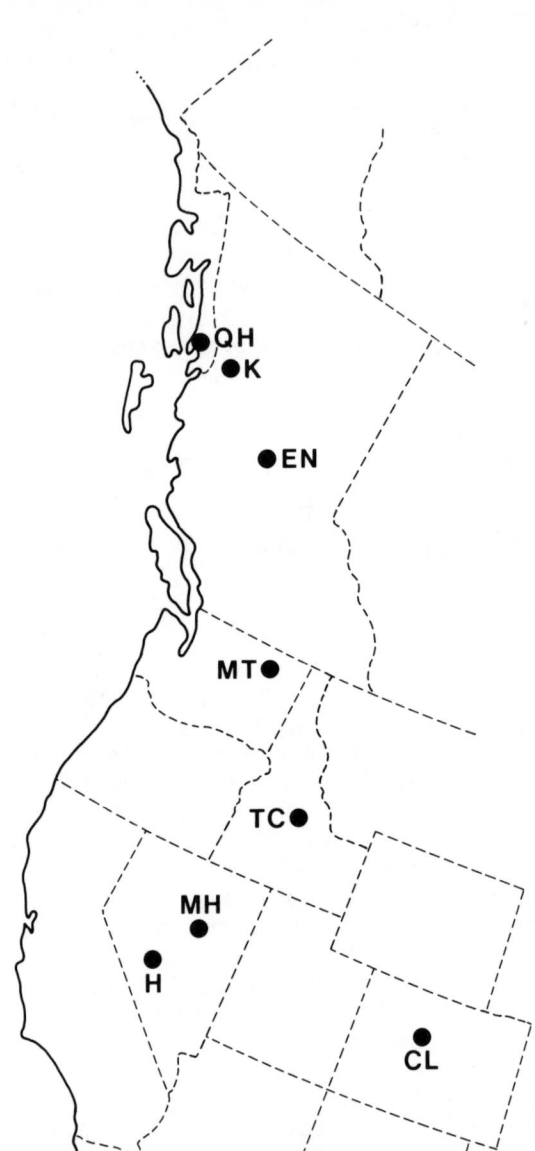

Figure 1. Map of western North America showing the location of Quartz Hill (QH), Kitsault (K), Endako (E), Mt. Tolman (MT), Thompson Creek (TC), Mt. Hope (MH), Hall (H), and Climax (CL).

ore grade molybdenite in the outcrop from which the sediment had been derived. Since then, Quartz Hill exploration has been primarily barge and helicopter supported.

The Quartz Hill deposit is related to an igneous stock that has intruded Jurassic and Cretaceous rocks (2). This stock was part of a much larger intrusive to the south, which is adjacent and somewhat lower in elevation. As the magma stopped moving, it began to crystallize from the outside in, as most intrusions do. This produces coarse-grained (less than 1 mm) igneous intrusive rock, not too far below the surface. As crystallization proceeds the pressure in the system increases. As it does, it tends to suppress the crystallization temperature of the constituent minerals, which means that it takes basically lower temperatures to get the magma to crystallize. Because the pressure builds up in the magma chamber itself, the strength of the overlying material is overcome and a crack develops to the surface, allowing the pressure in the system to be relieved, at least temporarily. As the pressure suddenly decreases in the magma, the temperature of cystallization for the constituent minerals is suddenly raised above the existing magmatic temperatures and thus the magma crystallizes instantaneously. It then produces two different textures, a very coarse texture from the original crystallization and a very fine-grained texture (less than 0.05 mm) from rapid crystallization (quenching), both shown in Fig. 2.

The depressurization or degassing of the magma chamber, a temporary event, tends to produce a breccia zone which is a zone of cracking and breaking of the constituent rocks. The still mobile portion of the magma fills in the cracks and quenches. This gives you some idea of the events that started about 27 million years ago (3).

The quenching or degassing event is probably coincidental with the separation of water from the magma. In other words, the magma undergoes a certain set of physical principles when its crystals are forming directly from the magma. When you get an aqueous phase separating from the magma, the crystal end products are somewhat different. The water phase does a couple of things. It produces alteration of the unstable minerals already produced and it is also the carrier for the ore mineral in the deposit. The next significant phase is the development of a series of veinlets filled with quartz and molybdenite cutting across all of the previously formed rocks. Following this period of mineralization is the last phase, which involves the intrusion of a series of post mineral dikes. They vary in composition from zeolites, the mineral stilbite and some of its relatives, to lamprophyre dikes

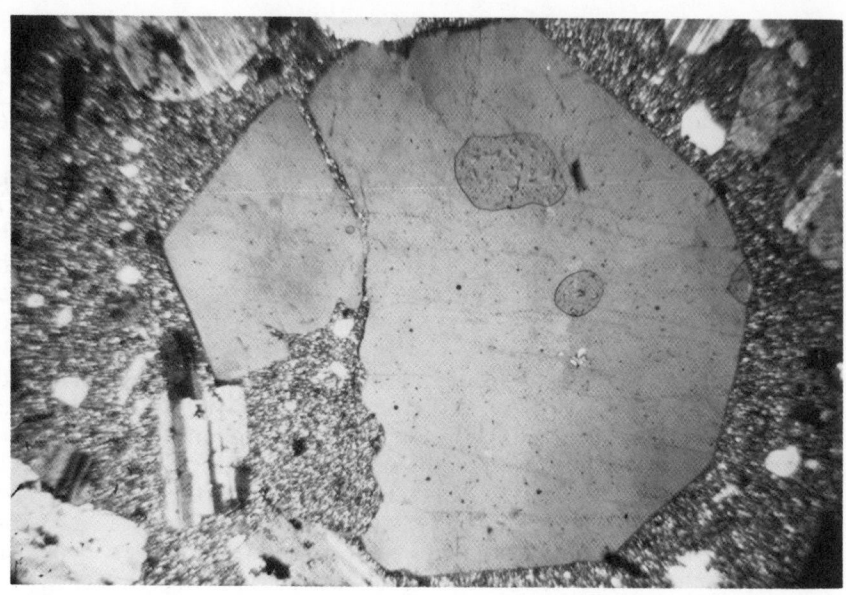

Figure 2. Photomicrograph of quartz latite porphyry which shows contrasting grain size of large quartz and feldspar phenocrysts and the very fine-grained matrix (< 0.03 mm). 24 X. Crossed polars.

(4) and some quartz latite dikes. The crystallization of a border zone and then the degassing of the intrusive is probably repeated several times because there is good evidence that the mineralization phase has been repeated more than once.

At the present time the top of the Quartz Hill stock is exposed at the surface over an area approximately 3-5 km across. The molybdenite veinlets are also exposed, indicating that the ore deposit has already come to equilibrium with the present environment. Ore grade samples are found in the local streams from boulder size through fine sand size. The mining process will not materially change the toxicity of the constituents of the Quartz Hill stock from the natural setting except that molybdenite will be extracted and thus reduced in quantity downstream. This is one aspect illustrating the uniqueness of Quartz Hill compared to active molybdenite mines or other developing deposits. Quartz Hill can also be favorably compared to other deposits when considering:

  ·quantity of ore;
  ·stripping ratio;
  ·lack of alteration products;

˙lack of other lithophile elements; and
˙restriction of mineralization to the intrusive.

The quantity and quality of ore for the deposits
located in Fig. 1 are shown diagrammatically in Fig. 3.
The quantity of ore and overburden is shown horizontally
and the quality of the ore (% MoS_2) is shown vertically.
From this you can see that Climax has a much greater
quality of ore than many deposits, but not nearly as much
quantity as some. Mt. Tolman has 0.09% MoS_2, but there is
quite a bit of it. The amount of overburden that must be
removed before you can mine the ore becomes a significant
factor. In the early phases of mining at Quartz Hill there
will be virtually no overburden but the later phases will
involve some stripping which averages 0.43:1 overall. This
compares to a stripping ratio of 5:1 at Thompson Creek and
2:1 at Kitsault. Two bars are shown on Fig. 3 for Quartz
Hill. They represent two different mining possibilities,
one making use of the greatest quantity of ore and the
other only the high grade ore. It is necessary to remember
that the quality of ore represents that part of the
processed material which you keep. In other words at
Quartz Hill, if you process 1,500,000,000 tons of ore
containing 0.125% MoS_2, you must discard 99.875% of the
material processed. In addition, you must discard the
overburden waste rock. You can see from Fig. 3 that Quartz
Hill represents the recovery of the greatest amount of
molybdenum and the mining of the least amount of rock to
get it, compared to the other properties shown.
The freshness of the rock and the restriction of the
ore to the intrusive at Quartz Hill is unusual compared to
other deposits. This affects the predictability of the
mining and milling process. Most molybdenite deposits are
characterized by strong alteration of the original
minerals. Altered rock is structurally weak and may be
mineralogically complex, including the enrichment of many
lithophile elements. The principal ore-bearing rocks in
the Quartz Hill deposit have a very simple mineralogy.
Quartz, K-feldspar, and albite-oligoclase comprise over
95% of the rock. Minor and accessory minerals include
biotite, chlorite, allanite, apatite, zircon, calcite,
dolomite, stilbite, fluorite, rutile, sphene, epidote,
clay minerals, sericite, molybdenite, pyrite, magnetite,
hematite, and ilmenite. The bulk of these minerals are
essentially insoluble silicates except for apatite
(calcium fluophosphate), calcite (calcium carbonate),
dolomite (calcium magnesium carbonate), fluorite (calcium
fluoride), rutile (titanium dioxide), molybdenite
(molybdenum disulfide), pyrite (iron disulfide),
magnetite/hematite (iron oxide), and ilmenite (iron

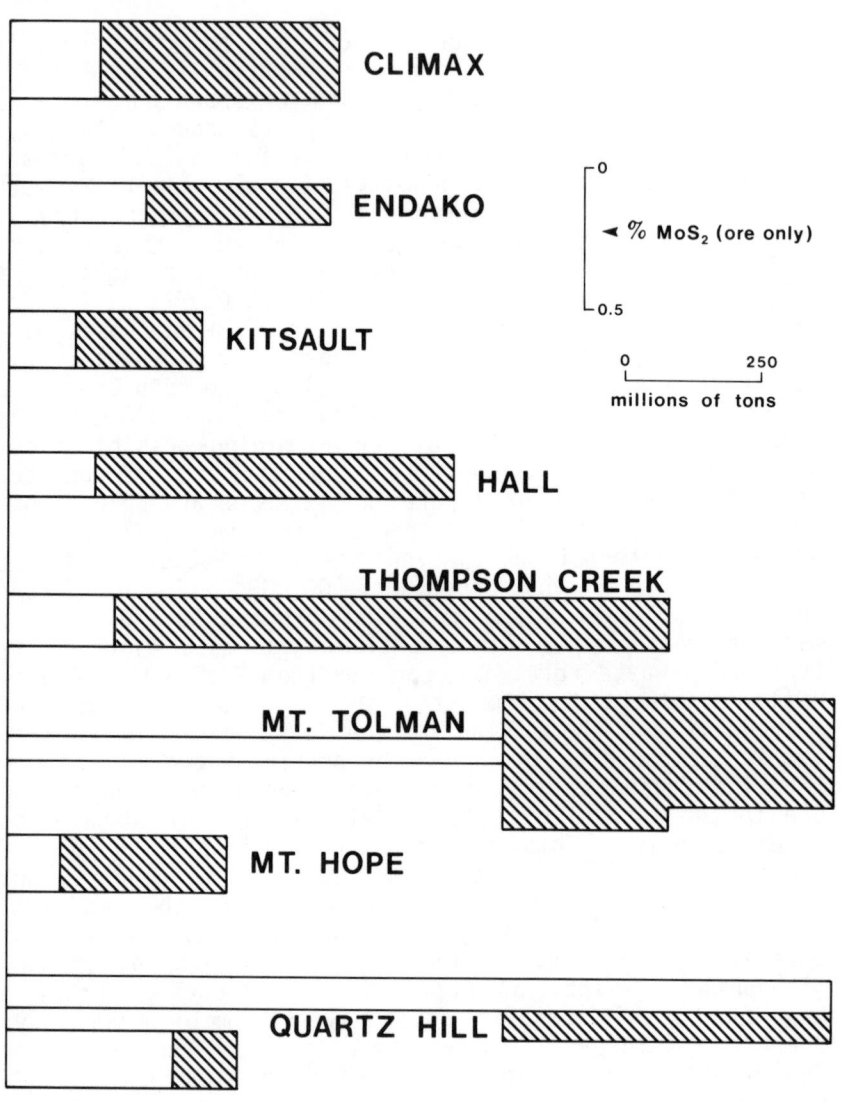

Figure 3. Quantity-quality of ore and stripping ratios for eight significant molybdenite deposits. Ore is shown as clear area and waste rock overburden is shown in pattern.

titanium oxide).
    The bulk composition of Quartz Hill as shown in Table
I sets it apart from an average igneous rock in the crust
and from other molybdenite deposits as well. It can be
shown from this that Quartz Hill is unique in that it has
a very high silicon dioxide content. It also has a low
iron content, and low magnesium and calcium contents. The

Table I. Comparative oxide and elemental percentages for
         Quartz Hill, an average igneous rock in the
         earth's crust and other molybdenite deposits.

	Thompson Creek	Climax	Kitsault	Quartz Hill	Igneous rock
SiO_2	67.10	76.10		78.08	59.14
Al_2O_3	17.01	11.9		11.52	15.34
Fe_2O_3		1.0		0.29	3.08
FeO	2.96	0.25		0.59	3.80
MgO	0.99	0.60		0.24	3.49
CaO	2.24	0.7		0.39	5.08
Na_2O	4.04	2.9		2.98	3.84
K_2O	2.65	5.0		5.09	3.13
Cu	0.1	0.1	0.01	0.0018	0.003
W			0.09	0.0003	0.0002
Pb			0.02	0.001	0.002
Zn				0.0005-0.002	0.0065-0.0094
Sn				0.0005	
Au				0.00002	
Ag				0.00002	
Mo			0.19	0.15	0.0001

trace element content of Quartz Hill is also unusual (5). The amount of copper in Quartz Hill rocks that have been examined to date is much less than in an average crustal igneous rock. It is also much lower than at Kitsault. The amount of tungsten is about that of an average igneous rock but is considerably less than at Kitsault. Quartz Hill is also low in lead content and very low in zinc, lower than an average crustal rock. The tin, gold, and silver quantities are all barely within the detectable limits of the analyses.

The mining prospects for Quartz Hill appear to be very high. From the standpoint of mining a useful metal for mankind, this deposit will yield the greatest amount of metal from the smallest amount of mined rock. In addition, the ore rock is quite uniform in both its physical and chemical properties, which simplifies the predictability of waste materials and metal recovery. The ore lacks significant amounts of objectionable elements, which should eliminate most environmental concerns. It is evident from these facts that most of the problems involving tailings disposal will be concerned with the grain size and volume of the tailings to be handled.

REFERENCES

1. Stephens, J.E. "A Large New Porphyry Molybdenum Discovery in Southeastern Alaska using Geology and Geochemistry", Geological Society of America, Denver, Annual Meeting, Abstracts with programs 8(8): 1121 (1976).
2. Stephens, J.E., L.E. Senter, P.R. Smith and J.R. Snook. "Geology of the U.S. Borax Quartz Hill Molybdenum Property, Southeast Alaska", Canadian Institute of Mining and Metallurgy, Second Annual Meeting, District 6, Victoria (1977).
3. Hudson, T., J.G. Smith and R.L. Elliot. "Petrology, Composition, and Age of Intrusive Rocks associated with the Quartz Hill Molybdenum Deposit, Southeastern Alaska", Can. J. Earth Sci. 16: 1805-1822 (1979).
4. Sutolov, A. International Molybdenum Encyclopaedia, 1778-1978 (Santiago, Chile: Intermet Publications, 1978), vol. 1, pp. 12-22.
5. Hudson, T., J.G. Arth and K.G. Muth. "Geochemistry of Intrusive Rocks associated with Molybdenite Deposits, Ketchikan Quadrangle, Southeastern Alaska", Economic Geology 76: 1225-1232 (1981).

DISCUSSION

Question: Are all these ore deposits simultaneous historically?

Answer: No, but they aren't that far apart.

Question: What is an approximate age compared to the initial material that they are flowing through?

Answer: Most of these are flowing through crustal material, which varies in age. Some of these go through paleozoic rocks, which would go back as much as 600 million years. Quartz Hill is fairly young at 27 million years. It is quite recent.

Question: not recorded.

Answer: This is another one of the things that makes Quartz Hill unique. It has so little in the way of alterations. It has great structural integrity and core recovery is really quite high. That probably will play a significant role in tailings disposal.

Question: How much of the original quantity of ore has gone?

Answer: Certainly some of it. We picked up 168 ppm downstream in the White Creek drainage. But I don't think much has been removed, on the basis of projecting the geology. It was not very far from the surface to start with. The shape of Quartz Hill itself is an indicator. Quartz is very resistant to erosion from glaciers so in part it has held up because of its resistance. It is also because of the shape of the intrusive itself. Most of the roof rocks have been removed. There isn't too much in the way of mineralization outside of the intrusive itself. It seems to be mostly restricted to the intrusive.

Question: Where is the other intrusive?

Answer: It is immediately to the south. The Travis Pluton is probably the mother Pluton for Quartz Hill. Quartz Hill is just a little finger off it. This seems to be the case with many molybdenite deposits, i.e. they represent a collecting pot for a larger intrusive. The molybdenite might be more dispersed if the main intrusive had maintained its overall dome shape. The volatiles, which always rise, tend to collect in this little upside-down cup. Since it is fairly restricted in area, this tends then to concentrate the goodies in a very small area. It makes it economic.

Question: not recorded.

Answer: When you have a high degree of alteration in one of these deposits, as you do at Climax for instance, you have lots of micas which represent flocculation problems during processing and other things of this sort involving particle settling. There isn't much of that sort of material in Quartz Hill. I am not in the tailings dispersal business, but I do know the mineralogy of Quartz Hill and you are looking at fresh fragments. It would be like comparing an average soil that has developed in Georgia or the middle southern part of the United States where tremendous thicknesses of soil are developed, and looking at the rock flour that has developed in the northern part of the U.S. and southern Canada. We have very fine ground up material. The material ground up by the glaciers is made up of mineral fragments, quartz, feldspar, and other mineral fragments. The soil, the fine-grained portion like that developed in an average soil in the middle part of the U.S., that has not been glaciated, is made up of clays and other kinds of material. They are all the same size, but their behavior is very different physically.

Question: What is the size of the ore?

Answer: The average size of the molybdenite goes from about 0.05-0.08 mm, so it is rather fine-grained.

Question: How deep is the molybdenite ore?

Answer: I know it goes down at least 1000 ft, but I am not too sure how much farther.

Question: I would like to register a small objection. I am not sure if I understand the connection between your presentation and the Symposium, which was billed as one on marine tailings disposal. I would have preferred to hear something, if any research has been done, on the grinding of the samples taken thus far, the constituent material of the tailings that we could possibly expect, etc. I appreciate the questions as they relate to tailings, but has there been any kind of research of that nature done?

Answer: The two adits that I showed in my presentation were for that purpose. As far as I know, U.S. Borax is studying that aspect. They have to find out how the ore mines and how it will mill before they can answer the question that you have raised. My role is to set some sort of a stage as to what it is that you are going to be involved with, whatever you do with it.

CHAPTER 9

THE CIRCULATION OF THE SMEATON BAY
AND BOCA DE QUADRA FJORD SYSTEMS

D.L. Nebert
    Institute of Marine Science
    University of Alaska
    Fairbanks, Alaska 99701
    U.S.A.

INTRODUCTION

The study area lies in the rugged mountainous region
east of Ketchikan, Alaska (Fig. 1). The circulation study
presented is a portion of a comprehensive program funded
by U.S. Borax & Chemical Corporation to determine
potential environmental impact relating to sea disposal of
mine tailings from the adjacent Quartz Hill molybdenum
mine. The fieldwork, which began in 1978, was originally
concentrated on the Boca de Quadra fjord system until 1980
when a decision was made to move into the Smeaton Bay
system. As a result, hydrographic data and current meter
data are now available for both fjord systems.

The main part of this presentation centers on the
work done in Smeaton Bay. This choice was made because the
relatively smaller size and less complicated basin
structure has made data collection and interpretation more
straightforward. The understanding gained in the simpler
Smeaton Bay system is then applied to the more complex
Boca de Quadra system in an attempt to explain
observations made there.

The Smeaton Bay system consists of one main basin
defined at the outer end by a 140 m sill at SB4 and at the
inner end by abrupt shoaling into the bifurcated arms of
Wilson and Bakewell (Figs. 2 and 3). Within this basin
there is a relatively deep sill, in the vicinity of SB2,
which serves to restrict bottom water movement to some
extent. Both arms shoal gradually from about 150 m, where
they join the basin, to near their respective heads where

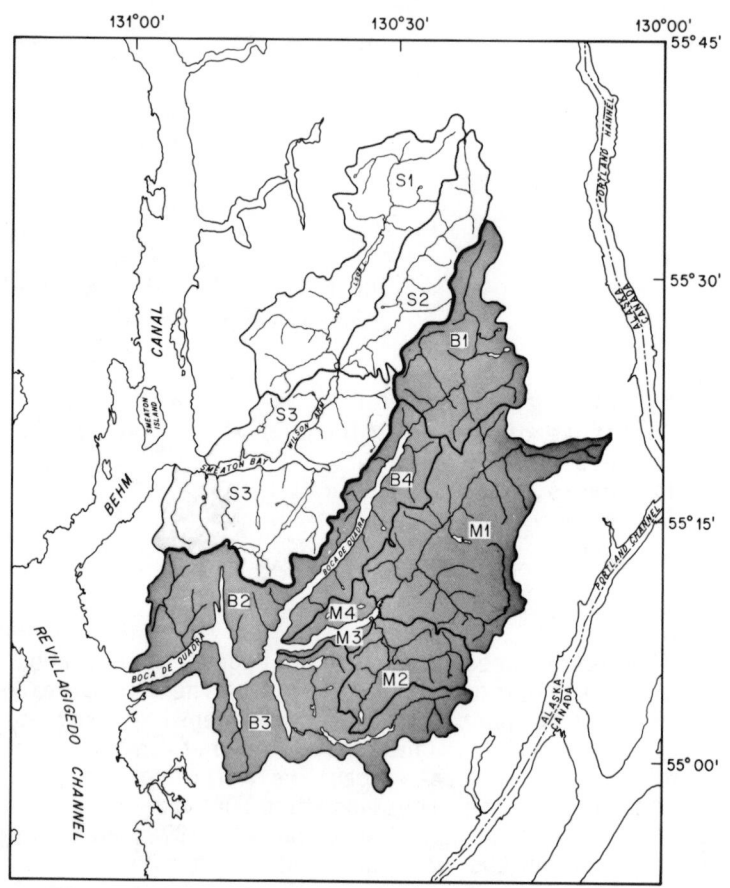

Figure 1. The study area, about 60 miles east of Ketchikan, Alaska, showing the Smeaton and Boca de Quadra watersheds.

they shoal rapidly.

The outer fjord joins Behm Canal, which joins Revillagigedo Channel, which in turn joins the eastern end of Dixon Entrance (see Section I Frontispiece). It is through this series of waterways that dense water from the Gulf of Alaska reaches the study area with a significant seasonal signal (1, 2). Because of the complex maze of water and wind channels in southeast Alaska, the source and route of waters appearing outside a given fjord must remain somewhat speculative.

The primary stations for basic hydrographic data

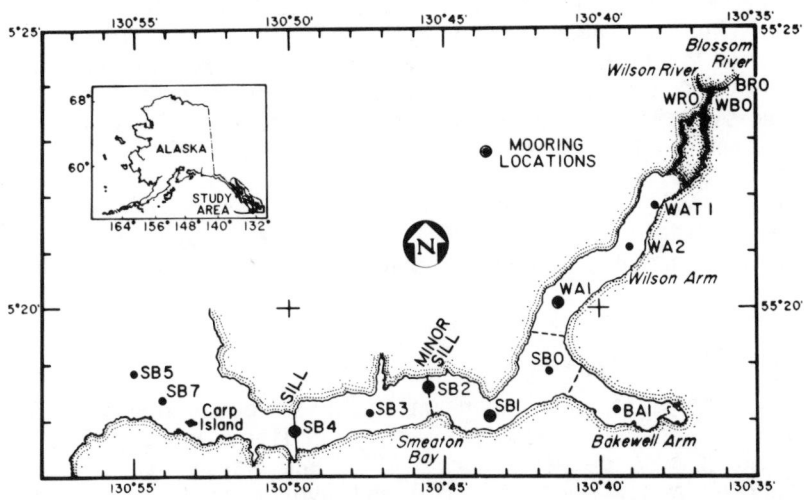

Figure 2. Major features of the Smeaton Bay fjord system. Note sill locations and the inner end of the main basin just up-fjord from SBO. STD and current meter stations (larger dots) are shown.

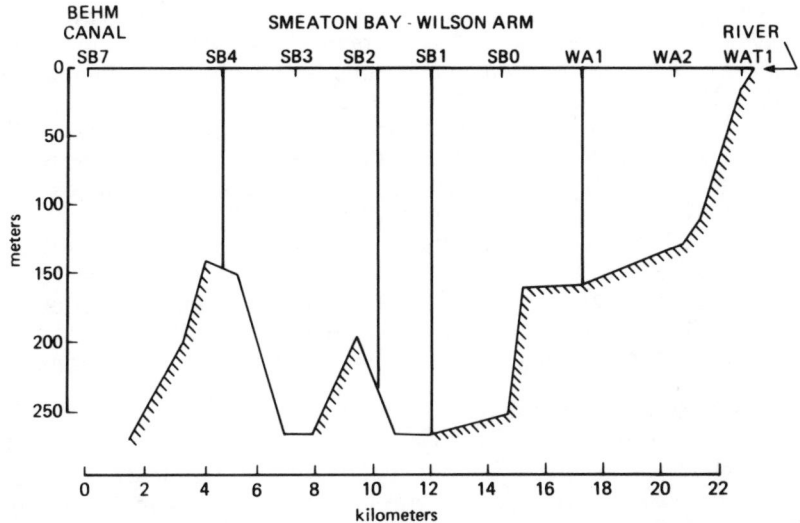

Figure 3. Cross section of Smeaton Bay-Wilson Arm. Vertical lines give approximate current meter mooring locations. Standard STD stations are indicated across the top.

collection are shown in Figs. 2 and 3. Emphasis has been placed upon collection of data outside as well as inside of the entrance sill. With a relatively simple fjord system such as Smeaton Bay, relatively fewer current meter moorings and hydrographic stations are required to define the system as compared to the more complex Boca de Quadra system (to be discussed in more detail later).

Substantial data collection and analysis have been done and are presently in progress for chemical, geochemical and physical oceanographic studies in both Smeaton and Boca de Quadra fjord systems (3, 4, 5, 6).

THE DATA

The data base upon which this report draws is large and is still growing. Twenty cruises were taken into Smeaton Bay between April 1979 and March 1982. Numerous current meter moorings, CTD and hydrographic casts provide a substantial data base upon which analysis is still underway.

An STD or CTD (salinity/temperature/depth or conductivity/temperature/depth) electronic profiling device has been used as the primary means of determining salinity and temperature profiles and hence computed density fields. From the computed density fields it is then possible to infer something about the fjord circulation. In an attempt to minimize potential drift errors associated with the profiling device, discreet samples are taken on each cruise for the purpose of computing offsets for both temperature and salinity. This method of "field calibration" has proven necessary in order to insure acceptable data quality. Thus, accuracy of the STD/CTD data is tied to the procedures for determining discreet temperature and salinity. With careful editing, the accuracy of our data is thought to be of the order $\pm$ 0.02 in both degrees Centigrade and parts per thousand salinity. This results in a reported accuracy for density of about $\pm$ 0.02 sigma-t. Sigma-t is a shorthand for density which is defined by the relationship:

$$sigma\text{-}t = (\, \rho - 1)*10^3$$

such that water with a density of ρ = 1.027 g/cm^3 has a sigma-t value of 27 sigma-t units. A more complete description of data acquisition and processing is available in (5).

The upper 5-10 m of the water column has been ignored for most of this discussion. This is done primarily because of difficulties encountered during sampling. The upper 5-10 m often contain strong gradients so that a small error in depth, or failure to sample synoptically, can cause serious problems with data interpretation. The study can be criticized on this point, but it is virtually impossible to sample synoptically in regions of high gradients without the use of numerous vessels, a luxury not available for this study. After examining the circulation presented along with supporting data, eliminating the upper layer appears not to be a serious problem.

The current meters used in this study are Aanderaa RCM4s. Every thirty minutes temperature, conductivity, pressure and direction are measured "instantaneously". The current speed is integrated over the thirty minute interval. For purposes of this discussion, only speed and direction are examined. The accuracy of these meters has been evaluated numerous times (7) and is considered sufficient for a study of this type. There are two problems that are associated with Aanderaa current meters which could effect the data discussed here:

1. Threshold speeds are on the order of 2 cm/sec.
2. There is a potential upward speed bias due to surface wave action.

The threshold problem may alter mean values over periods when currents are very low. This is more likely to be of academic interest than of practical concern since the data is being used to evaluate the circulation pattern as opposed to specifying whether a given mean current is exactly 2.7 cm/sec rather than some similar, but different value. The speed bias due to wave action has been avoided by mooring the current meters at depths of 10 m or greater. In a sheltered fjord environment where waves are generally small, this precaution is considered adequate.

The current meter data, after translation, are scanned by computer and anomalous values are flagged. The flagged data are then hand-edited by a person familiar with current meter data and potential problems. The recovery rate for the Smeaton Bay data is in excess of 95% for speed and direction. The rate for Boca de Quadra data is lower, due primarily to problems with premature battery failure. But it is still better than 80%.

The current meter data contain periodic signals as well as long-term (net) currents. The primary periodic influences in this region are due to tides and result in strong semi-diurnal and fortnightly tidal components. The resulting current meter record has semi-diurnal amplitudes that vary by more than a factor of three during the fortnightly period. These periodic signals must be removed to establish net flow. This can be done by using a tidal filter or by taking a mean over a dominant period, which is essentially a very simple form of filter. The current meter data have been averaged over 28 days, 14 days and 50 hours for this phase of the study. Only the 50 hour means are presented here for visual impact.

THE CIRCULATION OF SMEATON BAY

The density field, as computed from the CTD/STD data, suggests that Smeaton Bay circulation may be opposite to that of a classic positive estuary. A classic positive estuary is one in which excess low density water (river input) is found within the estuary, usually at the head, ostensibly driving the surface layer down-estuary. With river input at the head, constant density surfaces rise down-fjord as the saline water mixes with the river water. In Smeaton Bay, excluding the upper 5-10 m for the reasons already discussed, the constant density surfaces often tend to tilt upward toward the head. Fig. 4 presents a

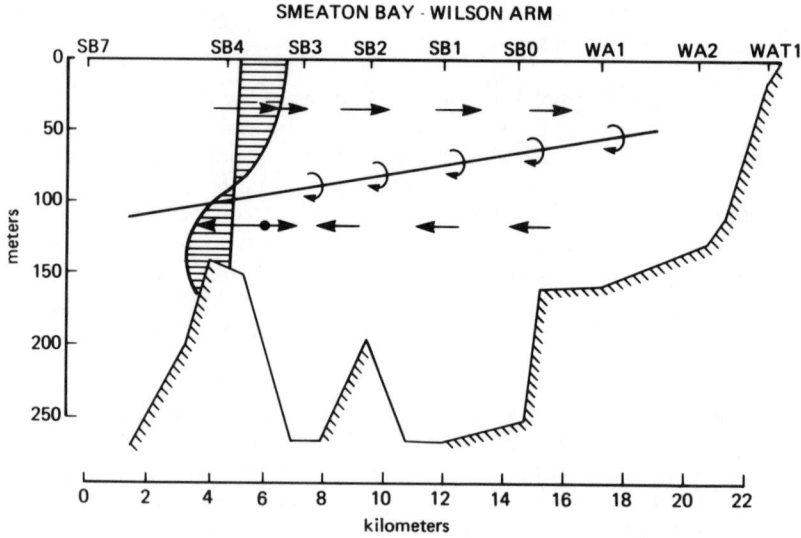

Figure 4. Schematic of density field and implied flow pattern. Constant density surfaces typically rise toward the head. Note that even though net flow is outward at the sill, bottom water renewal is possible as long as the tide (during flood) brings dense water in over the sill where it can sink.

schematic of the density field that might be found in
Smeaton Bay during much of the year. The resulting
(implied) circulation is as shown by the arrows, inward in
the upper waters and outward below. A relative current
profile is also suggested in Fig. 4.

It is not possible to say much about absolute
currents from a density field. A steady state cannot be
assumed; a cruise only yields a "snapshot" in time and
often the data span a period of a few days. In general,
sampling has not been adequate to resolve short-term
temporal variability such as that which might result from
internal wave action. Also, the observed density field
does not necessarily reflect the effects of baroclinic
flow such as those from a storm related event.

The vertical motion within Smeaton Bay associated
with the suggested flow pattern (as surmised from the
density field) would be downward. Materials added to the
upper waters would (on the average) be mixed downward into
the outward flowing layer. It must be stressed that while
this downward mixing might be expected based upon the mean
density field, individual events with upward transport are
possible and likely do occur. A flow pattern like the one
proposed would have a certain advantage with respect to
marine tailings disposal in that it might be less likely
that a "tailings cloud" would be carried into the upper
fjord waters.

Since analyses of the density field yield neither
absolute currents nor quantitative information on
persistence of the surmised flow pattern, it is helpful to
examine available current meter data.

The placement of the current meter moorings is
important in evaluating fjord circulation. With a simple
silled fjord, maximum currents can be expected at the
sill. Also, since the density field tends to be
horizontally stratified, net flows observed in a sill
region might be expected to extend horizontally into the
adjacent fjord basin(s). Thus, optimum mooring placement
would include one at the entrance sill to define the net
inflow/outflow and at least one within the fjord to
determine the extent to which the circulation extends into
the adjacent basin. With this in mind, moorings were
placed at the entrance sill at SB4 and in the main basin
at SB1 (Fig. 3).

Before discussing the current meter results, it is
helpful to remember that the density field suggested net
inflow in the upper waters and net outflow below (Fig. 4).
Fig. 5 is a time-series stick plot of the along-axis
velocity component for current meters moored at SB4/1 (the
designation SB4/1 means the first mooring placed at the
SB4 location). Each stick is a 50 hour (ca. 2 day) mean of

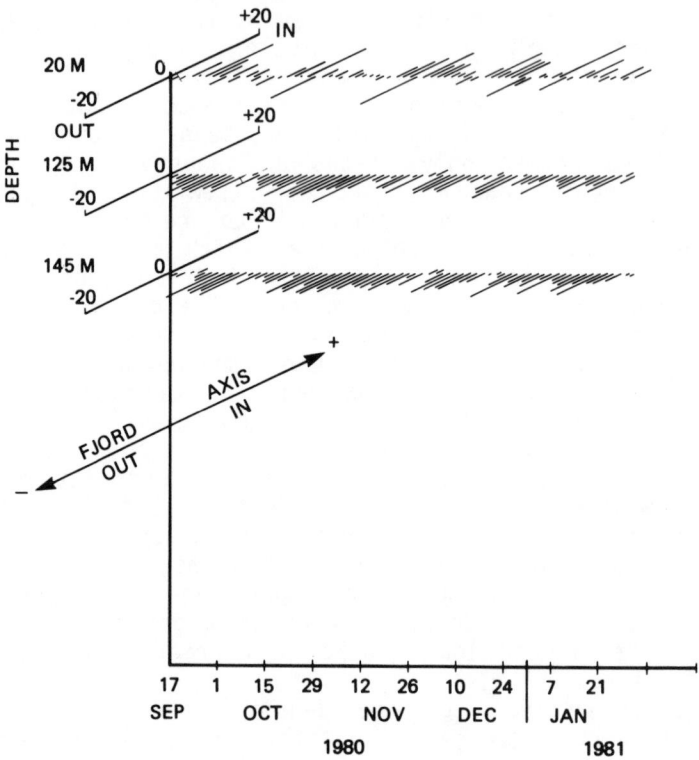

Figure 5. Time-series stick plot of means (50 hour) showing major axis flow at the Smeaton Bay outer sill. Velocities are in cm/sec. Current meter depths are indicated on the left margin and sill depth is about 140 m.

the 100 samples taken during the interval. Note the relative depths of the current meters on the vertical axis and the time-scale across the bottom marked off with 14 day time ticks. Recall that the upper current meter is not in the extreme surface layer.

The general conclusion drawn from an examination of Fig. 5 is that a net inflow occurs at 20 m and a net outflow occurs at both 125 m and 145 m. This is found during a period of expected high freshwater runoff, i.e. October 1980.

A similar look at the SB4/2 mooring (Fig. 6) shows a net inflow in the upper waters and net outflows below

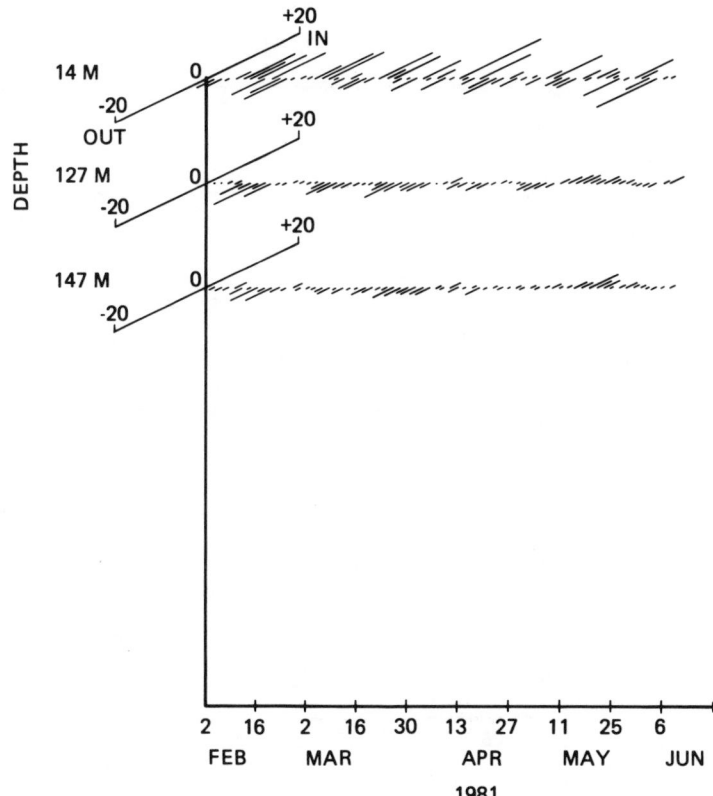

Figure 6. Time-series stick plot of means (50 hour)
showing major axis flow at the Smeaton Bay outer
sill. SB4/2 indicates the second mooring to be
placed at the SB4 location.

through April 1981. These reverse and show minor inflow
for the rest of the record. There is a noticeable
weakening of the deep outflow compared to the previous
mooring (SB4/1, Fig. 5).
    The subsequent mooring, SB4/3, shows an increasing
deep outflow near the sill (Fig. 7). The SB4/3 mooring was
placed in water deeper than was planned. One might guess
on the basis of continuity, that the upper waters provide
a compensating inflow. Remember that although the
summer-fall period normally experiences greatest
freshwater input, upper water inflow is still postulated.

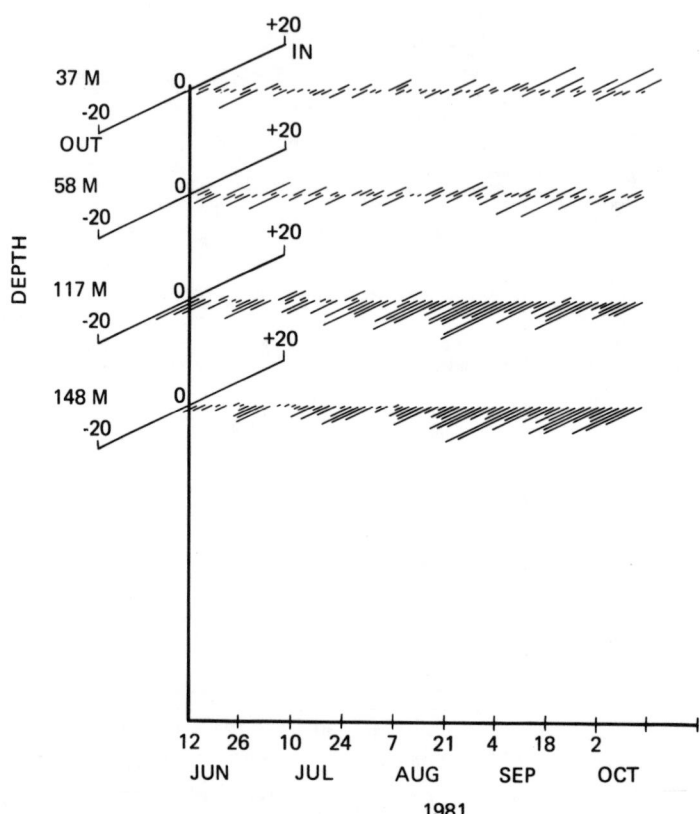

Figure 7. Time-series stick plot of means (50 hour)
showing major axis flow at the Smeaton Bay outer
sill. The third mooring at this location.

Examination of the SB1/3 mooring (Fig. 8) suggests
that the circulation observed at the entrance sill does
extend into the fjord basin. The upper two current meters
show minor inflow with a net zero flow at 65 m. There is a
definite outflow seen at 106 m (just above the sill) that
probably correlates with the outflow observed at the sill.
In addition, distinct bottom water renewal events are
evident on the 240 m record. Thus the basic features
observed at the sill are also seen within the fjord basin
although they are diminished in magnitude.
      The data above suggests that the general circulation
in Smeaton Bay is different from the estuarine

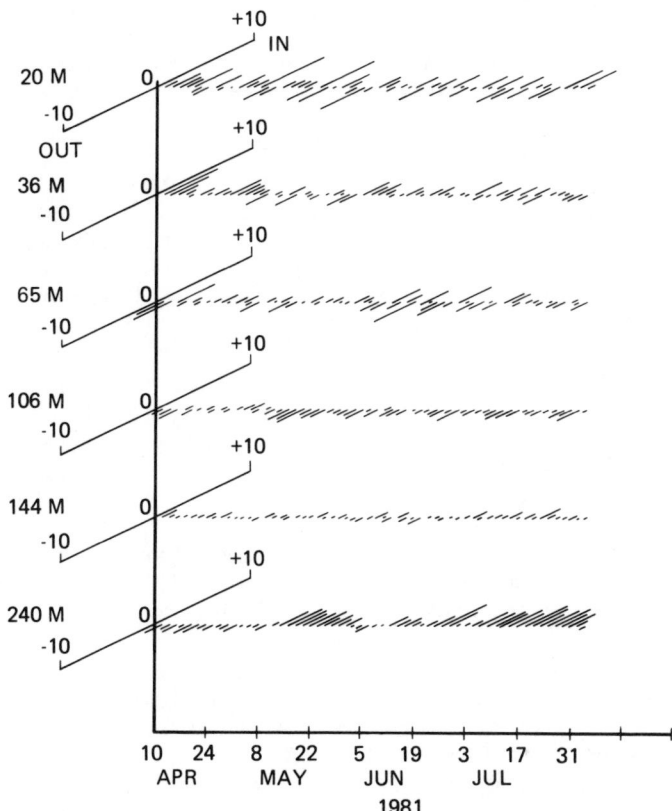

Figure 8. Time-series stick plot of means (50 hour) showing major axis flow within the Smeaton Bay basin. Note bottom water renewal commencing in May 1981 and net outflow at 106 m.

circulations described by other speakers at this symposium. Since this investigation is still in progress, some of the ideas put forward here may be modified, but I expect that ultimately some new ideas will result from this study.

POTENTIAL DRIVING FORCES

It is a well-known fact that fjord estuaries may be driven by freshwater input from within the fjord, a point discussed by participants at this symposium. By contrast, it has been reported to U.S. Borax that the study area fjords are driven backward, ostensibly by brackish water available outside the entrances (5, 6). The source of the brackish water is thought to be Nass and Skeena River water coupled with coastal British Columbia water moving northward through Hecate Strait (see Section I Frontispiece).

The water below sill depth is somewhat decoupled from the upper fjord water and tends to respond to dense coastal water which appears at the sill. Basin water replacement commences in spring and continues (probably intermittently) through fall, ceasing in September or October. The annual driving force for the deep circulation is attributed to the seasonal variations in the Gulf of Alaska which respond to weather patterns and coastal runoff. I have been considering the data for the last few weeks and have found a different way to describe the circulation just reported. My ideas must still be tested, but I would like to present them here. It is usually assumed that the surface circulation is mostly decoupled from the deep circulation and that the driving forces are mostly independent for top and bottom. Perhaps in Smeaton Bay there is a more direct coupling and, at least during part of the year, the deep water drives the above sill circulation.

It is known that relatively small horizontal density gradients can be associated with significant currents, at least in an open ocean regime. In Alaskan fjords, density gradients on the order of 0.1 to 0.2 sigma-t are normally sufficient to accomplish bottom water renewal.

With this in mind, let us examine a hypothetical situation. We will ignore other potentially important driving forces and just look at the density field (Fig. 9). We will begin with fall conditions, after deep water renewal has occurred and dense water fills the fjord to well above sill depth. Through fall and winter, the dense water outside drops away, setting up an outward current at the sill, coupled with an inflow above. This current pattern is maintained as long as the water inside the sill is greater than that outside. The denser water inside is at least partially maintained by the deep water below the sill which is being mixed by large amplitude tides which tend to vertically homogenize the water column. Thus, the deep water acts as stored potential energy for the suggested circulation. The energy of such a system would eventually run down or would be stopped by seasonally increasing water density outside the fjord. In Smeaton Bay the "reverse circulation" increases again in June. Some other driving force must be responsible since the stored dense water has been "consumed". That driving force might

be the appearance of brackish water outside the fjord due
to regional snowmelt.
     Note that even though a net outflow is observed at
the sill, it is possible to have bottom water renewal as
long as the flood flow reverses the current at the sill so
that water from outside can enter and sink. The subsequent
ebb tide will then preferentially remove water from above
sill depth due to energy considerations. Thus, it is not
necessary to observe a new inflow at the sill in order
that bottom water replacement may occur. In Smeaton Bay
this condition is observed (Figs. 7 and 8).
     In support of the proposed driving mechanisms, i.e.

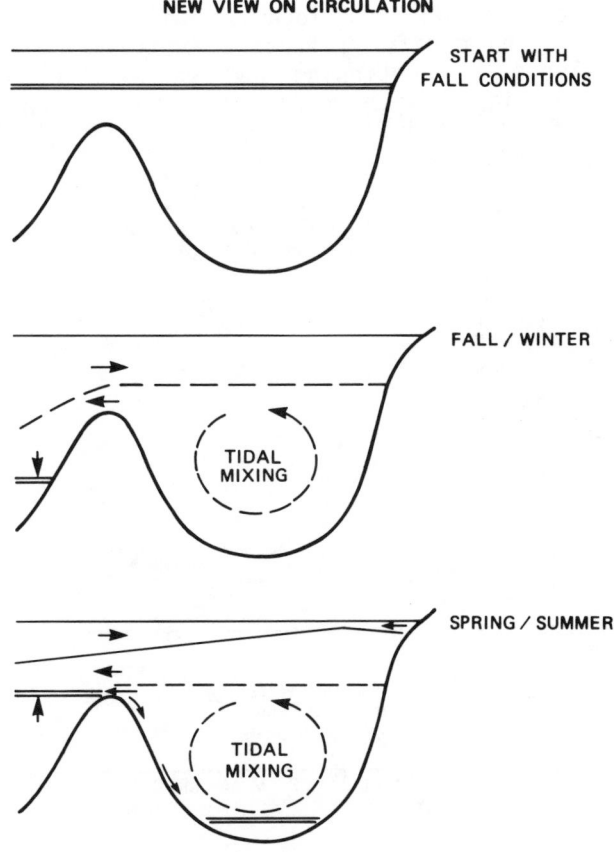

Figure 9. Hypothetical circulation which may explain some
          of the features observed in Smeaton Bay. Note
          that bottom water renewal can occur even though
          a net outflow is observed at the sill.

the fall/winter circulation is driven by dense water
within the fjord and the spring/summer circulation is
driven by brackish water outside the fjord, we have
density profiles across the sill that are in general
agreement. Also, the proposed circulation should run down
following a peak during summer or fall. Fig. 10 shows the
mean (14 day) of the bottom two meters at the Smeaton Bay
sill with time. This time-series is consistent with the
explanation presented. The "rundown" occurs as the dense
water is "consumed" and the increased outflow during
summer is attributed to an increase in brackish water
outside the fjord.

The proposed circulation does not operate in a
vacuum, but is influenced by many other factors such as
winds, local runoff and localized mixing. The proposed
circulation and modifying factors must now be evaluated.
Data analysis is continuing and we will shortly begin
testing our ideas with a numeric model which is being
developed for Smeaton Bay by Zygmunt Kowalik (Institute of
Marine Science).

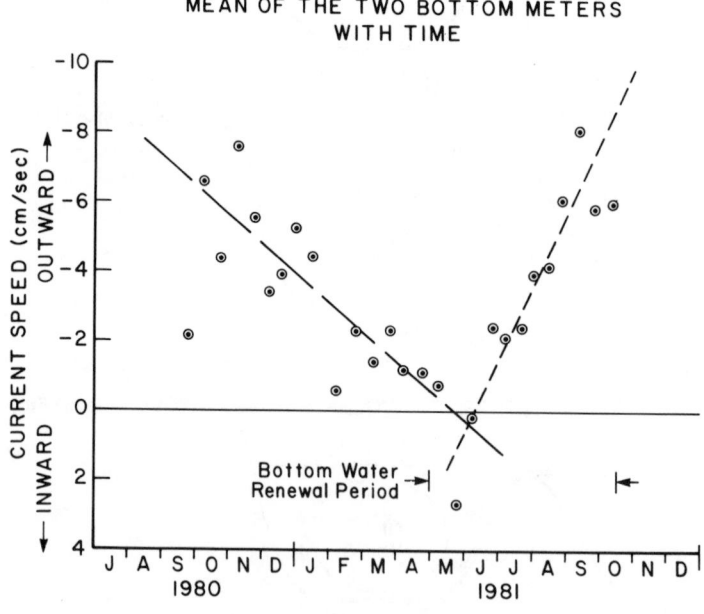

Figure 10. Major axis flow (14 day means) observed near
the sill in Smeaton Bay. It represents the mean
of two meters, one above and one just below
sill depth.

THE BOCA DE QUADRA FJORD SYSTEM

This discussion has been primarily about the Smeaton Bay fjord system because of its less complicated nature and because less data are required in order to define the circulation. Can the same findings be applied to the Boca de Quadra fjord system?

The answer, very briefly, is both yes and no. The Boca de Quadra system is much more complex in terms of shape, as can be seen from Figs. 11 and 12. In Fig. 12 the stylized density field shows many of the same features found in Smeaton Bay:

  ·uplift of density surfaces (mid depth) from sill toward head;
  ·seasonal inflow over the outer sill; and
  ·the seasonal buildup of brackish water at the extreme surface within the fjord.

However, the main features are not as well defined or as

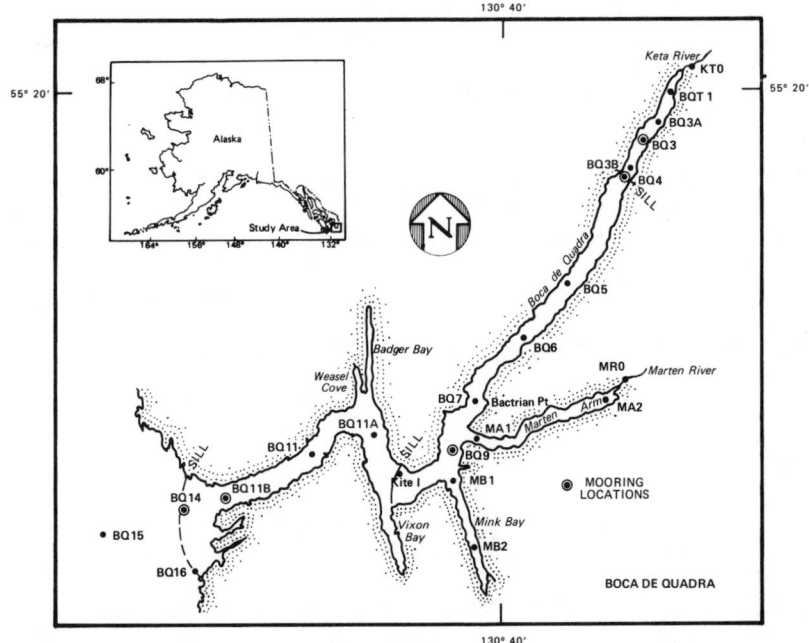

Figure 11. Map of Boca de Quadra showing a more complex configuration than that in Smeaton Bay (see Fig. 2). Station locations and current meter mooring locations are shown.

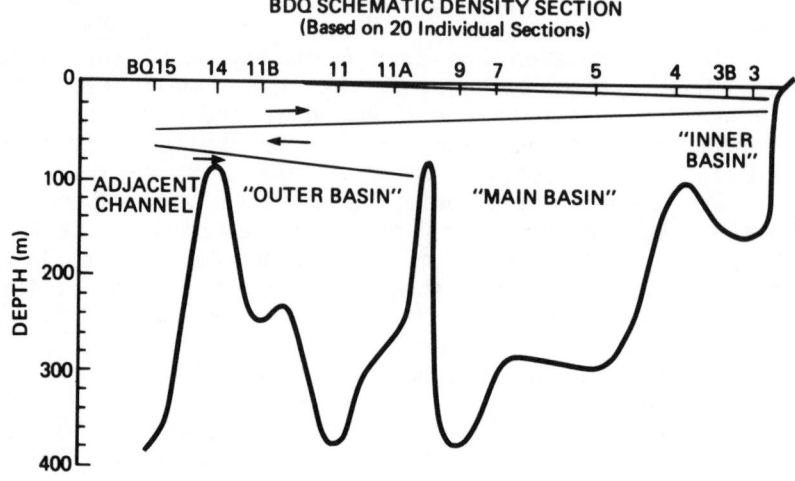

BDQ SCHEMATIC DENSITY SECTION
(Based on 20 Individual Sections)

Figure 12. Boca de Quadra schematic density section showing the various basins, station locations and the implied flow. Deep water replacement into the outer basin occurs over much of the year while the main basin remains somewhat isolated. The freshwater layer shown near the head is thought to be seasonal.

persistent as in Smeaton Bay. This may be largely due to the limited data set for the Boca de Quadra system which must be substantially larger to provide the same coverage we now have for the Smeaton system.

The field program is now to be redirected toward the Boca de Quadra system. This will enable us to gather data needed to further define the undoubtedly complex circulation there.

REFERENCES

1. Crean, P.B. "Physical Oceanography of Dixon Entrance, British Columbia", Fisheries Research Board of Canada Bull. 156 (1967).
2. Dodimead, A.J. "A General Review of the Oceanography of the Queen Charlotte Sound - Hecate Strait - Dixon Entrance Region", Canadian Manuscript Report of Fisheries and Aquatic Sciences No. 1574 (1980).
3. Burrell, D.C. "Marine Environmental Studies in Boca de Quadra and Smeaton Bay: Physical and Chemical", unpublished interim report for the period January-April 1980 to U.S. Borax & Chemical Corporation, Inst. Mar.

Sci., Univ. Alaska, Fairbanks (1980).
4. Burrell, D.C. and H.J. Niebauer. "Marine Environmental
   Studies in Boca de Quadra", unpublished report, Inst.
   Mar. Sci., Univ. Alaska, Fairbanks (1979).
5. Burrell, D.C., H.J. Niebauer and D.L. Nebert. "Marine
   Environmental Studies in Boca de Quadra and Smeaton
   Bay: Physical and Chemical, 1979", Inst. Mar. Sci.
   Rept. R80-1, Univ. Alaska, Fairbanks (1980).
6. Nebert, D.L. and D.C. Burrell. "Marine Environmental
   Studies in Boca de Quadra and Smeaton Bay: Physical and
   Chemnical, 1980", Inst. Mar. Sci. report prepared for
   U.S. Borax & Chemical Corporation (1981).
7. Beardsley, R.C., W.C. Boicourt, L.C. Huff, J.R.
   McCullough and J. Scott. "CMICE: A Near-Surface Current
   Meter Intercomparison Experiment", D̲.S̲.R̲. 28A (12):
   1577-1603 (1981).

ADDENDUM

Analysis of the above and new data is continuing.
This addendum expands on some of the ideas presented at
the symposium.

The idea that outflows observed at the sill might be
driven by dense water within the fjord has been checked
using two independent methods. Computed outflows are too
small by almost an order of magnitude to drive the deep
outflow observed (independent calculations by H. Gade and
myself). This suggests that the outflow is not due solely
to the deep density field.

Discussions at the symposium concentrated on the
paucity of current meter data and particularly analysis of
the extreme surface layer. D. Farmer and H. Gade (and
others) feel that the observed inflow at 10-30 m is a
return flow providing continuity for the surface,
estuarine outflow. I do not agree that the primary above
sill circulation is due to freshwater driven estuarine
flow. Many factors, including presently unreported drogue
studies, lead me to believe that the extreme surface layer
does not often experience a net seaward drift, even during
periods of high freshwater input.

H. Gade (personal communication) views the observed
deep net outflows at the sill with reservation, feeling
that they may be an artifact of data collection. This
reservation is always warranted, especially when apparent
results contradict established concepts. However, I feel
that the reported sill depth outflow is real since it can
also be seen at, or about at, sill depth in the adjacent
basins. This occurs in both Smeaton and Boca de Quadra
systems, although not at all times. Further investigation

will be needed to confirm or refute this interpretation.

Several people have expressed the opinion that a net outflow at sill depth is inconsistent with bottom water renewal. I reiterate that it is not inconsistent as long as current reversals (tidal) occur at the sill allowing water from outside the sill to enter and fall into the basin on flood tides. If this reversal occurs, the net flow at the sill may be (although not necessarily) independent of bottom water renewal.

DISCUSSION

Question: What is the length of the innermost basin in the Boca de Quadra system?

Answer: About 4.5 miles.

Question: If there is an outfall, approximately where might it be reasonable to locate it?

Answer: The engineers are going to have to tell us some things about how the tailings behave. What we know about the circulation now will allow us to do an intensive study at the innermost sill to examine what happens there. U.S. Borax, of course, is interested in putting tailings into this innermost basin if they will in fact stay out of the upper waters. We do not have sufficient data to say that they will or that they will not. There is a problem with continuity. If there is inflow at the upper waters and outflow in the bottom, then you can expect downward transport. We do know that during part of the year this reverses and there are inflows at the lower waters and outflows in the upper waters. In that condition, you have a potential uplifting of (deeper) denser water and if tailings were put in there, the question that has to be answered is how they are going to react to seawater. Are they going to sink and at what rate? I would expect that the vertical velocities associated with the continuity problem are very small and thus it really depends upon the tailings themselves. Certainly most of the tailings are going to drop out, like rocks.

Question: Given the fact that the tailings will be introduced for a long period of time with the methods chosen, wouldn't it be better to choose the deepest part of the fjord in which to introduce the tailings from the beginning?

Answer: Again, you are asking me a question that is not really oceanographic. Yesterday we heard Dr. Hay discuss channelling and the distribution of tailings via marine channels. I am sure that U.S. Borax' choice would be to fill in the innermost basin and then use the density channellings as a means of carrying the turbid, dense slurry over and into this main basin.

Question: Can you clarify the relationship between the variation in atmospheric pressure and the circulation?

Answer: What we see along the coast of Alaska in general, and I am sure this extends more or less along the B.C. coast, is that the density structure fluctuates in a vertical sense throughout the year. This is primarily due to atmospheric systems, storm systems and high and low pressure systems which tend to center themselves in the Gulf of Alaska. With a low pressure system set up you get a cyclonic wind system which tends to blow surface waters towards the coast, building up excess freshwater, depressing the dense water, and flushing it off the shelf. When the reverse atmospheric condition is prevalent, the opposite occurs. What we observe in the northern part of the gulf is actually a relaxation of the down-welling as opposed to an actual up-welling, although the up-welling can happen as well. Then, this dense water that was flushed off the shelf moves, when down-welling ceases, laterally several hundred km and perhaps becomes available to these fjords.

Question: Have you identified any major turbulences in the area of the Kite Island sill in Boca de Quadra?

Answer: No, we have no current meter data at that point. We are proposing to U.S. Borax that we now look at that sill. The projected peak tidal currents are only on the order of 7-14 cm/sec, assuming that it is a constant tide with depth. We don't expect to see any catastrophic mixing, although as the density gradient diminishes one might expect a hydraulic jump with lower velocities of water.

Question: not recorded.

Answer: As I have indicated, we have not addressed that

layer. There are some ancillary data. We have put drogues out there, primarily during times of high runoff. They moved up toward the river in spite of the fact that there was a wind blowing down and in spite of the fact that it was October in the particular year in which they had a high freshwater input. It was over a day, so it spanned the tides. I am not suggesting that the rivers don't have some influence, but we have not really addressed that particular feature. That is part of the problem that we are addressing with tailing disposal. If turbid tailings water makes it up into that upper layer, we have already lost the game. We must keep it out. We are interested in the circulation below because injection will hopefully be sufficiently far below to keep it out of the upper layer.

Question: not recorded.

Answer: Our highest current meter was in the order of 14 m deep to avoid wave contamination and also to avoid the problems of putting in a surface mooring from a small vessel. The hydrographic sections avoid the upper layer, the upper 5 m. Most of the freshwater input is put in down-fjord and would tend to retard circulation within that innermost basin.

The other point that I might make is that freshwater input into these basins is very variable. I particularly remember two days in October 1978 when first of all we had the lowest flow per day for the month and two days later, the highest. There was a 900% difference in the flow rates between those two days. These are very sharp pulses and they must certainly have some effect on the circulation in the span of a day or so. We have no evidence whether or not they have a significant effect for the whole period during October.

CHAPTER 10

THE PRE-IMPACT BIOGEOCHEMICAL ENVIRONMENT OF
BOCA DE QUADRA AND SMEATON BAY FJORDS, SOUTHEAST ALASKA*

D.C. Burrell
Institute of Marine Science
University of Alaska
Fairbanks, Alaska 99701
U.S.A.

INTRODUCTION

Boca de Quadra and Smeaton Bay are the two adjacent fjord-estuaries located nearest Quartz Hill (around $55^\circ20'N$), just north of the Alaska-British Columbia boundary (see Section I Frontispiece). One or both of these inlets are potential repositories for very large quantities of mill tailings, together with other associated industrial waste products, from the projected Quartz Hill molybdenum mine. This report reviews the current status of work in progress to describe and understand the way these estuarine systems function at the present time (i.e., prior to any large-scale development), on a seasonal and multi-year basis. The concern here is with the overall biogeochemical environment; other chapters included in this book consider aspects of the circulation, and benthic biology. The purpose of the overall program is:

1. To provide information required by the developers and the regulatory agencies so that potential impacts cause the least, and preferably reversible, environmental damage.
2. To attempt to predict the effects of various planned impacts.

---

*Contribution No. 487, Institute of Marine Science, University of Alaska.

Prior to this program, the oceanography of the fjord-estuaries in the southern part of S.E. Alaska had been characterized only in a very general fashion (1), and other than Alice Arm (see Section II and Section III Introductions), no fjord has been examined in detail anywhere in the vicinity. Fieldwork commenced with reconnaissance surveys in 1978 and 1979. It was apparent that close-interval seasonal detail would be required, and the information discussed here is derived primarily from observations made over consecutive periods, September 1979 - September 1980 in Boca de Quadra, and October 1980 - October 1981 within the Wilson Arm-Smeaton Bay system. Because of the marked seasonal variability, an approximately one-month sampling interval was maintained through these periods. Certain key localities (e.g., at the heads of the fjords and within the deep basins) have been monitored over a number of years, and some parameters are beginning to exhibit significant year-to-year variability. Natural multi-year ecological trends and cycling are well-known, e.g. see Gray (2), but are frequently poorly understood. It is essential to continue observations over a sufficient number of years to ensure that predictions are not based on observation of atypical conditions. Work in both fjords is therefore continuing, and the information provided here, and in more detail for some topics in (3), must be considered preliminary only, and subject to revision.

FJORDS AND MARINE POLLUTION

In order to assess the potential effects of pollution, it is instructive to consider, in a general fashion, the major processes occurring within any silled fjord. These are shown schematically in Fig. 1.

1. Water circulation is discussed in Chapter 9. As a gross simplification, transport in the upper layer and in the basin may be considered independently. Basin water is exchanged on a time-scale ranging from days to irregular periods in excess of one year. The near-surface waters are stratified through the summer and generally well mixed through the winter.
2. All estuaries by definition receive freshwater inflow. The rivers carry various types of sediment and chemicals in solution into the fjord.
3. Primary production of phytoplankton occurs in the surface waters in the summer. In many Alaskan fjords, including the examples considered here,

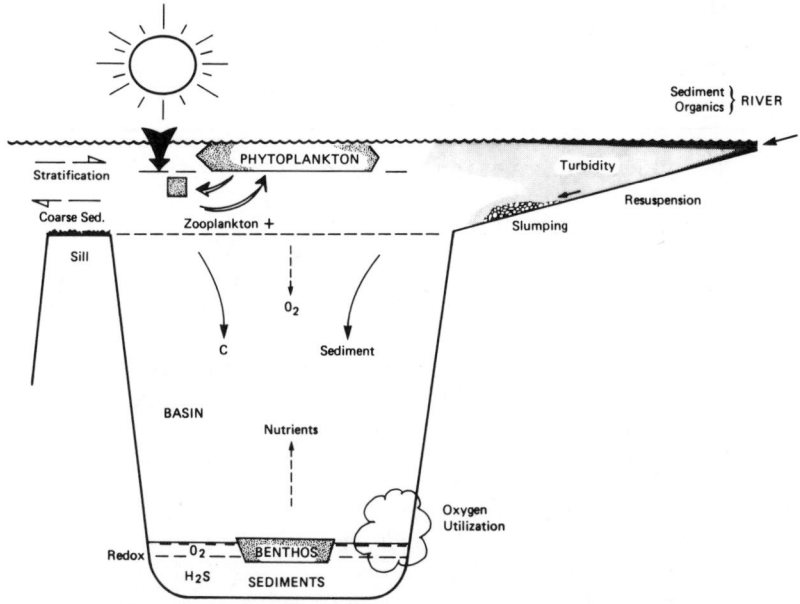

Figure 1. Schematic of fjord oceanographic processes.

this is the primary energy source to the pelagic predators, and indirectly or possibly directly to the benthos.
4. Biological activity influences the distribution of dissolved oxygen and nutrients throughout the fjord, including oxidation of organics in the deep basins and sediments. This may lead to complete depletion of available oxygen locally.
5. Sediments from various sources result in "turbidity", particularly in the surface zone. Sediment may also be transported along the sediment floor, and resuspension of previously deposited sediment may occur.

Only a few topics (in addition to those dealt with elsewhere in this volume) can be discussed in this report, concentrating on those processes potentially most at risk from the proposed industrial development (Table I). It should be noted that, although mine tailings disposal would constitute by far the greatest quantitative impact, equal or greater ecological damage could be generated by, for example, various types of more reactive organic waste from town-sites and transport. Table II lists some of the potential concerns; but, since it is the primary thrust of the book, this theme need not be further developed in this chapter.

Table I. Selected   fjord   oceanographic   processes   of
         particular interest in pollution studies.

---

1. Natural Sedimentation Patterns
   a. Seasonal riverine fluxes
   b. Major geochemical components
   c. Import, distribution, and reaction within the fjords
   d. Physical-chemical character of the bottom sediments

2. Carbon Flow
   a. Source and sinks, transport, reaction, and cycling
      of carbon and nutrients within the fjords
   b. The major biological components
   c. Seasonal effects

3. Oxygen Distributions
   a. Natural seasonal and yearly patterns of dissolved
      oxygen within the basins
   b. Redox character of the sediments

4. Heavy Metal Cycling
   a. Available concentrations in surficial sediments

TOPOGRAPHY

The environmental settings of the Boca de Quadra and
Smeaton Bay-Wilson Arm fjords have been discussed by
Burrell (3). The Smeaton Bay system is considerably
shorter (but has a greater mean width) than Boca de Quadra
(approximately 22 and 60 km in length, respectively). Both
are silled fjords; but Smeaton Bay contains a single
(though divided at depth) basin, some 240 m deep with
entrance sill height at around 140 m. Boca de Quadra is a
multi-basin system. However, Nebert and Burrell (4) have
shown that the outermost basin is well mixed year-round,
and the region of greatest practical interest is up-fjord
from the Kite Island sill (Fig. 2; sill height ca. 85 m),
especially the deep (ca. 350 m) central basin. Fig. 3
compares schematic longitudinal sections of the two fjord
systems on the same horizontal and (exaggerated) vertical
scales.

RIVER INFLOW

As discussed by Nebert and Burrell (4; see also

Table II. Mine tailings disposal in a fjord-estuary: potential concerns.

1. River
   a. Changes in natural seasonal sediment load and sedimentation patterns (e.g., disturbance of salmon spawning habitat)
   b. Effect on natural organic-inorganic fractionation of particulates
   c. Changes to short-term flow rates
   d. Changes in dissolved oxygen concentrations
   e. Introduction of exotic chemicals: organics and heavy metals (e.g., effect on salmon "homing" character; sublethal effects)

2. Fjord Surface Water
   a. Increased localized and seasonal turbidity (effect on primary production and pelagic organisms; aesthetic)
   b. Introduction of organics (e.g., from town-site operation and transportation)

3. Fjord Basin
   a. Increased particulate sediment
   b. Changes in oxygen consumption/nutrient regeneration patterns

4. Sediment
   a. Increased sedimentation rates. Hence:
      i. changes in physical/dynamic character of interface - sediment stability; cohesiveness, etc.
      ii. changes to geochemical regime at interface - porosity; redox boundary, etc.
   b. Increased bed-load transport - sediment instability; resuspension, etc.
   c. Changes in oxygen consumption/nutrient regeneration rates
   d. Changes to geometry of basins; altered circulation patterns, etc.
   e. Potential mobilization, biotic uptake of exotic chemicals (toxicity; sublethal effects, etc.)

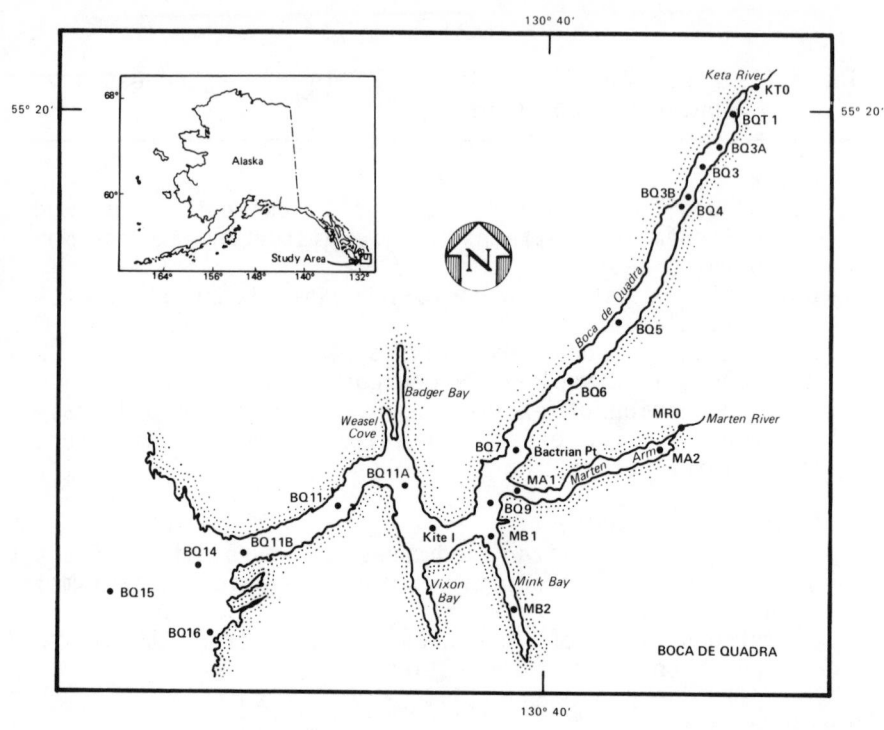

Figure 2. Boca de Quadra fjord system: location map.

Chapter 9), freshwater discharge does not appear to be driving the major circulation within these fjords; but riverine influx is probably the single most important influence on the biogeochemical regime. (The physical-chemical character of each of the major rivers is also of primary concern with regard to the economically important salmon runs, a topic which unfortunately is outside the scope of this chapter.) Fig. 4 shows the catchment areas for each of the principal rivers draining into the Smeaton Bay and Boca de Quadra systems. The integrated drainage areas for each fjord are not large relative to their surface areas (around 10x greater in both cases), and over half of the freshwater discharge is not via the major rivers, but peripherally around the margins. The two major rivers emptying into Boca de Quadra are the Keta at the head, and the Marten which impinges directly on the central basin via Marten Arm (Fig. 2). Volume flow from the Marten River is approximately 1.8x that of the Keta (Fig. 4: catchment areas M1 and B1 respectively).

The single most important freshwater influx into the Smeaton Bay system is at the head of Wilson Arm: the

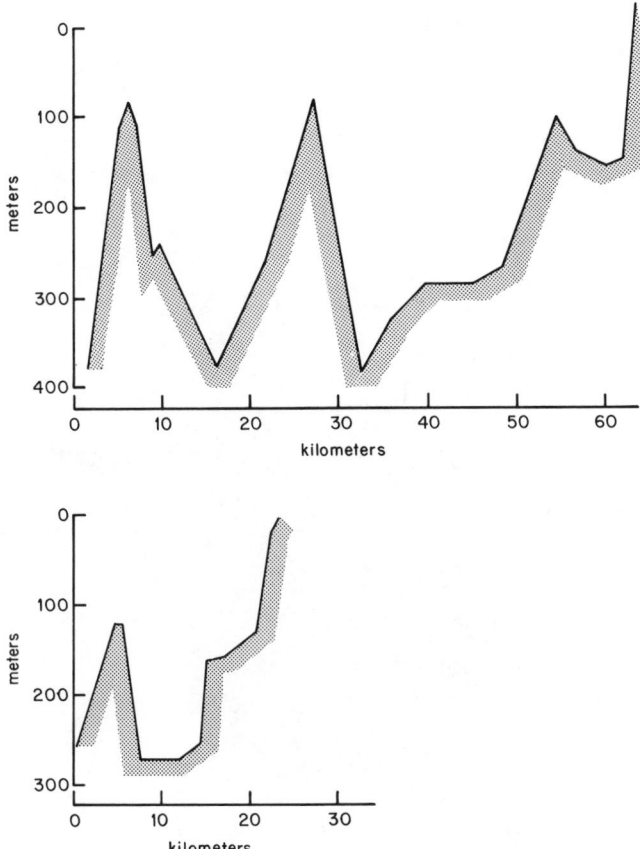

Figure 3. Schematic longitudinal profiles of Boca de Quadra (above) and Smeaton Bay (below) showing comparative sizes of basins.

combined Wilson and Blossom Rivers (Fig. 4: catchment areas S1 and S2). Although direct volume flow measurements are not yet available, it may be estimated that inflow into Wilson Arm is approximately 2.5x that into the head of Boca de Quadra. And, since the former fjord is much smaller, it is not surprising to find that the physical-chemical freshwater "signature" is typically more apparent here. Computed mean monthly freshwater discharge via the Wilson and Blossom Rivers into Wilson Arm is shown in Fig. 5. Such a synthetic construct has quite limited utility, especially because, since the residence times of (summer) precipitation in all the drainage basins surrounding these fjords are relatively short, the daily variations in volume discharge can be very large (Fig. 6). Individual characteristics of each river may also be

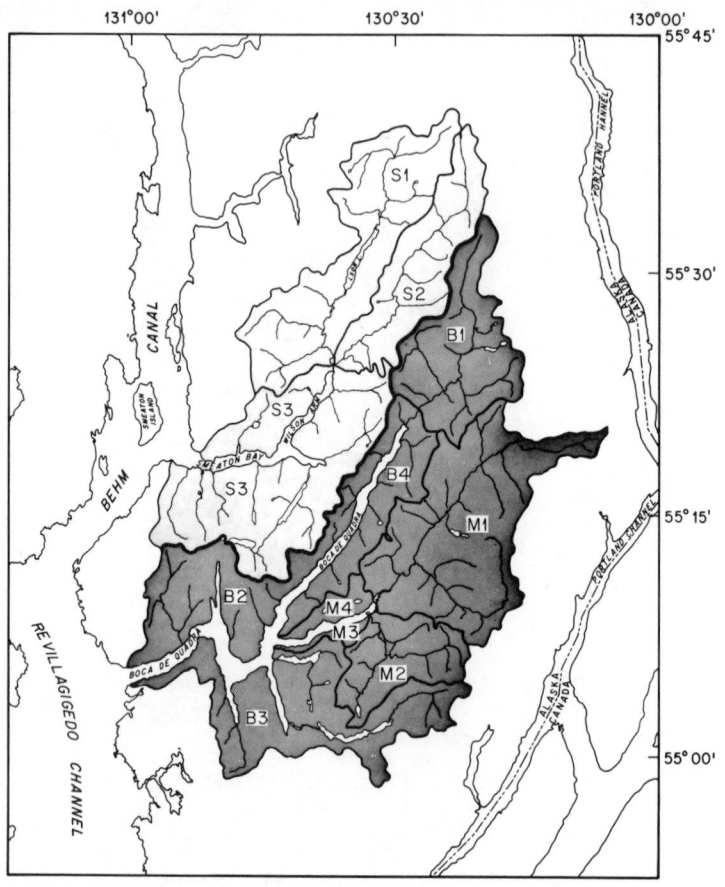

Figure 4. Boca de Quadra and Smeaton Bay catchment areas, from (4).

important. For example, short-term variations in the chemistry of Blossom River, which has a mean flow rate somewhat over half that of the Wilson, can be more extreme than that observed for the Wilson River (e.g., see Fig. 7). Possibly, this may be attributable to the intervention of Wilson Lake within the catchment area of the latter river. And there are major year-to-year variations, depending on rainfall patterns.

Nevertheless, the bimodal pattern of Fig. 5 is reproduced also from direct measurement of the Keta discharge, and is generally typical of this coastline (5). The basic pattern, representing maximum rainfall drainage in the fall and release of stored precipitation (snow) in the spring, has important geochemical implications. Fig. 7 shows the seasonal concentrations of total suspended

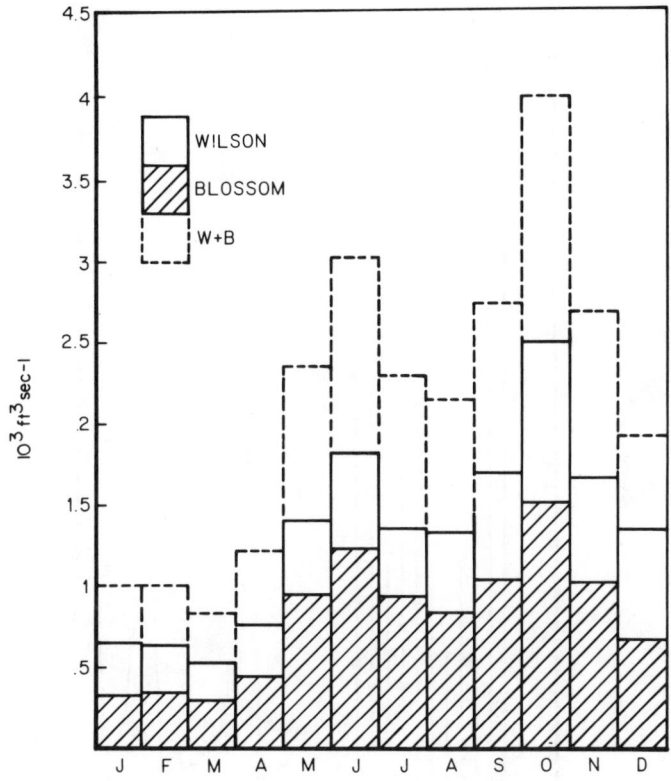

Figure 5. Synthetic monthly volume discharge rates from the Wilson and Blossom Rivers (computations courtesy L. Bartos, U.S. Forest Service).

sediment (> 0.4 μ m; mg/L) close to the mouths of the Wilson and Blossom Rivers, and at a standard confluence locality (Stations WB, WR, and WB, respectively). Over this specific one-year (October 1980-81) cycle, sediment load maxima in the integrated influx in January and early July follow periods of elevated discharge (Fig. 5; in 1980 maximum volume flow is believed to have occurred in December), and possibly represent erosional transport during periods of reduced dilution. Soluble SiO_4-Si contents, the product of alumino-silicate erosion, are also elevated at these times (Fig. 8).

Secondary peak loads are apparent (Fig. 7) in the spring, and in fall to early winter. The spring peak is paralleled by a particulate organic carbon (POC) concentration maximum in the combined river flow (Fig. 9);

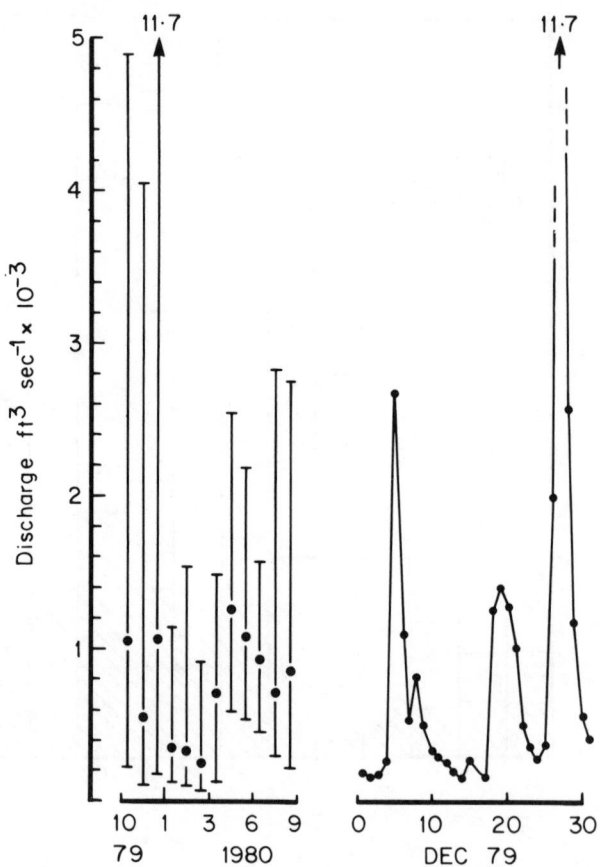

Figure 6. Variation in Keta River flow rates.
        Left: mean monthly rate and range, October 1979
        - September 1980.
        Right: mean daily rate, December 1979.

this may reflect early flushing of the low-lying (muskeg) terrain. Maximum POC concentrations occur, however, in late summer to early fall (September, 1980: Fig. 9), apparently slightly preceding the period of maximum (diluting) volume discharge. This pattern is characteristic also of the Keta and Marten discharge into Boca de Quadra. The particulate sediment at this time consists predominantly of organic material (> 80% POM), and the flux of terrigenous organic carbon into both fjords is very large. Fig. 10 shows computed inputs of carbon from the major rivers into Boca de Quadra for the

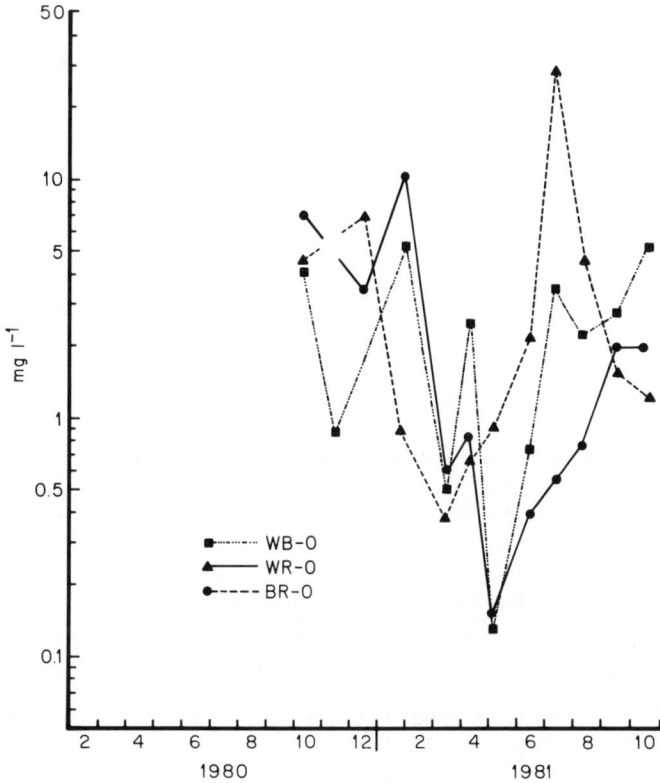

Figure 7. Time-series (October 1980 - October 1981) total suspended sediment load concentrations (mg/L) near the mouths of the Wilson and Blossom Rivers (Stations WR-0 and BR-0) and in the combined rivers (Station WB-0).

months in 1980 for which volume flow data are presently available.

Riverine concentrations of the inorganic nutrient species NO_3-N, NO_2-N and PO_4-P follow the seasonal POC trend described. There is a distinctive NH_4-N concentration maximum in late summer to early fall (Fig. 11). This may have some significance with regard to phytoplankton production at this time, but the flux is not thought to be large. In general, mean riverine concentrations are much lower than in the near-surface marine waters, and we calculate that nutrients supplied by the rivers are an insignificant source for primary production. In October and December 1980, for example, the surface waters of Wilson Arm were low in all the major

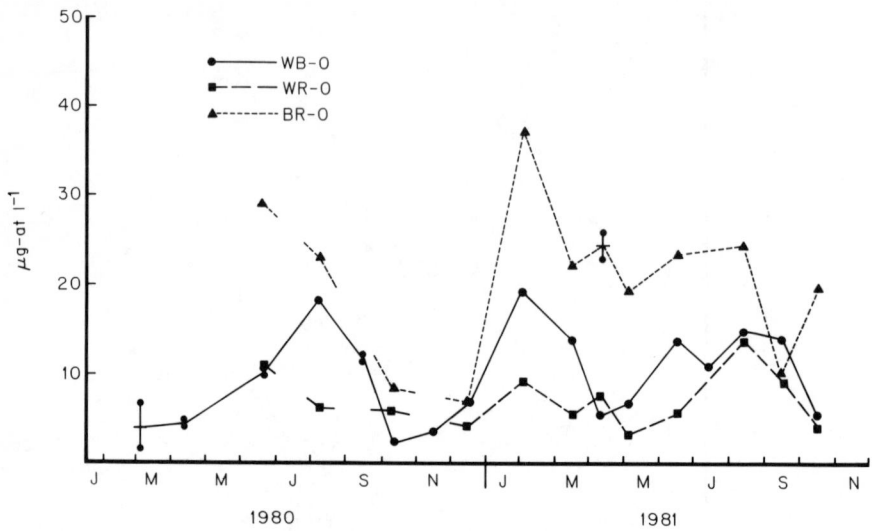

Figure 8. Seasonal concentrations of SiO_4-Si (μg-at/L) in the Wilson (WR-0), Blossom (BR-0) and combined rivers (WB-0), 1980-81.

inorganic nutrient species due, not to biological uptake, but to freshwater dilution.

SEDIMENT

It is apparent from the above discussion that one major source of sediment in the Smeaton Bay-Wilson Arm and Boca de Quadra fjords is terrigenous material which, at certain times of the year, is predominantly organic; Fig. 12 shows the resultant seasonal distribution at the head of Wilson Arm. Fig. 13 illustrates the time-series (1980) distribution of total suspended sediment within the deep central basin of Boca de Quadra. The spring maximum shown here is not riverine, but represents another source: the autochthonous spring (primarily diatomaceous) phytoplankton bloom. High concentrations of particulate material ("turbidity") are present only in the near-surface fjord waters through the summer (Figs. 12 and 13). This is partially due to reduced settling rates within this well stratified zone, but is also a reflection of high rates of consumption and decomposition of the labile organic component. Fig. 14 demonstrates the general summer depth pattern. It should be noted that except at the surface, and locally adjacent to the sediment floor,

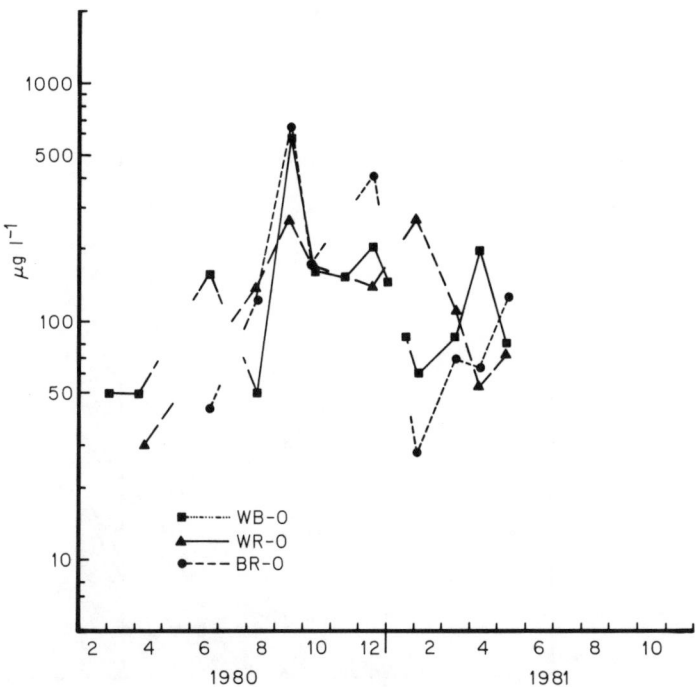

Figure 9. Time-series (March 1980 - May 1981) particulate organic carbon concentrations (POC: µg/L) near the mouths of the Wilson and Blossom Rivers (Stations WR-0 and BR-0) and in the combined rivers (Station WB-0).

the natural particulate sediment concentration range through most of the column is very low (<0.5 mg/L; see also Fig. 13). The near-bottom enhancement, attributable to resuspension (including the effects of bed-load transport), is only relatively well pronounced at the heads of the inlets (Fig. 15) through much of the year, and within the basins primarily at times during the summer season (Fig. 13). Nebert and Burrell (4; see also Chapter 9) have shown that replacement of the deep basin water occurs at this time, and this could constitute the resuspension mechanism within the basins. However, although not directly observed, sporadic movement of sediment along the fjord floor is known to occur (3), and to be especially prevalent at the heads of the primary inlets.

Irregular bed-load transport (turbidity flow, slumping) can locally transport very large quantities of

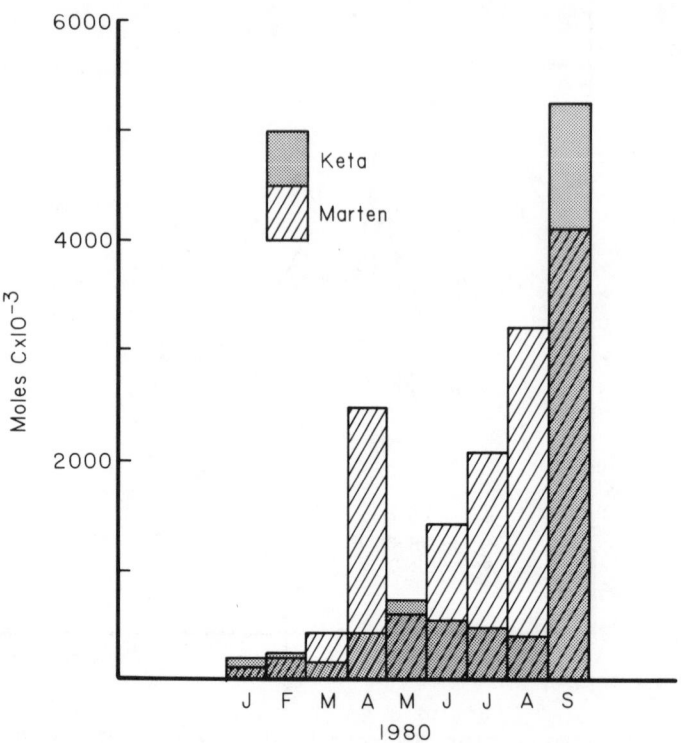

Figure 10. Computed mean monthly flux of carbon (moles C x 10^{-3} for each calendar month) via the Keta and Marten Rivers into the Boca de Quadra system, January - September 1980.

sediment: far in excess, and obliterating the effects, of "quiet" sedimentation downward through the column. The small innermost basin of Boca de Quadra protects the deep central basin from the effects of such flow originating from the head of the fjord, but there is no physical barrier between Wilson Arm and the Smeaton Bay basin (Fig. 3), and such bottom flow occurs along Wilson Arm, and within the basin. Thus Fig. 16 illustrates two depth profiles of unsupported Pb-210 from cores recovered from the Smeaton Bay basin in December 1980, and in October 1981. The later profile demonstrates the presence of contemporaneous, homogeneous sediment, greater than 10 cm in depth, which was absent at the time of the previous sampling.

The bottom sediments in the deep central basin of Boca de Quadra are:

  ·fine-grained: mean   grain   size   around   0.5   μ m,

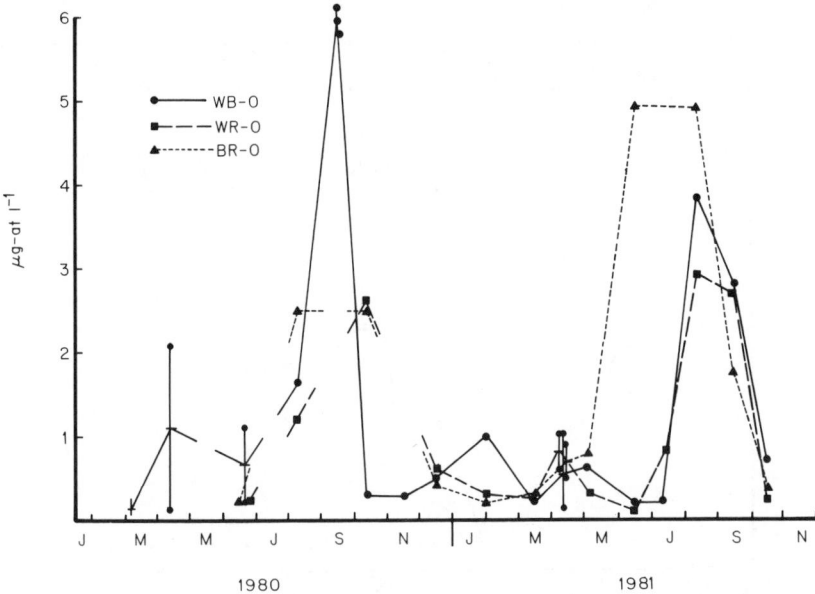

Figure 11. Seasonal concentrations of NH_4-N (µg-at/L) in the Wilson (WR-O), Blossom (BR-O) and combined rivers (WB-O), 1980-81.

(silt-clay boundary range);
˙compact: low porosity (40-50% water content);
˙organic rich: 4-5% POC range.

The sediment-water interface is sharp (as observed directly from a manned submersible within the innermost basin, and indirectly in the central basin from numerous core samples), and there is no evidence for significant bioturbation below the aerobic surface (ca. 1-2 cm) zone. Given these physical-chemical characteristics, a mechanically cohesive sediment would be expected, i.e., one not easily resuspended at moderate bottom current shear. The "quiet" sedimentation rate measured within the basins of both fjords is in the range 0.3-1.0 cm/year.

CARBON FLOW

Seasonal primary production in Boca de Quadra in 1980 (Fig. 17) followed the classic high-latitude pattern: an intense spring (diatom) bloom, and a very secondary fall peak. (Further north in Alaska, greatly increased turbidity tends to suppress the latter bloom; Fig. 17 includes comparative data from Resurrection Bay, S.

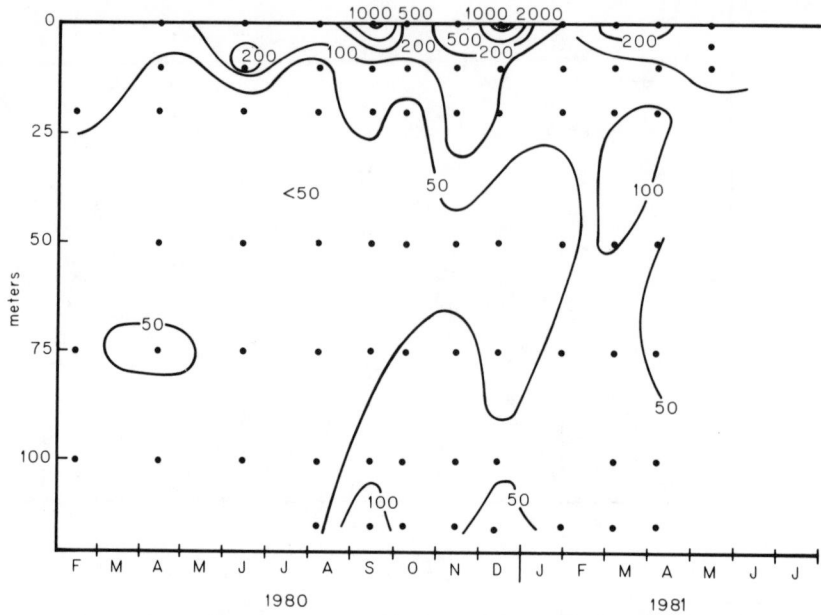

Figure 12. Time-series distribution of particulate organic carbon (POC: μg/L) at the head of Wilson Arm, February 1980 - May 1981.

Central Alaska.) The factor or factors that trigger the spring bloom at this locality, probably irradiation (6), are not presently known. Annual stratification of the surface waters intensifies around March - April (Fig. 18), and the timing and impact of the spring thaw appears to be locally important. Maximum carbon uptake rate and chlorophyll concentration values at the time of the primary bloom (see the sediment distribution data of Fig. 13), and through the summer season (e.g., Fig. 19) are typically sub-surface, in water greater than 25 °/oo salinity. Since this horizon appears generally to coincide with a pronounced T-S discontinuity, this characteristic vertical distribution is probably due to surface advection. Based on the few presently available data (Fig. 17), seasonal primary production in Smeaton Bay was much lower than in Boca de Quadra in 1980.

An important biological feature of these fjords is the apparently reduced importance of the pelagic component: the benthos (Fig. 20) is quantitatively more important (8), and hence mine tailings disposal must seriously disrupt the fjord ecology as it presently exists; such impact may, of course, be only temporary, on some presently indeterminate recovery time-scale. It is

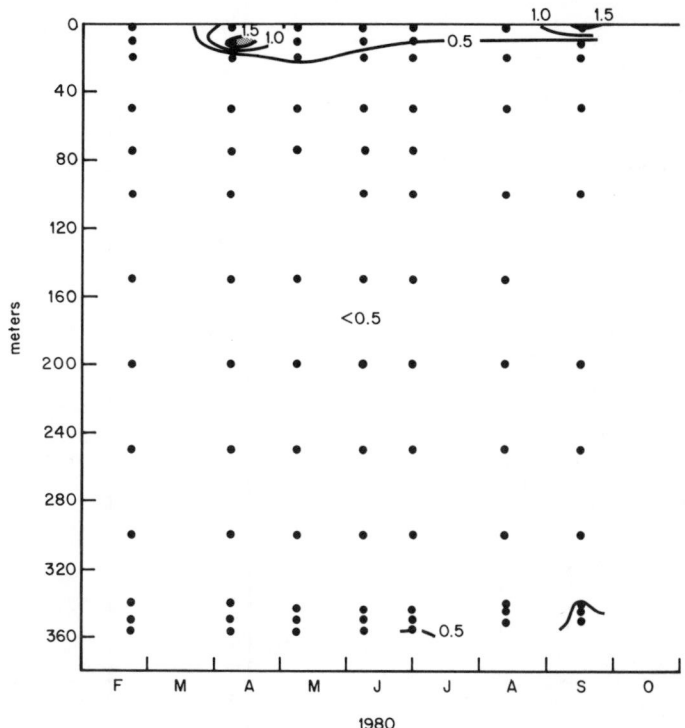

Figure 13. Time-series distribution of total suspended sediment load (> 0.4 μm; mg/L) within the central basin of Boca de Quadra (Station BQ-9: Fig. 2), February - September 1980.

believed (see below) that imported terrigenous carbon is largely refractory, and, although no definitive data are available, littoral productivity is also believed to be an insignificant food source to the benthos. Thus, although there is much consumption and recycling of phytoplankton in the euphotic zone, there is believed to be a relatively large transfer of primary produced carbon into the basins. This is a commonly observed pattern in high-latitude regions, e.g. (9). If the primary vernal diatom bloom occurs prior to near-surface stratification and buildup of the predator populations, as appeared to be the case in 1980 (Figs. 18 and 20), then a major downward transfer event could occur at this time (10). Later in the summer, productivity in the well stratified surface zone is nitrogen limited (Fig. 19).

Fig. 21 shows preliminary values for part of the carbon (energy) flow in Boca de Quadra in 1980. In the

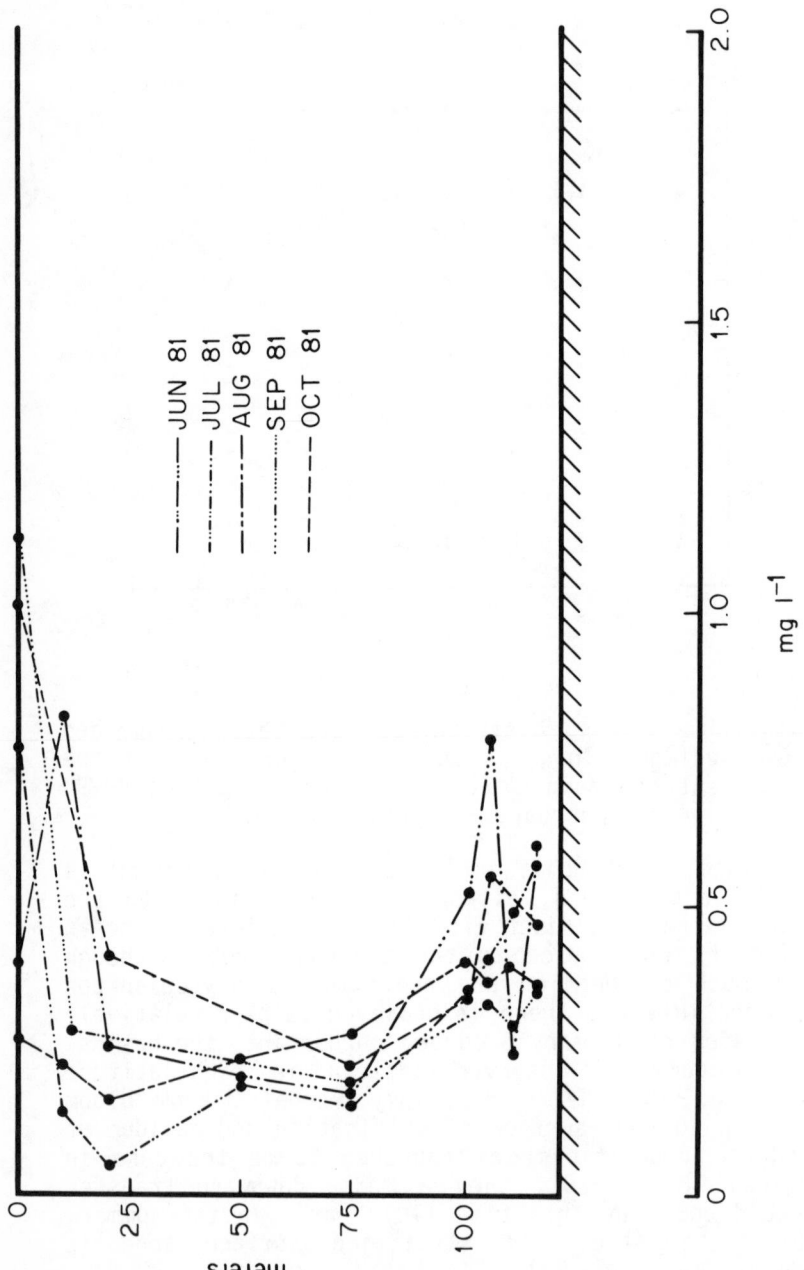

Figure 14. Time series vertical profiles of total suspended sediment load (>0.4 μm; mg/L) at the head of Wilson Arm, summer 1981.

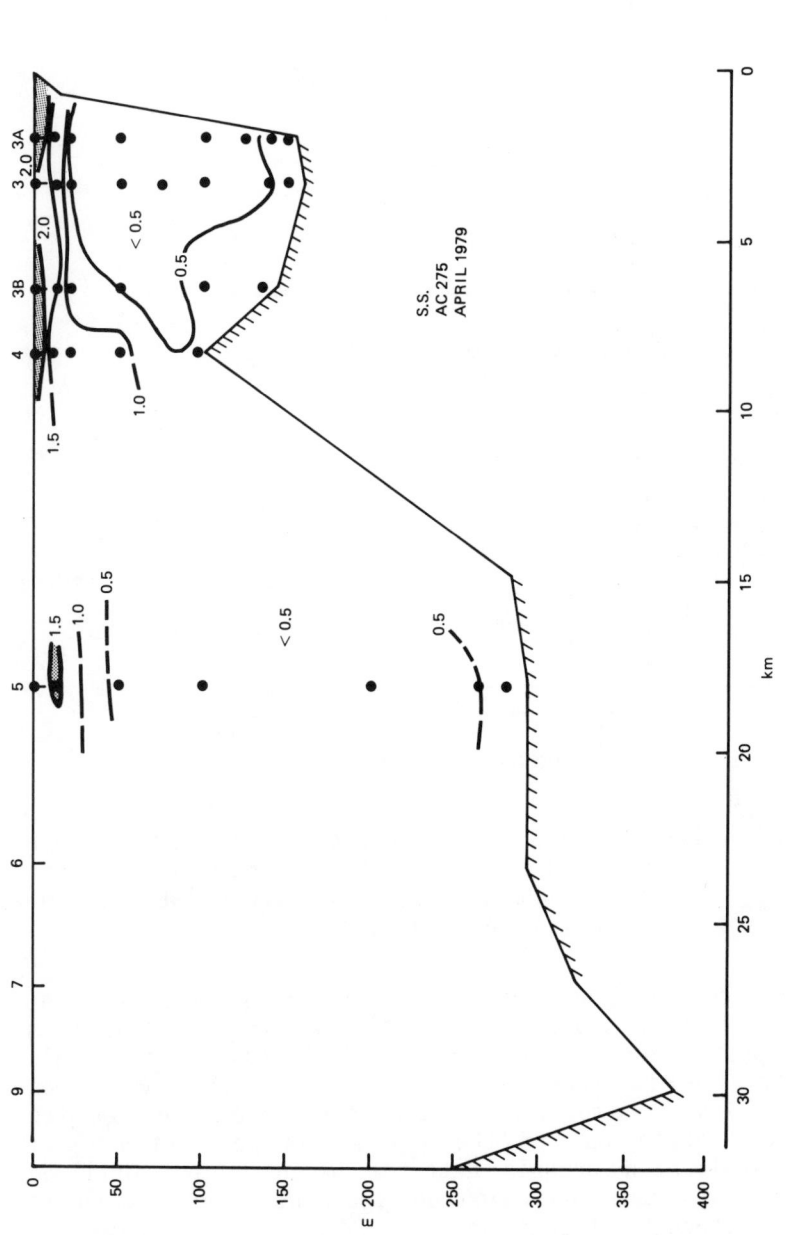

Figure 15. Longitudinal section showing distribution of total suspended sediment load (> 0.4µm; mg/L) within Boca de Quadra above Kite Island (Fig. 2), April 1979.

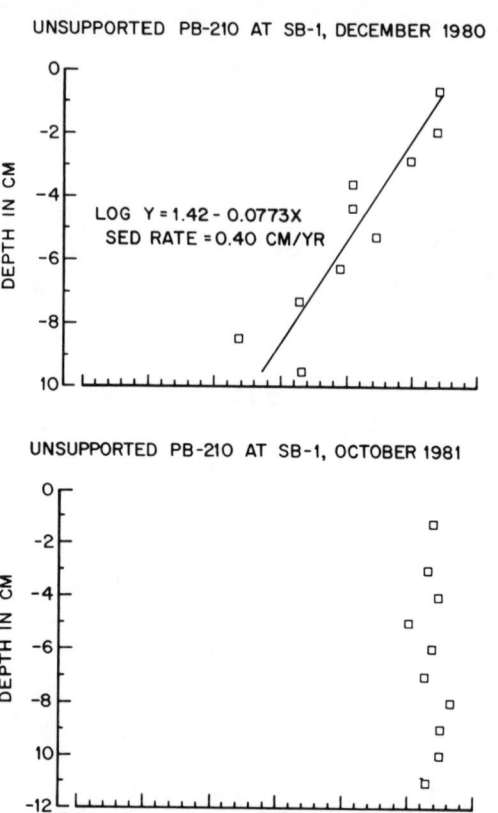

Figure 16. Log profiles of unsupported Pb-210 for sediment cores recovered from the Smeaton Bay basin: December 1980 (above) and October 1981 (below) (S. Sugai, unpublished data).

absence of knowledge of the quantitative importance of other potential sources, a relatively large leakage from the euphotic zone is indicated, and the estimated benthic productivity is then some 16% of the primary carbon uptake. Another important feature stems from the computed loss of carbon out of the system via sediment burial: i.e., sedimented organic detritus which is not utilized within the basin and surface sediments. This suggests that, although there is a large natural flux of terrigenous organic material into the fjords, and although the sediments are carbon rich, much of this detritus is refractory and apparently is not readily degraded during the relatively short residence time within the

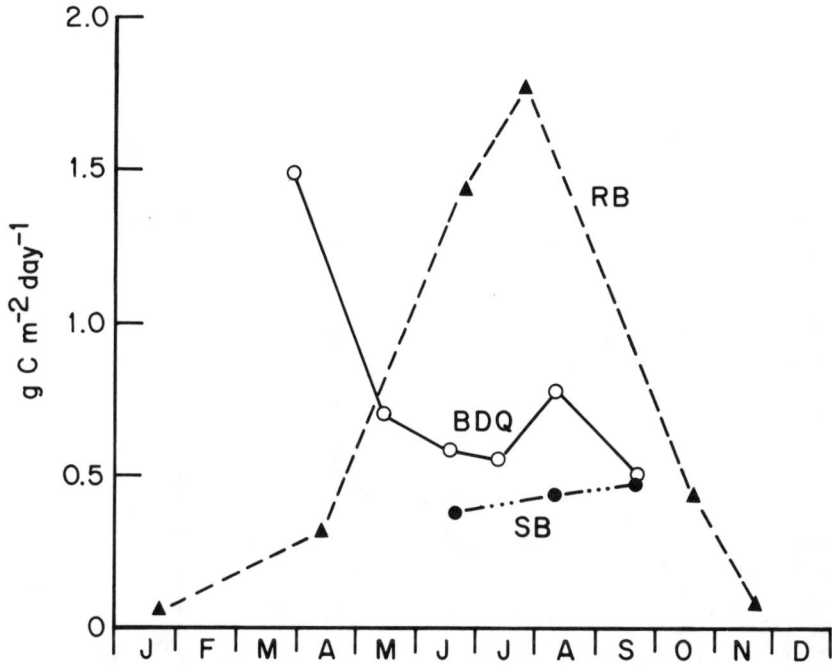

Figure 17. Integrated daily C-14 uptake rates (g C/m^2/day) in Boca de Quadra (BDQ) and Smeaton Bay (SB) through the 1980 season, and in Resurrection Bay, S. Central Alaska (RB) through the summer of 1974, from (7).

near-surface zone. An understanding of the relative importance of the rates and magnitudes of these processes is important with regard to the deep water distribution of dissolved oxygen.

DISSOLVED OXYGEN

The distribution of dissolved oxygen is always a factor of major concern in cases of marine pollution. And this is especially so in fjord basins because of seasonal restrictions on the free exchange of water; and hence limitations on the replenishment of dissolved oxygen to depth where it is being depleted by oxidation (respiration). Work in other Alaskan fjords (11) has demonstrated that waste dumping of organics can significantly deplete available deep basin oxygen, and that the effect is seasonal: i.e., the basin is vulnerable

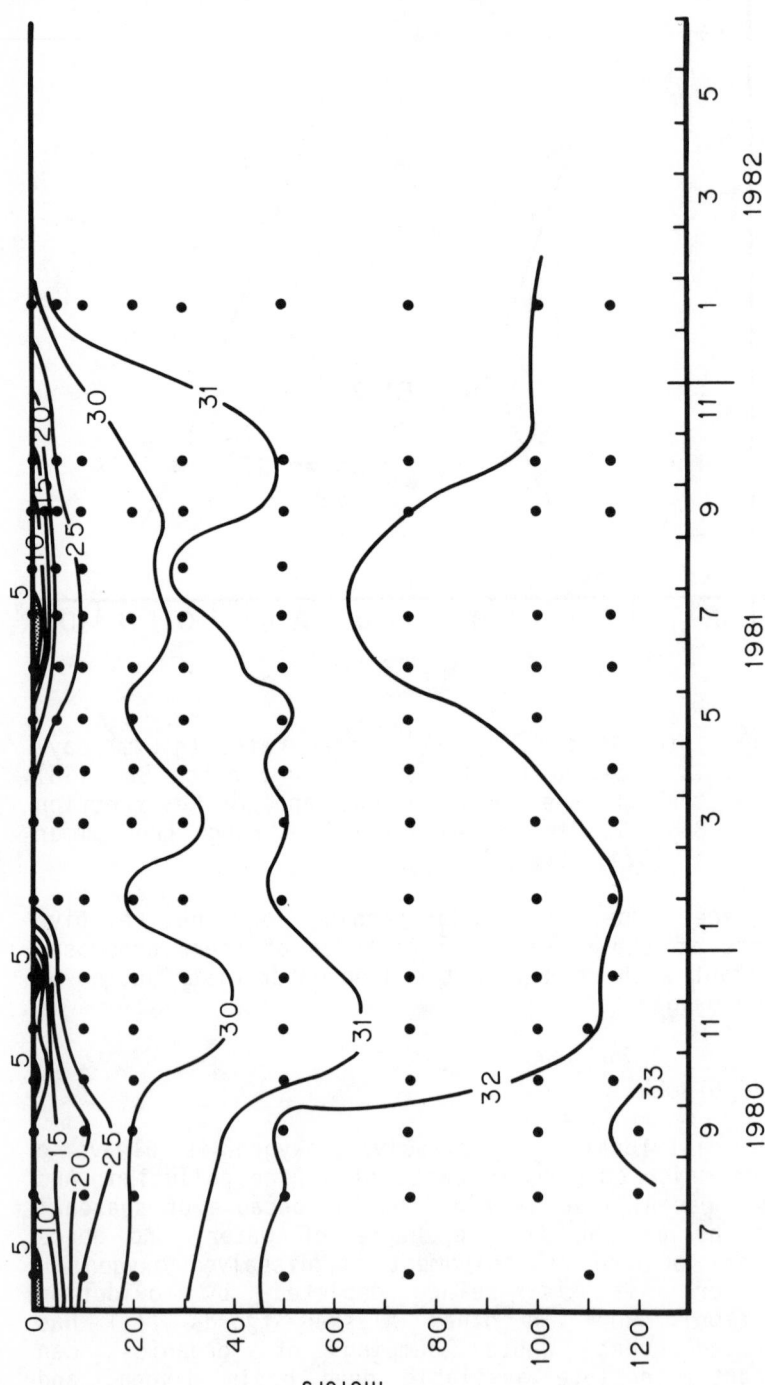

Figure 18. Time-series distribution of salinity (°/oo) at the head of Wilson Arm, June 1980 – January 1982.

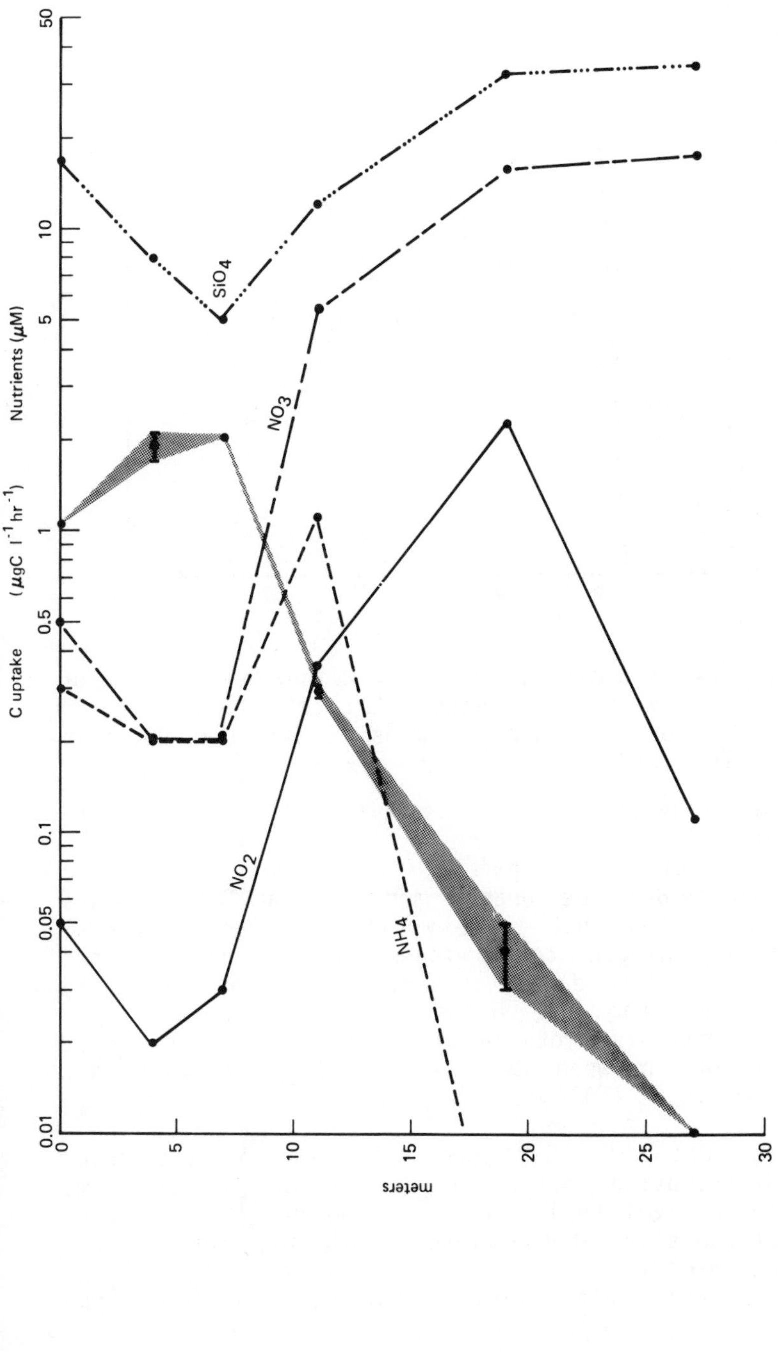

Figure 19. Vertical profiles of C-14 uptake rates (μg/L/hr) and nutrient species (μg-at/L) through the euphotic zone of the Smeaton Bay basin, August 1980.

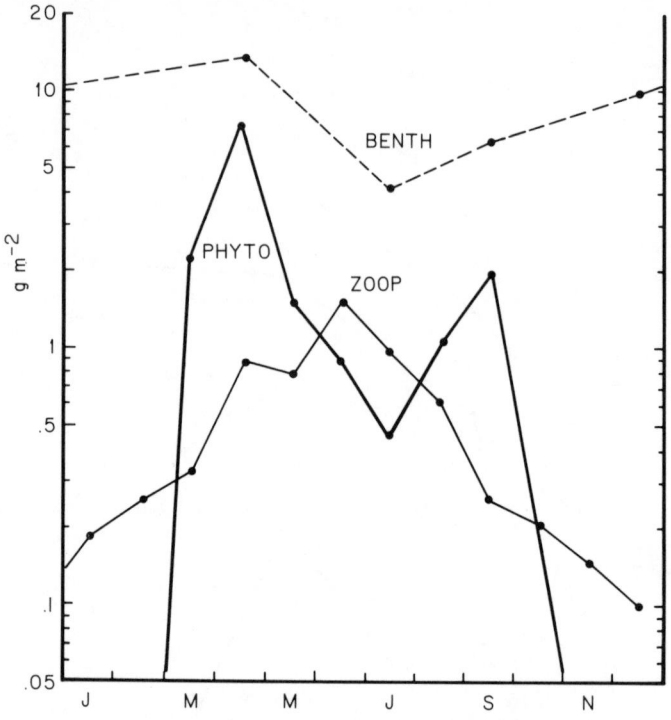

Figure 20. Mean standing stocks (dry weight g/m^2) of major
biological components within the central basin
of Boca de Quadra; composite year 1979-80: data
from (8).

at certain times of the year, usually at some time during
the winter.

Because of the apparent high input of organic
material into Boca de Quadra and Smeaton Bay, it was
initially expected that there would be a marked natural
depletion of oxygen deep within the basins. On the
evidence analyzed to date, however, this has not proved to
be the case. Fig. 22 illustrates time-series depth
profiles of dissolved oxygen within the central basin of
Boca de Quadra between two summer flushing seasons (4).
The sequence is as follows:

1. The October (1979) profile represents the depth
   distribution after completion of summer time
   (advective) replacements of the deep basin water,
   at the start of the winter "isolation" period.
2. December - April distributions show slight
   depletion with minimum concentrations at the

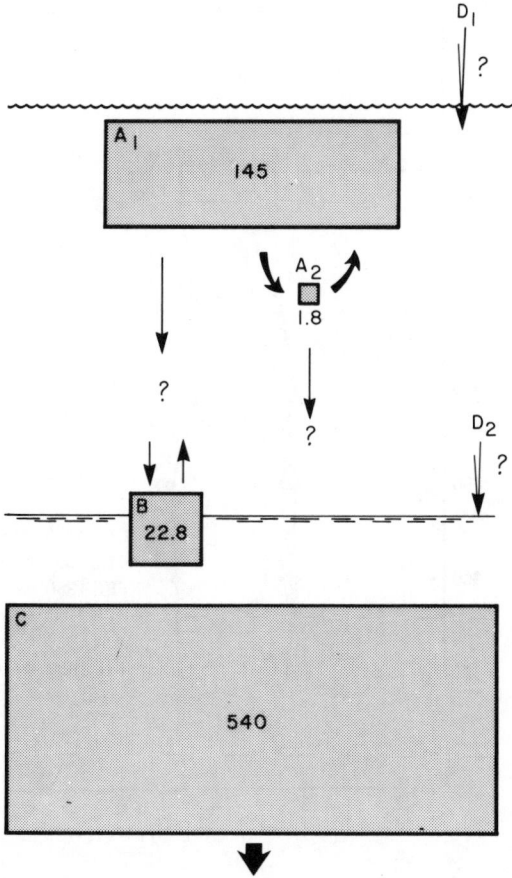

Figure 21. Schematic of partial carbon flow (g $C/m^2/year$) within the central basin of Boca de Quadra.
A_1 and A_2 - Phyto- and zooplankton production. Data from (8).
B - Benthic production. Data from (8).
C - C lost via burial.
D - Keta and Marten River flux.

bottom in April (1980).
3. In June a partial replacement brings higher oxygen-content water into the upper basin.
4. Complete flushing results in a homogeneous column in July.

Oxygen concentrations did not fall below 3 ml/L at the bottom of the basin through the 1979-80 winter. In the Smeaton Bay basin the "starting" concentration at the

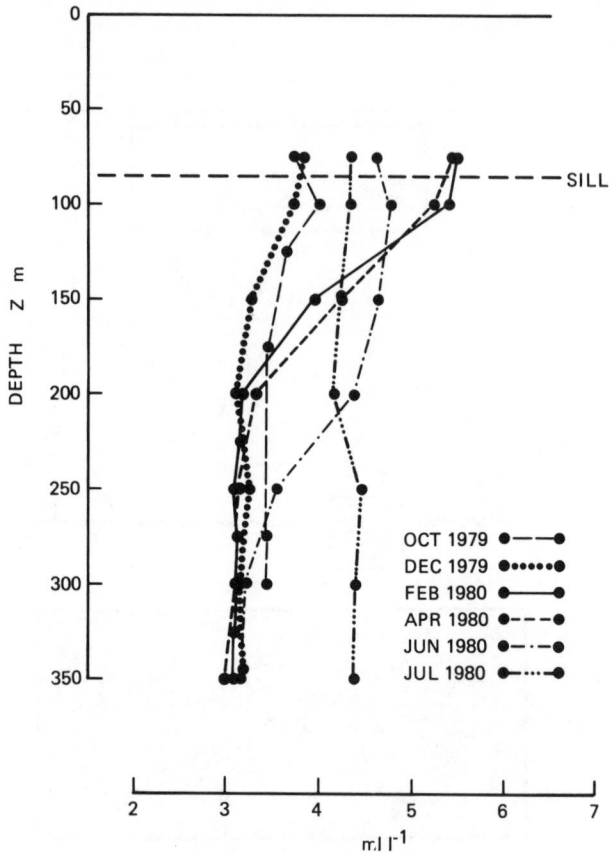

Figure 22. Time-series vertical profiles of dissolved oxygen (mL/L) within the central basin of Boca de Quadra (Station BQ-9: Fig. 2), October 1979 - July 1980.

beginning of the 1980-81 winter is less ($\sim$ 3.0 ml/L) than in Boca de Quadra, decreasing to $\sim$ 2.6 ml/L in December (Fig. 23), and increasing again through June. Although, at this time, significant advective transport of oxygen into the basins through the winter period cannot be ruled out even assuming a high diffusive flux through a near-homogeneous column it would appear that consumption rates at the bottom are relatively slight. This would suggest, a priori, relatively low benthic productivity. (Based on presently available information, a value of around 23 g/m^2 has been estimated for the macro- and meiofauna: Fig. 21.) It would then also follow that a

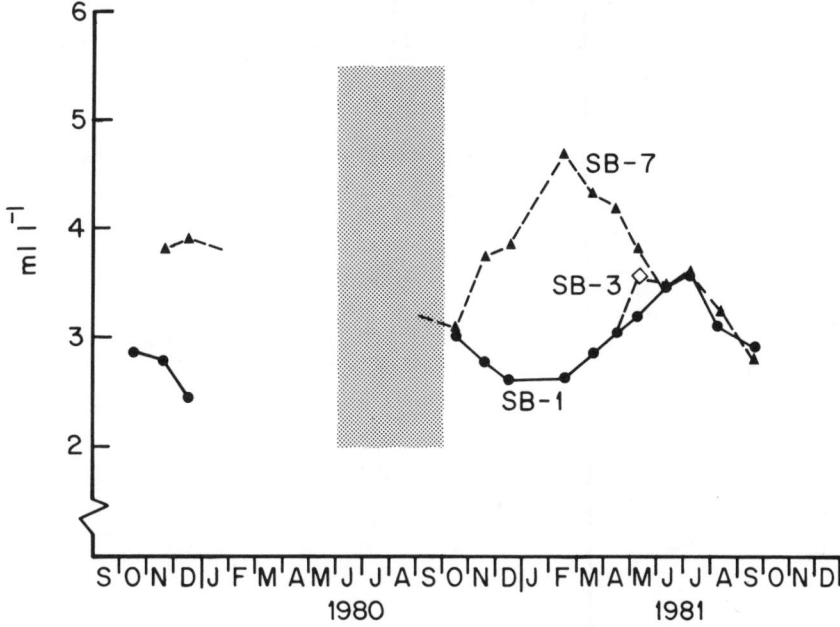

Figure 23. Time-series distribution of dissolved oxygen (mL/L) at the bottom of the Smeaton Bay basin (solid line) and outside the fjord at sill height (broken line). The shaded period is the time of advective replacement of the water within the basin.

large fraction, probably most, of the particulate organic material presently entering the fjords and accumulating on the bottom is refractory: it is estimated that in the order of 500 g C/m^2 is lost annually via sediment burial. It should, however, be noted that the artificial addition of small quantities of reactive carbon can locally generate significant changes. For example, in the vicinity of the mooring float which has been established for several years at the head of Wilson Arm, we have observed evolution of methane, indicating that anoxic conditions exist to the sediment interface.

The converse of oxygen utilization (degradation of organic matter) is regeneration of inorganic nutrients. Resupply of nutrient species from the sediments back into the euphotic zone is now generally recognized, e.g. (12), as a major source for estuarine and coastal productivity. This is particularly so in the type of fjords described where a major flux of carbon to the benthos has been

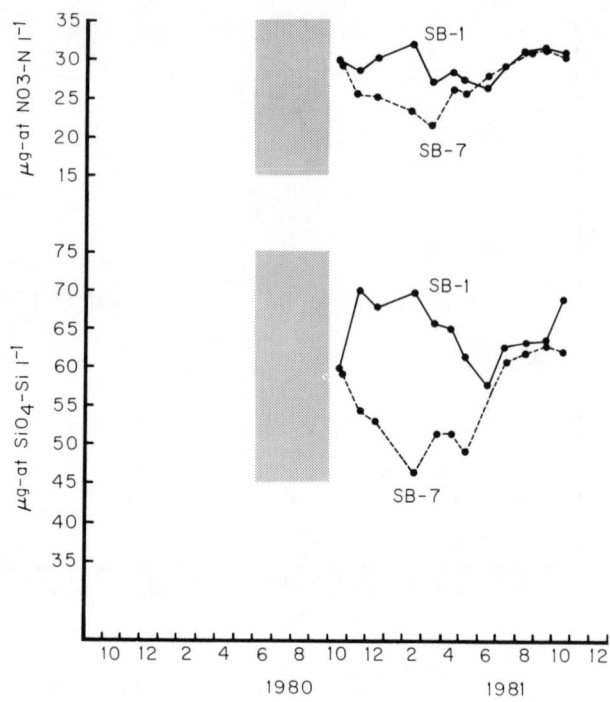

Figure 24. Time-series distribution of nutrients (μg-at/L) at the bottom of the Smeaton Bay basin (solid line) and outside the fjord at sill height (broken line). Above: NO_3-N. Below: SiO_4-Si. The shaded period is that of advective replacement of the water to depth within the basin.

advocated. Fig. 24 demonstrates regeneration of NO_3-N and SiO_4-Si within the basin (primarily from the basin sediments) through the 1980-81 winter "isolation" period. As noted previously the rivers are not believed to be a significant nutrient source for the fjord surface waters.

HEAVY METALS

In marine waters, the "heavy metals" (as conventionally defined by regulatory agencies) are predominantly partitioned onto or in solid (sediment) phases by various natural physical-chemical mechanisms: soluble concentrations are hence very low under all but the most atypical conditions. Since the sediments are the major repository, the most important pollutional concern is potential re-mobilization if the natural, near-surface sediment geochemical environment is altered. This is a subject about which there are still many uncertainties, but important factors include location of the redox boundary, and interstitial water sulfide and organic concentrations. The commonly observed pattern is for "available" (i.e., largely the fraction which is not incorporated into the alumino-silicate structures) heavy metal concentrations in the sediments to be correlated with POC contents, and with sediment grain fineness. This has been shown to be so for the Boca de Quadra basin sediments (3), and appears, from preliminary results (B. Spell, unpublished data), to be so for the Smeaton Bay sediments also. The bulk of the material proposed for disposal into these fjords consists of natural alumino-silicate material. Possible effects due to, for example, fine grinding leading to increased areas of unweathered mineral surfaces, and the potential addition of exotic processing chemicals, are best evaluated by long-term leaching experiments.

ACKNOWLEDGEMENTS

The work reported here has been commissioned and supported by the United States Borax & Chemical Corporation and the Pacific Coast Molybdenum Company, and represents the combined efforts of the Institute of Marine Science project team. In particular, the work of Barry Spell and Susan Sugai, and also that of Frank Flynn and Donna Weihs, is acknowledged.

REFERENCES

1. Pickard, G.L. "Some Oceanographic Characteristics of the Larger Inlets of Southeast Alaska", J. Fish. Res. Bd. Can. 24: 1475 (1967).
2. Gray, J.S. The Ecology of Marine Sediments (Cambridge: Cambridge University Press, 1981).
3. Burrell, D.C. "Marine Environmental Studies in Boca de Quadra and Smeaton Bay: Chemical and Geochemical, 1980", Report R82-2, Institute of Marine Science, University of Alaska (in press).
4. Nebert, D.L. and D.C. Burrell. "Marine Environmental Studies in Boca de Quadra and Smeaton Bay: Physical Oceanography, 1980", Institute of Marine Science, University of Alaska (1981).
5. Pickard, G.L. and B.R. Stanton. "Pacific Fjords: A Review of Their Water Characteristics", Report No. 34,

Department of Oceanography, University of British Columbia (1979).

6. Lenz, J. "Phytoplankton Standing Stock and Primary Production in the Western Baltic", Kieler Meeresforch., Sonderh. 5:29 (1981).

7. Heggie, D.T. and D.C. Burrell. "Hydrography, Nutrient Chemistry and Primary Productivity of Resurrection Bay, Alaska 1972-1975", Report No. R71-2, Institute of Marine Science, University of Alaska, Fairbanks (1977).

8. VTN Consolidated, Inc. "Boca de Quadra Baseline Report: Coastal and Marine Biology Program, Quartz Hill Molybdenum Project, Southeast Alaska", Report to U.S. Borax & Chemical Corporation (1980).

9. Peterson, G.H. and M.A. Curtis. "Differences in Energy Flow through Major Components of Subarctic, Temperate and Tropical Marine Ecosystems", Dana 1:53 (1980).

10. Smetacek, V. "Annual Cycle of Sedimentation in Relation to Plankton Ecology, Western Kiel Bight", Ophelia Suppl. 1:65 (1980).

11. Burrell, D.C. and D.T. Heggie. "Patterns of Oxygen Supply and Consumption in Two Alaskan Fjord Basins", 11th Int. Cong. Sediment., Hamilton, Ontario, August (1982).

12. Fisher, T.R., P.R. Carlson and R.T. Barber. "Sediment Nutrient Regeneration in Three North Carolina Estuaries. Est. Coast. Shelf. Science 14: 101 (1982).

DISCUSSION

Question: What is your P/B ratio for the benthos? What calculation do you use?

Answer: I used various figures for the infauna. I have estimated values for the meiofauna because I have even had to estimate their biomass. I used the arctic papers that were produced by the Danes fairly recently, mostly for comparative analysis of benthic productivity, comparing Greenland with temperate regions. The question was what value did I use to convert my standing stock of benthic biomass into a productivity figure. The questioner knows that I had to feed in a P/B ratio and he wanted to know exactly what I was using. We could be off here by a factor of two or three.

Question: Can you couple salmon into your calculations?

Answer: It would require much more detail looking at the

biological system which really isn't my job. What I showed
was a gross mean annual carbon flow. On that basis salmon
don't figure in. The biomass and everything else are not
large enough to come in the calculation. They are not
coupled into the pelagic system because the only way they
actually affect it is that the fry migrate out through the
system. It is of interest to note, perhaps, that the
primary productivity is lowest in that fjord which has the
greatest interest as far as salmon goes. This I suppose,
in a fashion, illustrates what de-coupling is. The primary
productivity is quite low, but the salmon really have
nothing to do with the primary productivity because they
depend on other factors, like suitability of spawning
beds, etc. But this is very rough and one has to go into
detail through the year to look at their feeding on the
zooplankton in the spring and it wasn't my job to do this.
I was looking at the carbon flow effect in the sea. If you
impact the system, one of the major components of the
impact will be on the benthic environment.

CHAPTER 11

COMPARATIVE BENTHIC ECOLOGY
OF TWO SOUTHEAST ALASKAN FJORDS,
BOCA DE QUADRA AND WILSON ARM/SMEATON BAY

R.L. Cimberg
VTN Oregon, Inc.,
25115 S.W. Parkway
Wilsonville, Oregon 97070
U.S.A.

Benthic ecological studies have been conducted since 1978 in the rocky intertidal, rocky subtidal, soft-bottom intertidal and soft-bottom subtidal habitats of both the Boca de Quadra and Wilson Arm/Smeaton Bay fjords. Studies have been conducted on the dominant organisms and communities and the factors that regulate their distributions.

The rocky intertidal is dominated by the barnacle Balanus glandula, the rockweed Fucus distichus and the mussel Mytilus edulis. The distribution and abundance of these organisms correspond with salinity values, tidal elevation and substrate angle. The abundance of the associated epifauna and flora varies with the tidal height, with season and between stations. The rocky subtidal is dominated by kelps (Agarum fimbriatum and Laminaria spp.), brachiopods, serpulid polychaetes and tunicates. Adjacent subtidal soft-bottom substrates are dominated by eelgrass, Dungeness crabs and flatfish. The soft-bottom intertidal region is inhabited, from high to low tides, by the following zones: herb/grass; sedge; rockweed/cobble; mussel/cobble; and eelgrass. The soft-bottom subtidal region is characterized by three to four major communities whose distributions correspond with depth, sediment grain size and organics.

The benthic biological systems of these two nearby fjords appear to have similar biotic communities regulated by the same environmental factors.

INTRODUCTION

VTN Environmental Sciences has been involved in marine biological studies for the Quartz Hill Molybdenum Project since 1978. These studies are part of the comprehensive Environmental Baseline Data Collection Program instituted by U.S. Borax & Chemical Corporation on behalf of the Pacific Coast Molybdenum Company.

The marine biological studies encompass the plankton, fish (including shellfish) and benthos. This chapter presents results from one component only, the distributional investigations of animals (other than fish and crabs) and attached plants in Boca de Quadra and Wilson Arm/Smeaton Bay. These results allow a description of the benthic communities occupying the sides and troughs of the two fjords, and provides a basis for predicting potential impacts from tailings disposal to the fjords. Benthic communities represent a major biological component of the energy flow and biogeochemical cycling processes in marine ecosystems. The balance of VTN's distributional and biological investigations have been or will be reported elsewhere (e.g. 1, 2, 3). The relevant physical and chemical oceanographic investigations leading to understanding the causes for such distributional patterns, and permitting predictions of impact on the benthic and other biological communities, are being conducted in both fjords by the University of Alaska, and are reported in Chapters 9 and 10.

The benthic environment is inhabited by organisms which live either on or in the substrate. Consequently a habitat approach was used for the benthic study. The benthos in each fjord was categorized into four habitats based on the nature of the substrate (rocky or soft-bottom) and exposure to tides (intertidal or subtidal). These habitats are:

˙the rocky intertidal;
˙the rocky subtidal;
˙the soft-bottom intertidal; and
˙the soft-bottom subtidal.

Each habitat was occupied by different organisms and usually required different sampling techniques. Therefore, each habitat is discussed separately in the following sections.

METHODS

The following section briefly describes the

scientific methods used to collect organisms from the four benthic habitats in Boca de Quadra and Wilson Arm/Smeaton Bay. The methodology presented below is summarized; a detailed discussion has been presented elsewhere (1, 2, 3).

The comprehensive baseline study was conducted in Boca de Quadra in 1979/80 (1) with continued monitoring in 1981 (3). The Wilson Arm/Smeaton Bay baseline study was conducted in 1980/81 (2). Monitoring in both fjords is continuing in 1982. During both baseline programs, samples were generally collected quarterly, while sampling for the monitoring studies is being performed during the winter and summer oceanographic periods.

The benthic habitats in both fjords have been sampled extensively at numerous stations. The reader is referred to the VTN reports for the 13 maps showing the locations of the sampling stations (1, 2, 3).

Rocky Intertidal Habitats

Qualitative surveys of intertidal zonational patterns were conducted at reconnaissance stations throughout both fjord systems. Zonational patterns were determined at low tides by recording the dominant species based on estimated cover at each station. Tidal heights for each zone were determined by placing a stadia rod in front of the fjord wall. An observer, standing in a boat 25 yards away, viewed both the intertidal zones and the stadia rod through a pair of binoculars and recorded the elevation corresponding with the breaks between each zone. The relative tidal heights were later adjusted to MLLW (mean lower low water). The substrate angle of each zone was measured with an inclinometer.

Quantitative seasonal data on the biota associated with each zone was determined from quadrats at permanent stations in both fjord systems. At each station, zones of the visually dominant organisms were determined. Three to six replicate quadrats (30 x 50 cm) were randomly established within each zone. During each sampling period, each quadrat was sampled using two procedures. Procedure one involved photographing the quadrats, once with drift algae present and once with drift algae removed, to record the overstory organisms. Slides were analyzed to determine the percent cover of each overstory species. Each slide was viewed under a dissecting microscope at 60X. A plastic sheet, having a field of 100 random dots, was placed over the photographic slide and the number of dots intercepted by each species was counted. These abundance values were transformed to a percent cover for each species.

The second method involved harvesting small,

non-permanent quadrats (15 x 25 cm) adjacent to each permanent quadrat. All organisms found within the non-permanent quadrat were removed during each sampling period. Only those algal plants whose holdfasts were located within the quadrat were removed. The samples were fixed in 10% formalin in the field and returned to the laboratory for analysis. Organisms from each harvested sample were separated, identified to species or the lowest practical level and verified by taxonomic experts. Invertebrates were counted to determine densities per 375 cm^2. Algae were blotted and weighed to the nearest 0.1 g to calculate wet weight biomass per 375 cm^2.

Rocky Subtidal Habitats

Reconnaissance surveys of habitat types and dominant biota were conducted at stations throughout both fjord systems. Observations of the substrate and dominant biota along transect lines were recorded on underwater tape recorders or slates. Photographs and voucher specimens of the dominant species were taken.

Detailed quantitative surveys of the biota at selected sites were conducted in 45 x 10 m study areas. Three different methods were used to sample different sized organisms: small quadrats for the small organisms; large quadrats for the large echinoderms; and a point intercept device for the large algae.

For the small organisms, three random, 30 x 50 cm quadrats were established and marked at depths corresponding with the different biotic assemblages. These quadrats were sampled seasonally using a combination of photographs and field notes, similar to the methods of Bonsack (4). Voucher specimens were collected from outside the permanent quadrats to verify field identifications.

In the laboratory, voucher specimens were identified and/or verified by taxonomic consultants. Abundance values were determined from a combination of field counts and laboratory assessments of photographic slides. These slides were viewed at 60X under a dissecting microscope. Densities (number of organisms/0.15 m^2) were determined for individual plants and solitary animals. Percent cover was determined from slides for all species, including colonial and encrusting forms, in the same manner as described for rocky intertidal organisms.

Large quadrats, 1.5 x 1.5 m, were established for sampling echinoderms, including seastars, sea urchins, sea cucumbers and crinoids. These quadrats were established along six parallel transects, each extending from 0 to 30 ft below MLLW. Density values for each species in each quadrat were recorded on underwater tape recorders or slates.

Abundance values for the large algae, such as Agarum fimbriatum, Laminaria spp., Constantinea subulifera, and Ahnfeltia plicata, were determined at 2.5 ft intervals, between 0 and 30 ft, using a point intercept method along the six parallel transects. A 5 ft bar with 9 equidistant points was held along the wall and the number of points intercepted by each species was counted, and later converted to a percentage.

Soft-bottom Intertidal Habitats

Reconnaisance surveys, using high resolution aerial observations and photographs of the Wilson, Bakewell and Keta River estuaries, were conducted to delineate the major zones based on the dominant biota and/or substrate. The dominant species and their associated habitats were later verified with ground truth observations and samples.

Quantitative surveys of the biota were conducted along transects using stratified random procedures. Each transect was divided into major habitat types. Sample sites within each habitat were randomly chosen along the transect. At each location, three random infaunal samples were taken with a core that measured 15 cm in diameter by 12 cm in depth. All vegetation present on the surface of the core was removed and rinsed. The rinse was poured through a 0.5 mm screen five or more times until no additional animals were collected on the screen. These organisms were placed in labeled plastic bags and fixed with 10% buffered formalin. Each plant species was blotted and weighed. The remainder of the core was transferred to buckets, elutriated with water and decanted through a 0.5 mm screen. This process was repeated five or more times until no additional organisms were collected on the screen. The unscreeened sediment was examined to remove any remaining organisms and discarded. All organisms were transferred to small plastic bags and preserved. Organisms were sorted to species, verified by taxonomic specialists and counted (animals) or weighed (plants).

Soft-bottom Subtidal Habitats

Samples were taken on a grid of stations in both fjord systems to ensure that variability in biota, both along and across the fjords, was investigated. Stations were selected to provide a large enough sample size to detect seasonal and spatial changes within each community.

In addition, five replicates were taken in at least one station within each community to compare variability within and between stations.

Benthic grabs were collected aboard the M/V Redoubt using a $0.1 m^2$ Smith-McIntyre grab for coarse sediments and a $0.1 m^2$ Van Veen grab for silt and clay sediments. Unsuccessful grabs (those which were not full of sediment) were discarded and retaken until a complete and satisfactory sample was obtained. The samples were sieved through 1.0 and 0.5 mm mesh screens, placed in plastic jars, stained with rose bengal and fixed in 10% buffered formalin before being transferred to the laboratory.

The contents of each grab were sieved through 0.5 and 1.0 mm screens in the laboratory. The 1.0 mm fraction was sorted under a dissecting microscope into taxonomic categories (polychaetes, molluscs, benthic crustaceans, echinoderms and other invertebrates) and preserved in 70% alcohol. Wet weights were determined to the nearest 0.01 g for these taxonomic groups. All organisms were then identified to species or to the lowest practical taxonomic level and counted.

Sediment samples were collected at soft-bottom subtidal stations by inserting three plexiglass sediment tubes through the top of the undisturbed grab. The contents of these tubes were transferred to plastic bags and frozen for laboratory analysis.

Sediments were analyzed for grain size using standard methods (5) with the following modifications. The frozen samples were thawed and split longitudinally; 15 g of each sample were wet sieved through a 62 µm screen. The coarse inorganic fraction was analyzed using an Emery settling tube. The finer sediment fraction was analyzed following standard procedures for pipette analysis using sodium hexametaphosphate as a dispersant. Subsamples of 25 ml were withdrawn at standard time intervals to sample for sands, silts and clays.

Total organic carbon was determined by weight using two methods. The LECO induction furnace method was used to calculate total combustible organic carbon, whereas the hydrogen peroxide digestion procedure was used to determine total digestible organic carbon. Total organic material was also determined volumetrically by first measuring an initial volume, then sieving the sample through a 0.25 mm screen and finally measuring the volume of the organic material retained in the sieve.

Data Management/Analysis

Field and laboratory results were transferred to

computer coding sheets, keypunched and placed on magnetic tapes. Benthic community data were analyzed using classification methods. These procedures involved generating a site dendrogram showing similarities among samples, a species dendrogram showing similarities among species and a two-way table which reorganizes the original data matrix to show the association between species groups and site groups. These analyses were conducted only on species which were found in 10% or more of the samples. They are documented elsewhere (1, 2, 3).

Population data were examined using analyses of variance. Multiple analyses of variance were conducted on data to determine significant variability in abundance values with specific environmental parameters (e.g., depth, season, station). One-way analysis of variance was conducted to determine the significance in differences among means.

RESULTS AND DISCUSSION

Rocky Intertidal Habitats

The results of qualitative reconnaissance surveys indicated similar zonational patterns in both fjords. These zonational patterns varied both on a large scale, along the axis of the fjords, and on a small or local geographic scale.

Large-scale differences in zonational patterns consisted of a major distributional break in the dominant biota near the entrances of both fjords. The region up-fjord of these breaks was low in the number of visually dominant species and was characterized by the rockweed Fucus distichus, the barnacle Balanus glandula and the mussel Mytilus edulis.

The region down-fjord of the break had a higher number of species, and was characterized by the three dominants found up-fjord and several more open coast species such as the barnacles Semibalanus cariosus and Chthamalus dalli, and the seastar Pisaster ochraceus.

This distributional break in both fjords corresponded to differences in surface salinities. At stations up-fjord from the break, salinities were usually below 15 °/oo during the year, while salinities down-fjord from the break were usually above 15 °/oo. Apparently many species found outside the fjord were not able to survive in the fjord where salinities fell below 15 °/oo. This minimum salinity level may regulate the up-fjord distributional range of the more oceanic species.

The distribution of the three dominant organisms in this habitat was plotted with tidal elevation and substrate angle. The Fucus zone occupied upper intertidal areas (6-13 ft above MLLW) on gradually sloping substrates ($<65^{o}$) and was restricted to cracks, crevices and small ledges on steeper slopes ($>65^{o}$). The Balanus zone occupied low intertidal levels (0-9 ft above MLLW) on gradually sloping substrates ($<65^{o}$) and occupied the entire intertidal range (0-13 ft above MLLW) on more vertical slopes where Fucus is less abundant. The Mytilus zone was found at mid-intertidal levels (3-9 ft above MLLW) on steeply sloping shores ($>40^{o}$) and therefore overlapped with the distribution of Balanus. The size and abundance of Mytilus increased in certain habitats, such as on logs and headlands that project into the fjords. These habitats probably have less drag and swifter currents than other areas along the fjord walls. This distribution suggests that mussels in the fjords may favor habitats with faster currents providing greater food availability. These distributions and corresponding factors were similar in Boca de Quadra and Wilson Arm/Smeaton Bay.

Several other environmental factors appear to have a localized effect on these zonational patterns. Ice scour appears to have a dramatic effect, particularly in the middle basin of Boca de Quadra. Narrow levels (1-3 ft) of the intertidal were void of all macroscopic organisms for varying distances along the fjord walls. The numerous freshwater streams had a modifying effect on zonational patterns, depending on the size and velocity. In areas where small streams ran directly over the rocky intertidal, zones were depressed and found at lower than normal tidal levels. In faster flowing streams Mytilus and Balanus were present, but Fucus was not noted.

Classification analyses conducted on quantitative data from harvested quadrats in each fjord system indicate that quadrats were grouped primarily into those dominated by Fucus and those dominated by Balanus. Quadrats visually dominated by Mytilus were grouped with those visually dominated by Balanus, indicating that these quadrats have a similar group of associated species but differ in the size of the visually dominant species, Mytilus versus Balanus.

Organisms consistently associated with the high Fucus quadrats in both fjords included oligochaetes and mites. Species consistently associated with the low Balanus quadrats included Mytilus and the algae Odonthallia and Rhodoglossum. The other species did not show consistent associations with the dominant organisms in the two fjord systems. Such results suggest that these organisms do not have the same habitat requirements as the dominant species.

Rocky Subtidal Habitats

Results of reconnaissance SCUBA dives indicate the presence of three major habitat types based on substrate. These were rock, sand and a combination of rock and sand. Rocky substrates consisted of nearly vertical rock walls with the amount of sediment increasing as the substrate angle decreased. Sand dominated substrates occurred on more horizontal areas, particularly near major river and stream deltas. Mixed habitats of rock and sand occurred where vertical rock substrates were locally interspersed with horizontal sandy areas or where large sandy areas were occupied with large boulders. All three habitat types occurred in both fjord systems; however, vertical rock substrates were more prevalent in Boca de Quadra.

The distribution of organisms within these three habitats varied primarily with substrate type and depth, and secondarily with other factors such as water movement. Rock wall areas were dominated by algae at shallow depths and filter-feeding invertebrates at depths below 20 ft. The red algae Ahnfeltia plicata and Constantinea subulifera were dominant at depths between 0 and 10 ft, whereas the kelps Laminaria saccharina, L. groenlandica and Agarum fimbriatum were dominant at depths between 10 and 20 ft. At depths between 20 and 35 ft the tunicates Halocynthia aurantium and Ascidia callosa, serpulid tube worms, and the bryozoan Microporina articulata were dominant. Crinoids and glass sponges were found at depth below 35 ft.

The relative abundance of these organisms varied among stations. For example, at depths between 10 and 20 ft, variability in the abundance of the shallow red algae and kelps can be attributed, in some cases, to heavy sea urchin predation. At depths below 20 ft, the relative abundance of the filter-feeding invertebrates, such as brachiopods, serpulids and tunicates, varied apparently with water velocity.

Sand bottom substrates were inhabited by fewer organisms on the substrate surface. In many areas, the eelgrass Zostera marina occurred at depths between 0 and 5 ft and provided a habitat for other organisms including the crab Cancer magister. At depths between 5 and 20 ft flatfish, the cockle Clinocardium and its predator the sunstar Pycnopodia helianthoides were found. This habitat supported the largest number of fish species in the fjords.

Much of both fjords consisted of areas with mixed substrates of rock and sand. The species composition in these areas therefore corresponded both with the particular dominant substrate (rock or sand) as well as with depth. This resulted in a mosaic of both rock and sand zonational patterns in these areas.

Soft-bottom Intertidal Habitats

Aerial and ground truth surveys were conducted in the soft-bottom intertidal habitats of the Bakewell Arm, Wilson River and Keta River estuaries. These surveys indicated the presence of several zones based on the dominant biota and substrate. These zones, distributed from high to low tidal level, were: an herb/grass zone; a sedge zone; a rockweed/cobble zone; a mussel/cobble zone; a sand/mud zone; and an eelgrass zone.

The herb/grass zone was dominated by a variety of mixed grasses and herbaceous plants. The sedge zone was dominated by the sedge Carex spp. and silverweed Potentilla and other grasses. The rockweed Fucus distichus was often found at the base of this zone.

The rockweed/cobble zone was dominated by Fucus distichus attached to the cobble along with the barnacle Balanus glandula and the crab Hemigrapsus nudus. The rockweed/cobble zone essentially represents an extension of the same zone found in the rocky intertidal region. Additional unattached Fucus was found in drifts up to 2 ft high.

The mussel/cobble zone was dominated by the mussel Mytilus edulis attached to cobble. This zone was found only in the Bakewell Arm estuary and represents an extension of the mussel zone found in nearby rocky intertidal areas.

The sand/mud zone was void of any large characteristic biota on the sediment surface. The eelgrass Zostera marina zone was found on the north side of both the Wilson and Bakewell Arm estuaries. This zone was not found in the Keta River estuary at the head of Boca de Quadra.

Soft-bottom Subtidal Habitats

Cluster analysis was conducted on soft-bottom subtidal infaunal data from both fjord systems. Four site groups were identified from the Boca de Quadra results. These site groups corresponded with specific geographic areas and depths in the fjords.

Cluster analysis on infaunal data from Wilson Arm/Smeaton Bay resulted in the identification of seven site groups. Three of these groups corresponded with specific geographic areas and depths, and paralleled the distribution of site groups in Boca de Quadra. These four site groups are considered as separate communities (Table I) since they occupy distinct habitats and, as discussed later, are characterized by different species assemblages. The four site groups in Wilson Arm/Smeaton Bay which did not correspond with particular areas or depths, are not considered as separate communities. They appeared to be subject to recent environmental disturbances as discussed later.

The four communities identified in the two fjords were not only associated with specific geographic areas and depths, but also with specific environmental parameters, particularly sediment grain size and organic content (Table I). Community 1 occupied shallow depths and was found only on the Kite Island sill in Boca de Quadra. This area was characterized by coarse sediments and low organics. Community 2 also occupied shallow depths, but was found in the inner basin of Boca de Quadra as well as in the shallow regions of both Wilson and Bakewell Arms. These areas were characterized by finer sediments and high organics. Community 3 was found in the deeper part of the inner basin of Boca de Quadra and the deeper parts of Wilson and Bakewell Arms. The sediments in this community were generally finer and organics, by volume, less than in Community 2. Community 4 occupied the deepest parts of both fjords, the middle and outer basins of Boca de Quadra and the deep basin that comprises Smeaton Bay. These areas had the finest sediments and low organics.

Sediment grain size was the primary factor separating these communities, as has been reported in many near-shore benthic infaunal studies (6). The mode was a better indicator of differences in grain size among these communities than other parameters measured. The mode is a measure of the most abundant grain size and is not affected by small abundances of other sediment sizes, which apparently do not have a significant influence on the observed infaunal distributional patterns.

The volume of organic material was a secondary factor separating these communities. This organic material, which comprised up to 38% of the sediment by volume and included tree branches, leaves and needles, could be more of an indicator of structural habitat than food abundance. Total organic carbon, which is an indicator of the total amount of food available in the sediments, did not vary much between communities. Total organic carbon did not appear to affect the distribution of species, but could affect the abundance of certain organisms.

Oxygen did not appear to have a major effect on the distribution of infaunal communities, as has been reported in some fjords. Such fjords have bottom waters that are either intermittently or permanently anoxic (6). Oxygen

Table I. Summary of soft-bottom subtidal communities in Boca de Quadra (BDQ) and Wilson Arm/Smeaton Bay (WA/SB).

Community	Area	Depth(m)	Grain Size (mode - ∅)	Organics (volume %)	Characteristic Species
1	BDQ: Kite I. sill WA/SB: Not present	45-85 -	1.5 -	2.5 -	Hiatella arctica Melinna cristata
2	BDQ: Inner basin WA/SB: Bakewell Arm, Wilson Arm	35-95 16-93	3.0-4.5 3.0-4.5	9.7-27.0 13.1-38.6	Axinopsida serricata Aricidea nr. suecica Glycera capitata
3	BDQ: Inner basin WA/SB: Bakewell Arm, Wilson Arm	110-155 115-155	2.5-7.0 4.5-8.0	2.3-10.0 4.5-9.4	Thyasira gouldi Nepthys cornuta cornuta
4	BDQ: Middle, outer basins WA/SB: Smeaton Bay	110-300 241-265	8.0-9.0 7.0-9.0	0.0-7.3 0.7-10.2	Ancistrosyllis groenlandica Brisaster latifrons Aglaophamus malgreni

levels in both fjord systems were relatively high; no values below 3 mg/L have been reported. These fjords are exposed to annual mixing and can be considered as permanently oxygenated.

Two-way tables were generated to show which species assemblages were associated with each community in the two fjord systems. Some of the characteristic species for each community are listed in Table I. There was a small number of species in the site groups which were not considered with any of the four communities in Wilson Arm/Smeaton Bay. These site groups consisted of stations from different areas of the fjord without any consistent similarity in sediment characteristics. Apparently other factors are regulating the distribution of species in these samples. The small number of species and scattered distribution suggests that periodic disturbances due to physical factors, such as slumping by sand, and/or by biological activity such as predation, may be the key environmental factor(s) regulating the distribution of organisms in these samples.

SUMMARY

In summary, Boca de Quadra and Wilson Arm/Smeaton Bay appear similar in the dominant benthic organisms and communities as well as in the environmental factors which regulate their distributions. Salinity, tidal height, substrate angle and water movement are important environmental factors in the rocky intertidal. In the rocky subtidal, the important factors are depth, substrate type, light and water movement. Tidal exposure is an important factor in the soft-bottom intertidal, while sediment grain size and organic content are important factors distinguishing soft-bottom subtidal communities.

REFERENCES

1. VTN. "Boca de Quadra Baseline Report, Quartz Hill Molybdenum Project, Southeast Alaska", U.S. Borax & Chemical Corp., Los Angeles, CA (1980).
2. VTN. "Wilson Arm/Smeaton Bay Baseline Report, Quartz Hill Molybdenum Project, Southeast Alaska", U.S. Borax & Chemical Corp., Los Angeles, CA (1981)
3. VTN. "1981 Boca de Quadra Monitoring Study, Quartz Hill Molybdenum Project, Southeast Alaska", U.S. Borax & Chemical Corp., Los Angeles, CA (1982).
4. Bonsack, J.A. "Photographic Quantitative Sampling of Hard-bottom Benthic Communities", Bull. Mar. Sci.

    29(2): 242-252 (1979).
5. Folk, R.L. Petrology of Sedimentary Rocks (Austin, TX: Hemphill Pub. Co., 1974).
6. Pearson, T.A. "Macrobenthos of Fjords", in Fjord Oceanography, J.J. Freeland, D.M. Farmer, and C.D. Leving, Eds. (NY: Plenum Press, 1980), pp. 569-602.

DISCUSSION

Question: There were questions earlier about the way in which juvenile salmon are coupled to the marine production system. An earlier report of VTN informs us that juvenile salmon eat primarily Chironomids and Harpacticoids. You have shown that the Chironomids are located in the Fucus zones. There should be no problem in ensuring that there is no impact from the tailings disposal system in the Fucus zone. But where are the Harpacticoids?

Answer: Some of the Harpacticoids are found also in the rocky intertidal zone. However, the same organisms, both the Chironomids and the Harpacticoids, are also found on soft bottoms such as sandy beaches and mud flats. In the studies that are going on right now, we are looking at gut analyses of the salmon in these habitats, in the mud flats and also along the rock walls. We are comparing what is in the habitats and what is in the guts.

Question: You noted what you considered to be estuarian organisms out in the fjord near the surface. Did you find any trace at depth also?

Answer: Not really. We did salinity profiles and found avoidance of low salinities by many non-estuarian species in the shallow subtidal. The Echinoderms are a good indication of organisms that are avoiding freshwater. We found that freshwater runoff provides a barrier for intertidal organisms to avoid predation by starfish. Thus this freshwater input not only affects the intertidal organisms, but also has an impact on the shallow subtidal. But it is not a question of having organisms that are strictly estuarian, since many found in the estuaries are found in other areas, as much as having oceanic organisms that cannot go past a low salinity point.

J.L. Littlepage
Department of Biology
University of Victoria
Victoria, B.C.
Canada V8W 2Y2

I have always admired chairmen who can manage to read and write in the dark and listen to a paper while writing a summary of the others already given. I am afraid that I am not one of those people. However, there are a few observations that I would like to make.

Dr. Snook brought us back to reality by, in essence, pointing out that the product of a mill is not tailing but a metal product, in this case molybdenum. The tailing are the residue that one has to dispose of. In looking at the ore-body that is present at Quartz Hill, it seems to be, to a non-geologist, a very favorable ore-body from the point of view of mining in that it is seemingly a large, coherent body. It was described as being a fine-grained ore of a quartz and molybdenum complex. In the comparisons of the Quartz Hill property to others along the western United States and Canada it appears as a large, if not the largest, one.

Speaking as a biologist, there are two or three things that are particularly important. First of all, I think the minimal overburden is an environmental asset in this case. We have heard that the overburden at Rupert Inlet does create disposal problems even though it is not toxic waste material. I know from experience at Alice Arm that while the case there is not of the same magnitude, the disposal of overburden is still a problem. You are fortunately free of this at Quartz Hill.

Also, I was very happy to note that the waste tailing rock was very low in copper, zinc, lead, iron and the associated arsenics. This makes it an environmentally essentially simple material to dispose of. Most of the metals that one worries about in other molybdenum mines such as lead, cadmium and zinc don't seem to be a problem. In short, I really cannot think of a better way in which there could be a molybdenum deposit than that at Quartz Hill.

Moving from land into water, I was impressed by the difference that one can see in oceanographic features in such a short geographical distance. I am very familiar with the conditions 50 miles to the east of us. While a quick glance at the map indicates that there are probably similar types of estuaries here, we heard this morning that they may not be similar at all. The type of circulation is quite different. The dominant features that we find in Alice Arm, and Rupert Inlet, are not active in this situation.

One has to be impressed by the amount of data available in the Quartz Hill baseline study and the diligence with which it was collected. I think one can, however, make a great deal of cross-transfer from one study to another. While the details from elsewhere are different with regard to water movements, water depths, sediment loading, etc. the principles that we have learned in the disposal of tailings in one area can be readily transferred to another.

I would like to mention Dr. Cimberg's benthic ecology paper. Again, I am very impressed by the amount of effort that has been put into these studies, particularly when you realize that Quartz Hill is still in the exploratory phase. To have this much data available this early in the program is certainly a sign of progress. I thought that we had come a long way in the development of the Alice Arm program because we had a long start on it. I am very happy that we have progressed from earlier practices of mine first, environment second. I think it is very commendable that we have the details that we have here.

As far as the communities that Dr. Cimberg describes, I just want to make one comparison. I will draw upon my work in Alice Arm because of the high sediment loading there. Dr. Burrell indicated that .5-1 mg/L was a reasonable load in Wilson Arm and Boca de Quadra. We have in Alice Arm something of the order of a minimum of 1 mg/L during the winter, and in the summer, surface loadings which are up to 50-75 mg/L. The low sediment loading is reflected in the type of fauna that you have. In Alice Arm the fauna is responsive to sediment loading. The southeast Alaska inlets have a much more diverse fauna and this, I believe, can be traced primarily to this difference in sediment loading.

This points to the need for site-specific investigations and site-specific regulations.

Investigations, data and regulations which were promulgated for Alice Arm, while informative, cannot be transferred in mass from one area to another. If today's papers did nothing else, they indicated the differences between areas which superficially look very similar.

I would like to make one observation about the symposium as a whole. To me it has carried out its prime objective, i.e., getting information across to people. The details that we heard were very helpful. But one should also realize that these are on-going programs and many of the details, as those of any program of this nature, are still forthcoming. This did not distract from the symposium in any way because our main purpose here was not to get the type of numbers that you can get from books, but rather to get ideas from our colleagues.

R.J. Buchanan was one of two Canadian government scientists appointed to a panel to review the tailings disposal performance of a mine in British Columbia. He represented the B.C. Provincial Government where he is employed as Director of the Aquatic Studies Branch of the Ministry of Environment. He was born in Regina, Saskatchewan and received a B.Sc. in Aquatic Biology and a Ph.D in Oceanography from the University of British Columbia. He has served as a Senior Research Associate in the Department of Civil Engineering at the University of Washington. His present duties include functioning as internal advisor to waste management officials of the B.C. Ministry of Environment.

R.W. Burling is a Professor of Oceanography at the University of British Columbia where he researches in the discipline of physical oceanography, especially on problems in the wind generation of waves, and air-sea interactions. His degrees are from the University of New Zealand (B.Sc. in Physics and M.Sc. in Mathematics) and Imperial College, London University, England (Ph.D.). He recently served as a member of the Technical Panel appointed to review the environmental performance of the marine tailings disposal system at the molybdenum mine at Kitsault.

D.C. Burrell is Professor of Marine Science at the University of Alaska, where he has been responsible for baseline oceanographic studies in the Quartz Hill area. He obtained a B.Sc. in Geology and a Ph.D. in Geochemistry from the University of Nottingham, England. Since taking up his appointment at the University of Alaska he has conducted extensive oceanographic investigations in state waters, especially estuaries. He conducts research in the area of marine trace metal cycling; and sedimentation dynamics and seawater interactions. Dr. Burrell is the author of Atomic Spectrometric Analysis of Heavy Metal Pollutants in Water.

J.A. Caldwell, P. Eng., M.Sc. (Eng.), LLB is a geotechnical engineer and manager of Steffen Robertson and Kirsten, Vancouver, B.C. He works primarily in the design of tailings impoundments and mine waste dispsoal facilities. Impoundment projects for which he has been responsible include those for Greens Creek (Alaska), Mt. Tolman (northeast Washington) and Bald Mountain (Maine). He has designed numerous other impoundments for tailings from gold, platinum, gypsum and copper mines. He obtained his degrees at the University of the Witwatersrand where he lectured for five years in geotechnical engineering. He is the author of a number of papers on impoundment design and the land disposal of tailings.

R.L. Cimberg is a senior marine ecologist with VTN's environmental sciences division. He has been director of the marine benthic studies program for the Quartz Hill project during the past two years. Dr. Cimberg received his doctorate in marine ecology from the University of Southern California and the Santa Catalina Marine Science Center. Before joining VTN, he taught marine ecology at the University of California and at the California State University. Dr. Cimberg has led field expeditions at project sites ranging from Alaska to Baja California. He has investigated rocky intertidal, estuary, mangrove, and subtidal habitats involving populational, community and physiological studies. He has also conducted laboratory investigations on the effects of various environmental factors on invertebrate life histories.

D.M. Farmer graduated from McGill University with a B.Sc. and M.Sc. in Physics. His doctoral research was at the Department of Physics and Institute of Oceanography at the University of British Columbia on the experimental and theoretical analysis of wind effects in Alberni Inlet. He is presently employed as Head, Coastal Zone Oceanography, Institute of Ocean Sciences, Pat Bay, B.C. His area of research is the physics of lakes and coastal waters, and fjord oceanography, especially the structure of tidal flow through constrictions. The Coastal Zone group has been involved in studies of fjords subject to mine tailings disposal, in particular Rupert Inlet and Alice Arm-Observatory Inlet.

A. Hay initiated seabed studies on tailings dispersal in the late 1970s as a graduate student in the Department of Oceanography, University of British Columbia, and subsequently received his Ph.D. in Physics with a thesis on the topic. He was born in Tillsonburg, Ontario, and obtained his B.Sc. and M.Sc. in Physics at the University of Western Ontario. He is now Assistant Professor in the Department of Physics, Memorial University of Newfoundland, where he continues his investigations on turbidity currents by means of acoustic remote sensing and other techniques.

W.J. Kuit has been employed by Cominco Ltd. since graduation in chemistry in 1966. He has conducted a range of chemical and metallurgical research assignments which, since 1972, were mainly related to process development for effluent control from mining and metallurgical operations. His experience has also ranged from external consulting activities to the field supervision of effluent treatment operations. He is currently Assistant Manager, Environmental Control, and is located at Trail, B.C. His professional environmental management experience includes the Polaris mine in the Canadian arctic.

J.L. Littlepage is Associate Professor of Biology at the University of Victoria in British Columbia and has been an oceanographic consultant to Climax Molybdenum Corp. and AMAX of Canada on the Kitsault project. He has also been involved in the design and implementation of baseline studies at other potentially marine discharging mines. He is a graduate of San Diego State and Stanford Universities where he was trained as a biological oceanographer. He has deep-sea experience in the Antarctic as an oceanographer, and aquacultural experience in the brine-shrimp industry.

D. Nebert received a bachelor's degree in Physics in 1965 from Portland State University. Graduate studies in Physical Oceanography were pursued at both the University of Washington and the University of Alaska, culminating in an M.S. degree in 1972. Mr. Nebert has been involved with various facets of oceanographic data from collection through final analysis. He has been responsible for physical oceanographic data quality and in this capacity has established data collection and processing procedures. Projects on which Mr. Nebert has worked are the Environmental Studies at Port Valdez, the OCSEAP Outer Continental Shelf Study, and the present Marine Environmental Studies in Boca de Quadra and Smeaton Bay.

C.A. Pelletier was in charge of the marine environmental laboratory and monitoring program at Island Copper Mine from 1970 through 1981, and since 1974 responsible for environmental/reclamation work there and at other sites under development. He was trained as a chemist and has a B.Sc. from the University of

Saskatchewan, and has worked as a metallurgist and process chemist. He is now an independent consultant of the Rescan Group, which provides planning and environmental services.

G.W. Poling is Professor and Head of the Department of Mining and Mineral Process Engineering, University of British Columbia. His training was received at the University of Alberta where he graduated with a B.Sc., M.Sc., and Ph.D. from the Department of Mining and Metallurgy. He has been employed previously as a mining engineer in Zambia, and in the research laboratories of Texaco. He was one of the original university scientists and engineers who in 1972 formed the environmental monitoring supervisory group for Island Copper Mine. He has published extensively in his professional area, and contributed to many environmental reports and published papers derived from mining activities.

J.R. Snook is a petrologist employed as Professor of Geology at Eastern Washington University, where he has been Chairman of the Department of Geology. He has degrees from Oregon State University in Business (B.S.) and Geology (M.S.), and a Ph.D. in Geology from the University of Washington. He has worked as a research, production and exploration geologist for a number of companies, and acts as a consultant to others.

J.D. Welsh is Vice-President of Steffen Robertson and Kirsten in Denver. He has a B.S. in Civil Engineering from the University of Missouri and an M.S. in Civil Engineering from Colorado State University. Mr. Welsh was for many years a construction engineer at the Climax mine tailings impoundments in Colorado. He has designed and supervised the construction of numerous tailings impoundments.

D.R. Young is a Senior Chemical Oceanographer with the Marine Services Group of Dames & Moore. He graduated with a B.A. in Physics from Pomona College and began ocean pollution investigations at Scripps Institution of Oceanography, studying the distribution and marine bioaccumulation of trace level radionuclides. After receiving his Ph.D. in Chemical and Biological Oceanography in 1970, he joined a government sponsored regional marine pollution agency, the Southern California Coastal Water Research Project, to found the Chemistry Department. During the next ten years, he directed numerous research projects involving inputs, distributional processes, and bioaccumulation of contaminants in the Southern California Bight.